ESSENTIALS OF
TERRORISM
CONCEPTS AND CONTROVERSIES

GUS MARTIN

California State University, Dominguez Hills

SAGE Publications

Los Angeles • London • New Delhi • Singapore

For information:

Sage Publications, Inc.
2455 Teller Road
Thousand Oaks, California 91320
E-mail: order@sagepub.com

Sage Publications India Pvt. Ltd.
B 1/I 1 Mohan Cooperative Industrial Area
Mathura Road, New Delhi 110 044
India

Sage Publications Ltd.
1 Oliver's Yard
55 City Road
London EC1Y 1SP
United Kingdom

Sage Publications Asia-Pacific Pte. Ltd
33 Pekin Street #02-01
Far East Square
Singapore 048763

Printed in the United States of America

Library of Congress Cataloging-in-Publication Data

Martin, C. Gus.
Essentials of terrorism : concepts and controversies / C. Gus Martin.
    p. cm.
Includes bibliographical references and index.
ISBN 978-1-4129-5313-9 (pbk.)
  1. Terrorism. I. Title.

HV6431.M366 2008
363.325—dc22   2007023519

This book is printed on acid-free paper.

07 08 09 10 11 10 9 8 7 6 5 4 3 2 1

Acquisitions Editor:	Jerry Westby
Editorial Assistant:	Kim Suarez
Associate Editor:	Elise Smith
Production Editor:	Catherine M. Chilton
Copy Editor:	Helen Glenn Court
Typesetter:	C&M Digitals (P) Ltd.
Proofreader:	William H. Stoddard
Indexer:	Molly Hall
Cover Designer:	Edgar Abarca
Marketing Manager:	Jennifer Reed

ESSENTIALS OF TERRORISM

Titles of Related Interest

From SAGE Publications

Risk Balance and Security, by Erin Van Brunschot and Leslie W. Kennedy

Understanding Terrorism (Second Edition), by Clarence Augustus Martin

The New Era of Terrorism: Selected Readings, by Clarence Augustus Martin

Terrorism in Perspective (Second Edition), by Sue Mahan and Pamala Griset

Communicating Terror: The Rhetorical Dimensions of Terrorism, by Joseph Tuman

Criminology: An Interdisciplinary Approach, by Anthony Walsh and Lee Ellis

Handbook of Transnational Crime and Justice, by Philip Reichel

Introduction to Criminology: Theories, Methods, and Criminal Behavior (Sixth Edition), by Frank Hagan

Crimes of Hate: Selected Readings, by Phyllis B. Gerstenfeld and Diana R. Grant

Violence: The Enduring Problem, by Alex Alvarez and Ronet Bachman

Brief Contents

About the Author   x

Preface   xi

Acknowledgments   xvi

Part I: Understanding Terrorism: A Conceptual Review   1

  1. Defining Terrorism   2

  2. Historical Perspectives and Ideological Origins   22

  3. Causes of Terrorist Violence   44

Part II: Terrorist Environments   65

  4. Terrorism by the State   66

  5. Terrorism by Dissidents   90

  6. Religious Terrorism   111

  7. International Terrorism   133

  8. Domestic Terrorism in the United States   154

Part III: The Terrorist Battleground   179

  9. Terrorist Violence and the Role of the Media   180

  10. Tactics and Targets of Terrorists   197

  11. Counterterrorism and the War on Terrorism   219

  12. Future Trends and Projections   249

Appendix A: Map References   272

Appendix B: National Intelligence Estimate: The Terrorist Threat to the US Homeland 285

Appendix C: Historical Examples   http://www.sagepub.com/martinessstudy/

Photo Credits   292

Glossary   293

Notes   311

Index   318

Detailed Table of Contents

About the Author   x

Preface   xi

Acknowledgments   xvi

Part I: Understanding Terrorism: A Conceptual Review   1

1. **Defining Terrorism**   2
   Understanding Political Extremism   3
   Formal and Informal Definitions   6
   Terrorists or Freedom Fighters?   11
   Chapter Summary   16
   Key Terms and Concepts   20
   Terrorism on the Web   20
   Web Exercise   20
   Recommended Readings   21

2. **Historical Perspectives and Ideological Origins**   22
   Historical Perspectives on Terrorism   22
   Ideological Origins of Terrorism   26
   September 11, 2001 and the New Era of Terrorism   37
   Chapter Summary   41
   Key Terms and Concepts   42
   Terrorism on the Web   43
   Web Exercise   43
   Recommended Readings   43

3. **Causes of Terrorist Violence**   44
   Political Violence as the Fruit of Injustice   45
   Political Violence as Strategic Choice   51
   The Morality of Political Violence   54
   Chapter Summary   60
   Key Terms and Concepts   63
   Terrorism on the Web   63
   Web Exercise   63
   Recommended Readings   63

Part II: Terrorist Environments   **65**

4. **Terrorism by the State**   **66**
   Perspectives on Terrorism by Governments   66
   Domestic Terrorism by the State   72
   Terrorism as Foreign Policy   81
   Chapter Summary   86
   Key Terms and Concepts   88
   Terrorism on the Web   89
   Web Exercise   89
   Recommended Readings   89

5. **Terrorism by Dissidents**   **90**
   Perspectives on Violent Dissent   92
   The Practice of Dissident Terrorism   96
   Dissidents and the New Terrorism   104
   Chapter Summary   107
   Key Terms and Concepts   109
   Terrorism on the Web   110
   Web Exercise   110
   Recommended Readings   110

6. **Religious Terrorism**   **111**
   Historical Perspectives on Religious Violence   111
   The Practice of Religious Terrorism   120
   Trends and Projections   129
   Chapter Summary   130
   Key Terms and Concepts   131
   Terrorism on the Web   132
   Web Exercise   132
   Recommended Readings   132

7. **International Terrorism**   **133**
   Understanding International Terrorism   134
   International Terrorist Networks   142
   The International Dimension of the New Terrorism   145
   Chapter Summary   151
   Key Terms and Concepts   152
   Terrorism on the Web   152
   Web Exercise   152
   Recommended Readings   153

8. **Domestic Terrorism in the United States**   **154**
   Extremism in America   157
   Leftist Terrorism in the United States   162

Right-Wing Terrorism in the United States 168

Chapter Summary 174

Key Terms and Concepts 175

Terrorism on the Web 176

Web Exercise 176

Recommended Readings 177

Part III: The Terrorist Battleground **179**

9. **Terrorist Violence and the Role of the Media** **180**

Understanding the Role of the Media 180

Mass Communications and the War for Information 185

Chapter Summary 193

Key Terms and Concepts 195

Terrorism on the Web 195

Web Exercise 195

Recommended Readings 196

10. **Tactics and Targets of Terrorists** **197**

Understanding Terrorist Objectives 198

The Terrorists' Arsenal 201

Terrorist Targets 209

Chapter Summary 215

Key Terms and Concepts 217

Terrorism on the Web 218

Web Exercise 218

Recommended Readings 218

11. **Counterterrorism and the War on Terrorism** **219**

The Use of Force 221

Operations Other Than War 229

Legalistic Responses 237

Chapter Summary 245

Key Terms and Concepts 247

Terrorism on the Web 248

Web Exercise 248

Recommended Readings 248

12. **Future Trends and Projections** **249**

What Does the Future Hold? 249

Revisiting the War on Terrorism 260

Chapter Summary 269

Key Terms and Concepts 271

Terrorism on the Web 271

Recommended Readings 271

Appendix A: Map References   272

Appendix B: National Intelligence Estimate:
   The Terrorist Threat to the US Homeland   285

Appendix C: Historical Examples   http://www.sagepub.com/martinessstudy/

Photo Credits   292

Glossary   293

Notes   311

Index   318

About the Author

C. Augustus "Gus" Martin is Assistant Vice President for Faculty Affairs at California State University, Dominguez Hills, where he is also the former chair of the Department of Public Administration & Public Policy. He has served on the faculty of the Graduate School of Public and International Affairs, University of Pittsburgh, where he was an Administration of Justice professor. His research and professional interests are terrorism and extremism, juvenile justice, administration of justice, and fair housing. He has served as a panelist for university and community symposia and interviews on the subjects of administration of justice, terrorism, and fair housing. He has also been a consultant to government and private agencies. Prior to joining academia, he was a "floor" legislative assistant to Congressman Charles B. Rangel of New York and Special Counsel to the Attorney General of the U.S. Virgin Islands. In addition, he served as managing attorney for the Fair Housing Partnership of Greater Pittsburgh, where he was also director of a program created under a federal consent decree to desegregate public and assisted housing.

Dedication

*This book is dedicated to the students of California State University,
Dominguez Hills, who continue to impart to me the meaning
of Docendo Discimus—we learn by teaching.*

Preface

Welcome to the first edition of *Essentials of Terrorism: Concepts and Controversies*. This book is designed as a foundational textbook to enhance the educational quality of the study of terrorism. It is a concise textbook for students and professionals who wish to explore the phenomenon of modern terrorist violence, and who wish to use additional resources adapted for their specific instructional needs. Readers will acquire a solid foundation for understanding the nature of terrorism in a manner best suited for their classroom or professional program.

This book is also designed to be a resource for university students and professionals who require fundamental expertise in understanding terrorist violence. The content of *Essentials of Terrorism: Concepts and Controversies* is directed to academic and professional courses whose subject areas have been selected for their specific educational program, including the study of terrorism, homeland security, international security, criminal justice administration, political conflict, armed conflict, and social environments. It can be incorporated into a variety of classes and seminars covering security studies, the administration of justice, the sociology of terrorism, conflict resolution, political theory, and other instruction in the social sciences. The intended level of instruction is undergraduate and master's students, as well as professionals with a need to understand terrorism.

No prerequisites are specifically recommended, but grounding within one of the following disciplines would be helpful: political science, government, administration of justice, sociology, history, or philosophy.

As will become readily apparent to instructors and students, *Essentials of Terrorism: Concepts and Controversies* is an ideal anchor textbook for investigating the many aspects of terrorism, political violence, and homeland security. That it is easily adapted to these subjects means that instructors will be able to design a variety of instructional packages around it. In this way, *Essentials of Terrorism* is a versatile resource.

Course Overview and Pedagogy

Essentials of Terrorism: Concepts and Controversies introduces readers to terrorism in the contemporary era, focusing on the post–Second World War period as its primary emphasis. It is a review of nations, movements, and individuals who have engaged in what many people would define as terrorist violence. It is also a review of the many kinds of terrorism that have existed in the postwar era. Very important, a serious exploration will be made of the underlying causes of terrorism—for example, extremist ideologies, religious intolerance, and traumatic episodes in the lives of nations and people.

The pedagogical approach of *Essentials of Terrorism: Concepts and Controversies* is designed to stimulate critical thinking in readers. Students, professionals, and instructors will find that each chapter follows a sequence of instruction that builds on previous chapters and thus incrementally enhances the reader's knowledge of each topic. Chapters incorporate the following features:

♦ *Chapter Introduction.* Each chapter is introduced by an overview of the subject under investigation. This provides the perspective to incorporate each chapter's topic into the broader themes of the textbook.

♦ *Chapter Perspectives.* Chapters incorporate focused presentations of perspectives that explore people, events, organizations, and movements relevant to the subject matter of each chapter.

♦ *Chapter Summary.* A concluding discussion recapitulates the main themes of each chapter and introduce the subject matter of the chapter that follows.

♦ *Discussion Boxes.* Discussion Boxes present provocative information and pose challenging questions to stimulate critical thinking and further debate.

♦ *Key Terms and Concepts.* Important terms and ideas introduced in each chapter are listed for review and discussion. These are further explored and defined in the book's Glossary.

♦ *Terrorism on the Web.*

♦ *Web Exercises.* Internet exercises at the ends of chapters have been designed for students, professionals, and instructors to explore and discuss information found on the Internet.

♦ *Recommended Readings.* Suggested readings listed at the end of each chapter provide either further information on or avenues of research into each topic.

Chapter Guide

This volume is organized into three thematic units, each consisting of several chapters. Several Appendices and a Glossary are included after the substantive chapters.

Part I. Understanding Terrorism: A Conceptual Review

The first section of the book is a comprehensive discussion of definitions of terrorism and the root causes of violent political extremism. Readers develop comprehensive, contextual, and critical skills for defining terrorism and for understanding the many causes of terrorist behavior.

Chapter 1. Defining Terrorism

This chapter investigates the reasons underlying why certain groups, movements, and individuals are labeled as terrorists or freedom fighters. It compares and contrasts radical and reactionary ideological tendencies, and defines and investigates the characteristics of extremism. Terrorism is discussed at length by sampling official definitions, reviewing the American context, and summarizing several types of terrorism. Readers are introduced to several perspectives of terrorism that pose problems for definitional issues.

Chapter 2. Historical Perspectives and Ideological Origins

This chapter explores the historical and ideological origins of modern terrorism. Historical perspectives are discussed within the contexts of conceptual themes used throughout the book. Ideological foundations for modern terrorist violence are also discussed at some length. The causes of left-wing and right-wing terrorism are identified, as are the qualities of ideological violence. Because both ideological poles were inextricably entwined during the 20th century, and adherents continue to be active in the 21st century, it is important for readers to grasp the importance of the ideologies of class struggle, national liberation, order, and race. This chapter also discusses regional examples of ideological terrorism.

Chapter 3. Causes of Terrorist Violence

Readers become familiar with central factors in the personal and group histories of individuals and groups who become associated with terrorism. The motives of extremists and several explanations of terrorism are explored, including acts of political will, sociological explanations, and psychological explanations. An important discussion probes the degree to which a fresh generation of new terrorists is being forged in reaction to how the West and its allies have conducted the post–September 11, 2001, war on terrorism.

Part II. Terrorist Environments

Part II educates readers about the many manifestations of terrorism by helping them develop skills to critically assess and understand historical and modern examples of political violence. In particular, state- and dissident-initiated terrorism are discussed, compared, and contrasted. Readers are also guided through how to distinguish between religious and international terrorism. Domestic terrorism in the United States is also explored.

Chapter 4. Terrorism by the State

This chapter investigates state-initiated repression and terror. A detailed discussion explores terrorism as foreign policy and terrorism as domestic policy. Important examples of state terrorism include the Khmer Rouge in Cambodia and the link between Janjaweed fighters and the government in Sudan.

Chapter 5. Terrorism by Dissidents

This chapter critically evaluates terrorism emanating from dissident movements. Several typologies and the morality of the New Terrorism, are investigated. Finally, a detailed discussion explores antistate dissident terrorism and communal terrorism. Important examples include the modern use of child soldiers by extremists and Chechen terrorism against Russia.

Chapter 6. Religious Terrorism

This chapter evaluates the historical and modern origins and quality of religious terrorism. The goal is to engender critical discussion on the subject and to develop a contextual perspective for it. Because religious terrorism has become so prominent, it is important for readers to investigate different manifestations and to understand the contexts of regional case studies.

Chapter 7. International Terrorism

This chapter discusses recent and historical examples of international terrorism, defines what is meant by international terrorism, and explores the reasons for terrorist spillovers. Both the phenomenon of international terrorist networks and the concept of stateless revolutionaries are discussed. In this regard, readers evaluate newly emerging threats from groups and networks that have adapted the Al Qaeda example as a model.

Chapter 8. Terrorism in the United States

This chapter presents an overview of terrorism in postwar America. It probes the background to political violence and presents a detailed discussion of leftist and rightist terrorism in the United States. The chapter also evaluates international terrorism and prospects for violence from rightist, leftist, and religious extremists today.

Part III. The Terrorist Battleground

Part III discusses the nuts and bolts of the terrorist trade, including the informational war waged between adversaries and the role of the mass media. Readers investigate how the applications of the concepts of propaganda by deed and armed propaganda have been historically common to extremist violence. With the availability of high-yield weaponry in the arsenals of terrorists and the globalization of information, it is important for readers to grasp the significance of the terrorist trade in the modern world.

Chapter 9. Terrorist Violence and the Role of the Media

This chapter investigates and evaluates the centrality of the media and mass communications in the modern era of political violence. It first discusses the nature of mass communications and reporting within the context of terrorist environments. It also investigates the war of manipulation for favorable media coverage. In particular, readers assess the manipulation of information technologies and the media by modern terrorists. A discussion is also presented on the efficacy of regulating the media.

Chapter 10. Tactics and Targets of Terrorists

This chapter investigates the methodology of terrorism. Terrorist objectives, methods, and targets are analyzed at length, as is the question of whether terrorism is effective. Recent data and examples identify new challenges in the new era of terrorism, including examples of the use of the Internet to post incidents and communiqués.

Chapter 11. Counterterrorism and the "War on Terrorism"

This chapter explores counterterrorist options and security measures. Several categories of responses are assessed: the use of force, repressive operations other than war, conciliatory operations other than war, and legalistic responses. Contemporary controversies, such as the status and treatment of captured suspects, are explored.

Chapter 12. Future Trends and Predictions

Readers are challenged to critically assess trends and other factors that can be used to project the near future of terrorism. In particular, this chapter presents fresh discussions and data on the near-term projections for the future of terrorism. New issues and likely scenarios are offered for the near future of ideological terrorism, religious terrorism, international terrorism, political violence against women, and criminal terrorism.

Ancillaries

This new *Essentials* edition comes with a variety of ancillaries to supplement instructor course preparation and student learning.

Instructor's Resource CD

This CD offers the instructor a variety of resources with which to supplement the book material, including PowerPoint lecture slides, Web resources, activity and lecture suggestions, guidance regarding the discussion boxes, and more. Also included is an updated electronic test bank using Brownstone's Diploma test bank software, which consists of multiple choice questions with answers and page references, as well as short answer and essay questions for each chapter.

Web-Based Student Study Site

http://www.sagepub.com/martinessstudy/

This Web-based student study site provides a variety of additional resources to enhance students' understanding of the book content. Taking their learning one step further, the site includes comprehensive study materials, such as introductions to the chapter articles, updated chapter articles, recommended Web sites, flash cards, and Web exercises.

Acknowledgments

I am indebted for the support and encouragement of many people in bringing this venture to completion, with special appreciation given to the very professional and expert attention given to this project by the editorial group at Sage. Without their patient professionalism and constructive criticism, this project would not have incorporated the comprehensiveness and completeness that was its underlying objective from the beginning.

Thanks are extended to colleagues who shared their expert advice and suggestions for crafting this volume. Deep appreciation is also given to the panel of peer reviewers assembled by the very able editors and staff of Sage Publications during several rounds of review. The insightful, constructive comments and critical analysis of the following reviewers was truly invaluable.

Joseph T. McCann, PsyD, JD
Binghamton University

Lee Ayers Schlosser
Southern Oregon University

Barry J. Balleck
Georgia Southern University

Pinky S. Wassenberg
University of Illinois, Springfield

Thomas E. Baker, Lt. Col.
MP USAR (ret.)
University of Scranton

Daniel P. Ford
Cameron University

Jennifer Kunz
West Texas A&M University

James M. Lutz
Indiana University–Purdue University Fort Wayne

Jerome Randall
University of Central Florida

Ronald D. Server
Prairie View A&M University

Finally, I thank my wife and children for their constant support, encouragement, and humor during the course of this project.

PART I

Understanding Terrorism

A Conceptual Review

Part I Photo   The Pentagon on the morning of September 11, 2001.

1

Defining Terrorism

Terrorism, however defined, has always challenged the stability of societies and the peace of mind of everyday people. In the modern era, the impact of terrorism—that is, its ability to terrorize—is not limited to the locales or regions where the terrorists strike. In the age of television, the Internet, satellite communications, and global news coverage, graphic images of terrorist incidents are broadcast instantaneously into the homes of hundreds of millions of people. Terrorist groups understand the power of these images, and manipulate them to their advantage as much as they can. Terrorist states also fully appreciate the power of instantaneous information, and thus try to control the spin on reports of their behavior. In many respects, the beginning of the 21st century is an era of globalized terrorism.

Some acts of political violence are clearly acts of terrorism. Most people would agree that politically motivated bombings of marketplaces, massacres of enemy civilians, and routine government use of torture are terrorist acts. However, as we begin our study of terrorism, it is important to appreciate that we will encounter many definitional gray areas. Depending on which side of the ideological, racial, religious, or national fence one sits, political violence can be interpreted either as acts of unmitigated terrorist barbarity or as freedom fighting and national liberation. These areas will be explored in the chapters that follow.

This chapter investigates definitional issues in the study of terrorism. Readers will probe the nuances of these issues and learn that the truism "one person's terrorist is another person's freedom fighter" is a significant factor in the definitional debate. It must be remembered that this debate occurs within a practical and real-life framework—in other words, a nontheoretical reality that some political, religious, or ethno-nationalist beliefs and behaviors are so reprehensible that they cannot be considered to be mere differences in opinion. Some violent incidents are *malum in se* acts of terrorist violence. For example, the New Terrorism of today is characterized by the threat of weapons of mass destruction, indiscriminate targeting, and intentionally high casualty rates—such as the attacks of September 11, 2001, in the United States; of March 11, 2004, in Spain; of July 7, 2005, in Great Britain; of July 23, 2005, in Saudi Arabia; and repeated attacks in Baghdad. Using these weapons and tactics against civilians is indefensible, no matter what cause is championed by those who use them. The New Terrorism is discussed in detail in Chapter 2.

Photo 1.1 The U.S. Rewards for Justice poster for Osama bin Laden, founder of the Al Qaeda terrorist organization.

The discussion in this chapter will review the following:

◆ Understanding political extremism
◆ Formal and informal definitions
◆ Terrorists or freedom fighters?

Understanding Political Extremism

An important step toward defining terrorism is to develop an understanding of the sources of terrorism. To identify them, one must first understand the important role of extremism as a primary feature of all terrorist behavior.

Behind each incident of terrorist violence is some deeply held belief system that has motivated the perpetrators. Such systems are, at their core, extremist systems characterized by intolerance. One must keep in mind, however, that though terrorism is a violent expression of these beliefs, it is by no means the only possible manifestation of extremism. On a scale of activist behavior, extremists can engage in such benign expressions as sponsoring debates or publishing newspapers. They might also engage in vandalism and other disruptions of the normal routines of their enemies. Our focus in this and subsequent chapters will be on violent extremist behavior that many people would define as acts of terrorism. First, we must briefly investigate the general characteristics of the extremist foundations of terrorism.

Defining Extremism

Extremism is a quality that is "radical in opinion, especially in political matters; ultra; advanced."[1] It is characterized by intolerance toward opposing interests and divergent opinions, and is the primary catalyst and motivation for terrorist behavior. Extremists who cross the line to become terrorists always develop noble arguments to rationalize and justify their acts of violence toward nations, people, religions, or other interests.

It is important to understand that extremism is a radical expression of political values. Both the content of one's beliefs and the style in which one expresses them are fundamental to extremism. Laird Wilcox summed up this quality as follows:

> Extremism is more an issue of style than of content. . . . Most people can hold radical or unorthodox views and still entertain them in a more or less reasonable, rational, and nondogmatic manner. On the other hand, I have met people whose views are fairly close to the political mainstream but were presented in a shrill, uncompromising, bullying, and distinctly authoritarian manner.[2]

Extremism is a precursor to terrorism—it is an overarching belief system terrorists use to justify their violent behavior. It is characterized by what a person's beliefs are as well as how a person expresses his or her beliefs. Thus, no matter how offensive or reprehensible one's thoughts or words are, they are not by themselves acts of terrorism. Only those who violently act out their extremist beliefs are terrorists.

Common Characteristics of Violent Extremists

Scholars and other experts have identified common characteristics exhibited by violent extremists. These characteristics are expressed in different ways, depending on a movement's particular belief system. The following commonalities are summaries of traits these experts have identified, but are by no means an exhaustive inventory.[3]

Intolerance. Intolerance is the hallmark of extremist belief systems and terrorist behavior. The cause is considered to be absolutely just and good, and those who disagree with it (or some aspect of it) are cast as the opposition. Terrorists affix their opponents with certain negative or derisive labels to set them apart. These characterizations are often highly personalized, so that individuals are identified who symbolize the opposing belief system or cause. Thus, during the Cold War, the American president was referred to by the pro-U.S. camp as the leader of the free world and by Latin American Marxists as the embodiment of Yankee imperialism.[4]

Moral Absolutes. Extremists adopt moral absolutes, so that the distinction between good and evil is clear, as are the lines between the extremists and their opponents. The extremist's belief or cause is a morally correct vision of the world and is used to establish moral superiority over others. Violent extremists thus become morally and ethically pure elites who lead the oppressed masses to freedom. For example, religious terrorists generally believe that their one true faith is superior to all others and that any behavior committed in defense of the faith is fully justifiable.

Broad Conclusions. Extremist conclusions are made to simplify the goals of the cause and the nature of the opponents. These generalizations are not debatable and permit no

Photo 1.2 The IRA at work. A photograph of members of the Provisional Irish Republican Army as they prepare for a mission.

exceptions. Evidence for them is rooted in a belief system rather than based on objective data. For example, ethno-nationalists frequently categorize all members of their opponent group as having certain broadly negative traits.

New Language and Conspiratorial Beliefs. Language and conspiracies are created to demonize the enemy and set the terrorists apart from those not part of their belief system. Extremists thus become an elite with a hidden agenda and targets of that agenda. For example, some American far right conspiracy proponents express their anti-Semitic beliefs by using coded references to international bankers or a Zionist occupied government (ZOG). Neo-Nazi rightists degrade members of non-European races by referring to them as mud people.

The World of the Extremist

Extremists have a very different—and at times fantastic—worldview compared with nonextremists. They set themselves apart as protectors of a truth or as the true heirs of a legacy. They frequently believe that secret and quasi-mystical forces are arrayed against them and that these forces are the cause of worldwide calamities. One conspiracy theory widely believed among Islamic extremists in the aftermath of September 11, 2001, for example, was that Israeli agents were behind the attacks, that 4,000 Jews received telephone calls to evacuate the World Trade Center in New York, and thus that no Jews were among the victims of the attack.

As in the past, religion is often an underlying impetus for extremist activity. When extremists adopt a religious belief system, their worldview becomes one of a struggle between supernatural forces of good and evil. They view themselves as living a righteous life that fits with their interpretation of God's will. Those who do not conform to the belief system are opposed to the one true faith. Those who live according to it are chosen, and those who do not are not chosen. These interpretations of behavior include elements of the underlying social or political environment. For example, one student at a Pakistani religious school explained that "Osama [bin Laden] wants to keep Islam pure from the pollution of the infidels. . . . He believes Islam is the way for all the world. He wants to bring Islam to all the world."[5]

Extremists have a very clear sense of mission, purpose, and righteousness. They create a worldview that sets them apart from society. Extremist beliefs and terrorist behaviors are thus very logical to those who accept the belief system, but illogical to those who reject it. Chapter Perspective 1.1 illustrates the rigid intolerance of one belief system.

CHAPTER PERSPECTIVE 1.1

We the Klan Believe

We believe in the eternal separation of the church and state:

    Roman Catholicism teaches the union of church and state with the church controlling the state. . . .

    Every Roman Catholic holds allegiance to the Pope of Rome, and Catholicism teaches that this allegiance is superior to his allegiance to his country. . . .

    We believe in white supremacy:

The Klan believes that America is a white man's country, and should be governed by white men. Yet the Klan is not anti-Negro, it is the Negro's friend. The Klan is eternally opposed to the mixing of the white and colored races. Our creed: Let the white man remain white, the black man black, the yellow man yellow, the brown man brown, and the red man red. God drew the color line. . . .

The Klan believes in England for Englishmen, France for Frenchman, Italy for Italians, and America for Americans. . . . The Klan is not anti-Catholic, anti-Jew, anti-Negro, anti-foreign, the Klan is pro-Protestant and pro-American. . . .

We the Klan will never allow out [sic] blood bought liberties to be crucified on a Roman cross: and we will not yield to the integration of white and Negro races in our schools or any where else. . . .

SOURCE: Excerpts from "The Principle of the United Klans of America," (Tuscaloosa, AL: Office of the Imperial Wizard, 1974), in *Extremism in America: A Reader,* ed. Lyman Tower Sargent (New York: New York University Press, 1995), 139.

Formal and Informal Definitions

There is some consensus among experts—but no unanimity—on what kind of violence constitutes an act of terrorism. Governments, individual agencies within governments, and private agencies have each developed, adopted, and designed their own definitions, and academic experts have proposed and analyzed dozens of definitional constructs. This lack of unanimity, which exists throughout the public and private sectors, is an accepted reality in the study of political violence.

Terrorism would not, from a layperson's point of view, seem to be a difficult concept to define. Most people likely hold an instinctive understanding that terrorism is

- ◆ Politically motivated violence
- ◆ Usually directed against soft targets (i.e., civilian and administrative government targets)
- ◆ With an intention to affect (terrorize) a target audience

This instinctive understanding would also hold that terrorism is a criminal, unfair, or otherwise illegitimate use of force. Laypersons might presume that this is an easily understood concept, but defining terrorism is not that simple. Experts have for some time grappled with designing (and agreeing on) clear definitions of terrorism; the issue is, in fact, at the center of an ongoing debate. The result is a remarkable variety of approaches and definitions. Walter Laqueur

Photo 1.3 Ilich Ramirez Sanchez, also known as Carlos the Jackal. He was personally responsible for the deaths of scores of victims.

noted that "more than a hundred definitions have been offered," including several of his own.[6] Even within the U.S. government, different agencies apply several definitions.

A significant amount of intellectual energy has been devoted to identifying formal elements of terrorism, as illustrated by Alex Smid's surveys, which identified more than 100 definitions.[7] Establishing formal definitions can, of course, be complicated by the perspectives of the participants in a terrorist incident, who instinctively differentiate freedom fighters from terrorists, regardless of formal definitions. Another complication is that most definitions focus on political violence perpetrated by dissident groups, even though many governments have practiced terrorism as both domestic and foreign policy.

Guerrilla Warfare. One important observation must be kept in mind and understood at the outset: Terrorism is not synonymous with guerrilla warfare. The term *guerrilla* (little war) was developed during the early 19th century when Napoleon's army fought a long, brutal, and ultimately unsuccessful war in Spain. Unlike the Napoleonic campaigns elsewhere in Europe, which involved conventional armies fighting set-piece battles in accordance with rules of engagement, the war in Spain was a classic unconventional conflict. The Spanish people, as opposed to the Spanish army, rose in rebellion and resisted the invading French army. They liberated large areas of the Spanish countryside. After years of costly fighting—in which atrocities were common on both sides—the French were driven out. Thus, in contrast to *terrorists*, the term *guerrilla fighters* refers to

> a numerically larger group of armed individuals, who operate as a military unit, attack enemy military forces, and seize and hold territory (even if only ephemerally during the daylight hours), while also exercising some form of sovereignty or control over a defined geographical area and its population.[8]

Dozens, if not scores, of examples of guerrilla warfare exist in the modern era. They exhibit the classic strategy of hit-and-run warfare. Many examples also exist of successful guerrilla campaigns against numerically and technologically superior adversaries.

A Sampling of Formal Definitions

The effort to formally define terrorism is critical, because government antiterrorist policy calculations must be based on criteria that determine whether a violent incident is an act of terrorism. Governments and policy makers must piece together the elements of terrorist behavior and demarcate the factors that distinguish terrorism from other forms of conflict.

In Europe, countries that endured terrorist campaigns have written official definitions of terrorism. The British have defined terrorism as "the use or threat, for the purpose of advancing a political, religious or ideological cause, of action which involves serious violence against any person or property."[9] In Germany, terrorism has been described as an "enduringly conducted struggle for political goals, which are intended to be achieved by means of assaults on the life and property of other persons, especially by means of severe crimes."[10] The European Interior Ministers note that "terrorism is . . . the use, or the threatened use, by a cohesive group of persons of violence (short of warfare) to effect political aims."[11]

Scholars have also tried their hand at defining terrorism. For example, Ted Gurr has described it as "the use of unexpected violence to intimidate or coerce people in the pursuit of political or social objectives."[12] J. P. Gibbs described it as "illegal violence or threatened violence against human or nonhuman objects," so long as that violence meets additional criteria such as secretive features and unconventional warfare.[13] Bruce Hoffman wrote:

> We come to appreciate that terrorism is ineluctably political in aims and motives; violent—or, equally important, threatens violence; designed to have far-reaching psychological repercussions beyond the immediate victim or target; conducted by an organization with an identifiable chain of command or conspiratorial structure (whose members wear no uniform or identifying insignia); and perpetrated by a subnational group or non-state entity.
>
> We may therefore now attempt to define terrorism as the deliberate creation and exploitation of fear through violence or the threat of violence in the pursuit of change.[14]

To further illustrate the range of definitions, Whittaker notes the following descriptions by terrorism experts:[15]

> Contributes the illegitimate use of force to achieve a political objective when innocent people are targeted (Walter Laqueur).
> A strategy of violence designed to promote desired outcomes by instilling fear in the public at large (Walter Reich).
> The use or threatened use of force designed to bring about political change (Brian Jenkins).

From this discussion, we can identify the common features of most formal definitions:

◆ The use of illegal force   ◆ Political motives
◆ Subnational actors   ◆ Attacks against soft civilian and passive military targets
◆ Unconventional methods   ◆ Acts aimed at purposefully affecting an audience

The emphasis, then, is on terrorists adopting specific types of motives, methods, and targets. One fact readily apparent from these formal definitions is that they focus on terrorist groups rather than terrorist states. As will be made abundantly clear in Chapter 4, state terrorism has been responsible for many more deaths and much more suffering than terrorism originating in small bands of terrorists.

Defining Terrorism in the United States

The United States has not adopted a single definition of terrorism as a matter of government policy, instead relying on definitions developed from time to time by government agencies. These definitions reflect the United States' traditional law enforcement approach to distinguishing terrorism from more common criminal behavior. The following definitions are a sample of the official approach.

The U.S. Department of Defense defines terrorism as "the unlawful use of, or threatened use, of force or violence against individuals or property to coerce and intimidate governments or societies, often to achieve political, religious, or ideological objectives."[16] The U.S. Code defines terrorism as illegal violence that attempts to "intimidate or coerce a civilian population; . . . influence the policy of a government by intimidation or coercion; or . . . affect the conduct of a government by assassination or kidnapping."[17] The **Federal Bureau of**

Investigation (FBI) has defined terrorism as "the unlawful use of force or violence against persons or property to intimidate or coerce a Government, the civilian population, or any segment thereof, in furtherance of political or social objectives."[18] For the State Department, terrorism is "premeditated, politically motivated violence perpetrated against noncombatant targets by subnational groups or clandestine agents, usually intended to influence an audience."[19]

Using these definitions, the following common elements can be used to construct a composite American definitional model:

> Terrorism is a premeditated and unlawful act in which groups or agents of some principal engage in a threatened or actual use of force or violence against human or property targets. These groups or agents engage in this behavior intending the purposeful intimidation of governments or people to affect policy or behavior with an underlying political objective.

These elements indicate a fairly narrow and legalistic approach. When they are assigned to individual suspects, the suspects may be labeled and detained as terrorists. Readers, in evaluating the practical policy implications of this approach, should bear in mind that labeling and detaining suspects as terrorists is not without controversy. Some post–September 11 counterterrorist practices have prompted strong debate. For example, when enemy soldiers are taken prisoner, they are traditionally afforded legal protections as prisoners of war. This is well recognized under international law. In the war on terrorism, many suspected terrorists have been designated as enemy combatants and not afforded the same legal status as prisoners of war. Such practices have been hotly debated among proponents and opponents. Chapter Perspective 1.2 discusses the ongoing problem.

CHAPTER PERSPECTIVE 1.2

The Problem of Labeling the Enemy in the New Era of Terrorism

When formulating counterterrorist policies, policy makers are challenged by two problems: first, the problem of defining terrorism, and second, the problem of labeling individual suspects. Although defining terrorism can be an exercise in semantics, and is often shaped by subjective political or cultural biases, certain fundamental elements are objective. In comparison, official designations (labels) used to confer special status on captured suspects have become controversial.

After September 11, it became clear to experts and the public that official designations and labels of individual suspected terrorists is a central legal, political, and security issue. The question of a suspect's official status when he or she is taken prisoner is central. It determines whether certain recognized legal or political protections are or are not observed.

According to the protocols of the third Geneva Convention, prisoners who are designated as prisoners of war, and who are brought to trial, must be afforded the same legal rights in the same courts as soldiers from the country holding them prisoner. Thus, prisoners of war held by the United States would be brought to trial in standard military courts under the Uniform Code of Military Justice, and would have the same rights and protections (such as the right to appeal) as all soldiers.

Suspected terrorists have not been designated as prisoners of war. Official and unofficial designations such as enemy combatants, unlawful combatants, and battlefield detainees have been used by American authorities to differentiate them from prisoners of war. The rationale is that suspected terrorists are not soldiers fighting for a sovereign nation, and are therefore not eligible for prisoner of war status. When hundreds of prisoners were detained at facilities such as the American base in Guantanamo Bay, Cuba, the United States argued that persons designated as enemy combatants were not subject to the Geneva Conventions. Thus, such individuals could be held indefinitely, detained in secret, transferred at will, and sent to allied countries for more coercive interrogations. Under enemy combatant status, conditions of confinement in Guantanamo Bay included open-air cells with wooden roofs and chainlink walls. In theory, each case was to be reviewed by special military tribunals, and innocent prisoners would be reclassified as nonenemy combatants and released.

Civil liberties and human rights groups disagreed with the special status conferred by the labeling system on prisoners. They argued that basic legal and humanitarian protections should be granted to prisoners regardless of their designation.

In one interesting development, the U.S. Department of Defense conferred protected persons status on members of the Iranian Mujahideen-e Khalq Organization (MKO), who were under guard in Iraq by the American military. The MKO is a Marxist movement opposed to the postrevolution regime in Iran. The group was regularly listed on the U.S. Department of State's list of terrorist organizations, and was responsible for killings of Americans and others in terrorist attacks.

Types of Terrorism

The basic elements of terrorist environments are uncomplicated, and experts and commentators generally agree on the forms of terrorism found in modern political environments. For example, the following environments have been described by academic experts:

- ◆ Barkan and Snowden describe vigilante, insurgent, transnational, and state terrorism.[20]
- ◆ Hoffman discusses ethno-nationalist/separatist, international, religious, and state-sponsored terrorism.[21]
- ◆ While undertaking the task of defining the New Terrorism, Laqueur contextualizes far rightist, religious, state, exotic, and criminal terrorism.[22]
- ◆ Other experts evaluate **narco-terrorism**, toxic terrorism, and netwar.[23]

We will explore all of these environments in later chapters within the following contexts:

State Terrorism. Terrorism from above committed by governments against perceived enemies. State terrorism can be directed externally against adversaries in the international domain or internally against domestic enemies.

Dissident Terrorism. Terrorism from below committed by nonstate movements and groups against governments, ethno-national groups, religious groups, and other perceived enemies.

Religious Terrorism. Terrorism motivated by an absolute belief that an otherworldly power has sanctioned—and commanded—the application of terrorist violence for the greater glory of the faith. Religious terrorism is usually conducted in defense of what believers consider the one true faith.

International Terrorism. Terrorism that spills over onto the world's stage. Targets are selected because of their value as symbols of international interests, either within the home country or across state boundaries.

Terrorists or Freedom Fighters?

It should now be clear that defining terrorism can be an exercise in semantics and context, driven by one's perspective and worldview. Absent definitional guidelines, these perspectives would be merely personal opinion and the subject of academic debate.

Perspective is a central consideration in defining terrorism. Those who oppose an extremist group's violent behavior—and who might be its targets—would naturally consider them terrorists. On the other hand, those who are being championed by the group—and on whose behalf the terrorist war is being fought—often see them as liberation fighters, even when they do not necessarily agree with the methods of the group. "The problem is that there exists no precise or widely accepted definition of terrorism."[24] We will consider several perspectives that illustrate this problem.

Perspective 1: Four Quotations

The term *terrorism* has acquired a decidedly pejorative meaning in the modern era, so that few if any states or groups who espouse political violence ever refer to themselves as terrorists. Nevertheless, these same states and groups can be unabashedly extremist in their beliefs or violent in their behavior. They often invoke—and manipulate—images of a malevolent threat or unjust conditions to justify their actions. The question is whether these justifications are morally satisfactory (and thereby validate extremist violence), or whether terrorism is inherently wrong.

Evaluating the following aphorisms critically will help to address difficult moral questions:

- "One person's terrorist is another person's freedom fighter."
- "One man willing to throw away his life is enough to terrorize a thousand."
- "Extremism in defense of liberty is no vice."
- "It became necessary to destroy the town to save it."[25]

"One Person's Terrorist Is Another Person's Freedom Fighter"

Who made this statement is not known; it most likely originated in one form or another in the remote historical past. The concept it embodies is, very simply, perspective. It is a concept that will be applied throughout our examination of terrorist groups, movements, and individuals.

As will become abundantly clear, terrorists never consider themselves the bad guys in their struggle for what they would define as freedom. They might admit that they have been forced by a powerful and ruthless opponent to adopt terrorist methods, but they see themselves as freedom fighters. Benefactors of terrorists always live with clean hands because they present their clients as plucky freedom fighters. Likewise, nations that use the technology of war to attack known civilian targets justify their sacrifice as incidental to the greater good of the cause.

"One Man Willing to Throw Away His Life Is Enough to Terrorize a Thousand"

This concept originated with the 19th-century Chinese military philosopher **Wu Ch'i**, who wrote,

> Now suppose there is a desperate bandit lurking in the fields and one thousand men set out in pursuit of him. The reason all look for him as they would a wolf is that each one fears that he will arise and harm him. This is the reason one man willing to throw away his life is enough to terrorize a thousand.[26]

These sentences are the likely source for the better-known aphorism, **"kill one man, terrorize a thousand."** Its authorship is undetermined but has been attributed to the leader of the Chinese Revolution, **Mao Zedong**, and to the Chinese military philosopher **Sun Tzu**. Both Wu Ch'i and Sun Tzu are often discussed in conjunction with each other, but Sun Tzu may be a mythical figure. Sun Tzu's book *The Art of War* has become a classic study of warfare. Regardless of who originated these phrases, their simplicity explains the value of a motivated individual who is willing to sacrifice himself or herself when committing an act of violence. They suggest that the selfless application of lethal force—in combination with correct timing, surgical precision, and an unambiguous purpose—is an invaluable weapon of war. It is also an obvious tactic for small, motivated groups that are vastly outnumbered and outgunned by a more powerful adversary.

"Extremism in Defense of Liberty Is No Vice"

Senator Barry M. Goldwater of Arizona made this statement during his bid for the presidency in 1964. His campaign theme was staunchly conservative and anti-Communist. However, because of the nation's rivalry with the Soviet Union at the time, every major candidate was overtly anti-Communist. Goldwater simply tried to outdo incumbent President Lyndon Johnson, his main rival, on the issue.[27]

This aphorism represents an uncompromising belief in the absolute righteousness of a cause. It defines a clear belief in good versus evil and a belief that the end justifies the means. If one simply substitutes *any cause* for the word *liberty,* one can fully understand how the expression lends itself to legitimizing uncompromising devotion to the cause. Terrorists use this reasoning to justify their belief that they are defending their championed interest (be it ideological, racial, religious, or national) against all perceived enemies—who are, of course, evil. Hence, the practice of ethnic cleansing was begun by Serb militias during the 1991–1995 war in Bosnia to forcibly remove Muslims and Croats from villages and towns. This was done in the name of Bosnian Serb security and historical claims to land occupied by others.[28] Bosnian and Croat paramilitaries later practiced ethnic cleansing to create their own ethnically pure enclaves.

"It Became Necessary to Destroy the Town to Save It"

This quotation has been attributed to a statement by an American officer during the war in Vietnam. When asked why a village thought to be occupied by the enemy had been destroyed, he allegedly replied that American soldiers had destroyed the village to save it.[29] The symbolic logic behind this statement is seductive: If the worst thing that can happen to a village is for it to be occupied by an enemy, then destroying it is a good thing. The village has been denied to the enemy, and it has been saved from the horrors of enemy occupation. The symbolism of the village can be replaced by any number of symbolic values.

Terrorists use this kind of reasoning to justify hardships that they impose not only on a perceived enemy but also on their own championed group. For example, in Chapter 5, readers will be introduced to nihilist dissident terrorists, who are content to wage "revolution for revolution's sake." They have no concrete plan for what kind of society will be built upon the rubble of the old one—their goal is simply to destroy an inherently evil system. To them, anything is better than the existing order. A historical example of this reasoning on an enormous scale is found in the great war between two totalitarian and terrorist states—Germany and the Soviet Union—from July 1941 to May 1945. Both sides used scorched-earth tactics as a matter of policy when their armies retreated, destroying towns, crops, roadways, bridges, factories, and other infrastructure as a way to deny resources to the enemy.

Perspective 2: Participants in a Terrorist Environment

Typically, the participants in a terrorist environment include the following actors, each of whom may advance different interpretations of an incident:[30]

The Terrorist. Terrorists are the perpetrators of a politically violent incident. For them, the violent incident is a justifiable act of war against an oppressive opponent. "Insofar as terrorists seek to attract attention, they target the enemy public or uncommitted bystanders."[31] In their minds, this is a legitimate tactic, because in their view they are always freedom fighters, never terrorists.

The Supporter. Supporters of terrorists are patrons, in essence persons who provide a supportive environment or apparatus. Supporters will generally refer to the terrorist participants as freedom fighters. Even if supporters disagree with the use or with the application of force in an incident, they will often rationalize its as the unfortunate consequence of a just war.

The Victim. Victims of political violence, and of warfare, will rarely sympathize with the perpetrators, regardless of the underlying motive. From their perspective, the perpetrators are little better than terrorists.

The Target. Targets are usually symbolic. They represent some feature of the enemy and can be either property targets or human targets. Like the victims, human targets will rarely sympathize with the perpetrators.

The Onlooker. Onlookers are the broad audience to the terrorist incident. They can be directly affected at the scene or indirectly by mass media. They may either sympathize with the perpetrators, revile them, or remain neutral. Depending on the onlooker's worldview, he or she might actually applaud an incident or an environment. Television is particularly effective for broadening the scope of who is an onlooker. This was evident during the live broadcasts of the attacks on the World Trade Center on September 11, 2001. The Internet has also become a primary medium for broadening the audience of terrorist acts, such as beheadings and bombings.

The Analyst. The analyst is an interpreter of the terrorist incident. Analysts are important participants because they create perspectives, interpret incidents, and label other participants. Analysts can include political leaders, media experts, and academic experts. Very often, the analyst will simply define for the other participants who is—or is not—a terrorist.

The same event can be interpreted a number of ways, causing participants to adopt biased spins on that event. The following factors illustrate this problem:

- ◆ *Political associations* of participants can create a sense of identification with either the target group or the defended group. This identification can be either favorable or unfavorable, depending on the political association.
- ◆ *Emotional responses* of participants after a terrorist incident can range from horror to joy. This response can shape a participant's opinion of the incident or the extremist's cause.
- ◆ *Labeling* participants can create either a positive or negative impression of an incident or cause. Labeling can range from positive symbolism on behalf of the terrorists to dehumanization of enemy participants (including civilians).
- ◆ *Symbolism* plays an important role in the terrorist's selection of targets. The targets can be inanimate objects that symbolize a government's power or human victims who symbolize an enemy people. Other participants sometimes make value judgments on the incident based on the symbolism of the target, thus asking whether the selected target was legitimate or illegitimate.

Perspective 3: Terrorism or Freedom Fighting?

Members of politically violent organizations tend to adopt the language of liberation, national identity, religious fervor, and even democracy. For example, ethno-nationalist and religious organizations such as Hamas (Islamic Resistance Movement) in Israel, Liberation Tigers of Tamil Eelam (LTTE) in Sri Lanka, and the Provisional Irish Republican Army (the Provos) in the United Kingdom all declare that they are armies fighting on behalf of an oppressed people, and are viewed by their supporters as freedom fighters. Conversely, many Israelis, Sinhalese, and British consider them terrorists.

Governments have also adopted authoritarian measures to counter domestic threats from perceived subversives. Similarly, they rationalize their behavior as a proportional response to an immediate threat. Numerous cases of such rationalization exist, such as when the Chilean and Argentine armed forces seized power during the 1970s and engaged in widespread violent repression of dissidents. In Argentina, an estimated 30,000 people disappeared during the so-called **Dirty War** the military government waged from 1976 to 1983.

Thus, from the perspective of many violent groups and governments, extremist beliefs and terrorist methods are both logical and necessary, and rational and justifiable. They become mainstreamed within the context of the worldview and political environment, which, in the minds of the extremists, offer no alternative to using violence to achieve freedom or maintain order. Conversely, those who oppose the political violence reject the justifications and disavow the moral proportion of the perceived political environment.

The Political Violence Matrix

As discussed earlier, experts have identified and analyzed many terrorist environments. As readers will learn in the chapters that follow, these include state, dissident, religious, ideological, and international frameworks. One distinguishing feature within each model is the relationship between the quality of force used by the terrorists and the characteristics of the victim of the attack. Figure 1.1 depicts how this relationship often defines the type of conflict.

A Definitional Dilemma: Combatants, Noncombatants, and the Use of Force

Definitional and ethical issues are not always clearly drawn when one uses terms such as combatant, noncombatant, discriminate force, or indiscriminate force.

Combatants and Noncombatants

The term *combatants* certainly refers to conventional or unconventional adversaries who engage in armed conflict as members of regular military or irregular guerrilla fighting units. The term *noncombatants* obviously includes civilians who have no connection to military or other security forces. There are, however, circumstances in which these lines become blurred. For example, in times of social unrest, civilians can become combatants. This has occurred

---

Figure 1.1   The Political Violence Matrix

When force (whether conventional or unconventional) is used against combatants, it occurs in a warfare environment. When force is used against noncombatants or passive military targets, it often characterizes a terrorist environment. Violent environments can be broadly summarized as follows:

- **Total war.** Force is indiscriminately applied to military targets of an enemy combatant to destroy them.
- **Total war/unrestricted terrorism.** Indiscriminate force is applied against noncombatants without restraint, by either a government or dissidents.
- **Limited war.** Discriminating force is used against a combatant, either to defeat the enemy or to achieve a more limited political goal.
- **State repression/restricted terrorism.** Discriminating force is directed against noncombatants either as a matter of domestic policy or as the selective use of terrorism by dissidents.

The figure summarizes factors to consider when evaluating the application of force against certain types of targets.

Indiscriminate force, combatant	• Total war (WWII Eastern Front)	• Limited war (Korean War)	Discriminate force, combatant
Indiscriminate force, noncombatant	• Total war (WWII bombing of cities) • Unrestricted terrorism (Rwandan genocide)	• State repression (Argentine "Dirty War") • Restricted terrorism (Italian Red Brigades)	Discriminate force, noncombatant

Photo 1.4   A mother and child lie dead after a cleansing sweep by paramilitaries in Bosnia.

repeatedly in societies where communal violence (e.g., civil war) breaks out between members of ethno-national, ideological, or religious groups. Similarly, noncombatants can include off-duty members of the military in nonwarfare environments.[32] They become targets because of their symbolic status.

Indiscriminate and Discriminate Force

Indiscriminate force is applied against a target without attempting to limit the level of force or the degree of destruction of the target. *Discriminate force* is a surgical use of limited force. Indiscriminate force is considered acceptable when used against combatants in a warfare environment, but regularly condemned when used in any nonwarfare environment, regardless of the characteristics of the victim.[33] There are, however, many circumstances in which adversaries define *warfare environment* differently. When weaker adversaries resort to unconventional methods (including terrorism), they justify them by defining them as necessary during a self-defined state of war. Discriminate force is considered moral when it is applied against specific targets with the intention to limit so-called **collateral damage**, or unintended destruction and casualties.

Chapter Summary

This chapter presented readers with the nature of terrorism and probed the definitional debates about the elements of these behaviors. Several fundamental concepts were identified that continue to influence the motives and behaviors of those who support or engage in political violence.

It is important to understand the elements that help define terrorism. Common characteristics of the extremist beliefs that underlie terrorist behavior include intolerance, moral absolutes, broad conclusions, and a new language that supports a particular belief system. Literally scores of definitions of terrorism have been offered by laypersons, academics, and

Photo 1.5   The war on terrorism. A U.S. Army soldier on patrol in Iraq.

policy professionals to describe the elements of terrorist violence. Many of these are value laden and can depend on one's perspective as an actor in a terrorist environment.

The role of perspective is significant in the definitional debate. Terrorists always declare that they are fighters who represent the interests of an oppressed group. They consider themselves freedom fighters and justify their violence as a proportional response to the object of their oppression. Their supporters will often mainstream the motives of those who violently champion their cause. In addition, the underlying principles of long-standing ideologies and philosophies continue to provide justifications for the support and use of political violence.

In the United States, official definitions have been adopted as a matter of policy. No single definition has been applied across all government agencies, but there is some commonality among their approaches. Commonalities include premeditation, unlawfulness, groups or agents, force or violence, human or property targets, intimidation, and a political objective.

DISCUSSION BOX

This chapter's Discussion Box is intended to stimulate critical debate about the role of perspective in labeling those who practice extremist behavior as freedom fighters or terrorists.

Cold War Revolutionaries

The Cold War between the United States and the Soviet Union lasted from the late 1940s until the fall of the Berlin Wall in 1989. During the roughly 40 years of rivalry, the two superpowers never entered into direct military conflict—at least conventionally.

(Continued)

Rather, they supported insurgent and government allies in the developing world (commonly referred to as the Third World),[a] who often entered into armed conflict. These conflicts could be ideological or communal in nature. Conflicts were often proxy wars, in which the Soviets or Americans sponsored rival insurgent groups (such as in Angola), or wars of national liberation (such as in Vietnam).

The following examples illustrate how Cuba became an important front in the Cold War between the United States and the Soviet Union.

The Cuban Revolution

The American influence in Cuba had been very strong since it granted the country independence in 1902 after defeating the Spanish in the Spanish-American War of 1898. The United States supported a succession of corrupt and repressive governments, the last of which was that of Fulgencio Batista. Batista's government was overthrown in 1959 by a guerrilla army led by Fidel Castro and **Ernesto "Che" Guevara**, an Argentine trained as a physician. Castro's insurgency had begun rather unremarkably, with significant defeats at the Moncada barracks in 1953 and a landing on the southeast coast of Cuba from Mexico in 1956 (when only 15 rebels survived to seek refuge in the Sierra Maestra mountains).

It was Batista's brutal reprisals against urban civilians that eventually drove many Cubans to support Castro's movement. When Batista's army was defeated and demoralized in a rural offensive against the rebels, Castro, his brother Raul, Guevara, and Camilo Cienfuegos launched a multifront campaign that ended in victory when their units converged in the capital of Havana in January 1959. The revolution had not been a Communist revolution, and the new Cuban government was not initially a Communist government. By early 1960, however, Cuba began to receive strong economic and military support from the Soviet Union. Castro and his followers soon declared the revolution to be a Communist one, and the Soviet-American Cold War opened a new and volatile front. American attempts to subvert Castro's regime included the Bay of Pigs invasion in April 1961 and several assassination attempts against Castro.[b] The Soviets and Americans came close to war during the Cuban Missile Crisis in October 1962.

Cubans in Africa

In the postwar era, dozens of anticolonial and communal insurgencies arose in Africa. During the 1970s, Africa became a central focus of the rivalry between Soviet- and Western-supported groups and governments. Thousands of Cuban soldiers were sent to several African countries on a mission that Fidel Castro justified as their internationalist duty. For example, in the 1970s, Cuba sent 20,000 soldiers to Angola, 17,000 to Ethiopia, 500 to Mozambique, 250 to Guinea-Bissau, 250 to Equatorial Guinea, and 125 to Libya.[c]

Angola

Portugal was the colonial ruler of this southern Africa country for more than 500 years. Beginning in 1961, guerillas began conducting raids in northern Angola, committing brutal atrocities that few can argue were not acts of terrorism. Three guerrilla movements eventually drove the Portuguese from Angola and declared independence in November 1975. These were the Front for the Liberation of Angola (FNLA), the National Union for the Total Independence of Angola (UNITA), and the Movement for the Liberation of Angola (MPLA).

In the civil war that broke out after the Portuguese withdrew, the United States and China supported the FNLA, the Soviets and Cubans supported the MPLA, and the United States and South Africa supported UNITA. The MPLA became the de facto government of Angola. Cuban soldiers were sent to support the MPLA government, the United States and South Africa sent aid to UNITA, and South African and British mercenaries fought with UNITA. The FNLA never achieved much success in the field. Direct foreign support was withdrawn as the Cold War and South African apartheid ended, although the conflict continued through the 1990s. The MPLA finally forced UNITA to end its insurgency when UNITA leader Jonas Savimbi was killed in February 2002.

Nicaragua

U.S. influence and intervention in Nicaragua were common during most of the 20th century. Its governments had been supported by the United States, and its National Guard (the Guardia) had been trained by the United States. These pro-American Nicaraguan governments had a long history of corruption and violent repression. Cuban-oriented Marxist guerrillas, the Sandinista National Liberation Front, overthrew the government of Anistasio Samoza in 1979 with Cuban and Soviet assistance.

During much of the next decade, the United States armed, trained, and supported anti-Sandinista guerrillas known as the **Contras** (counterrevolutionaries). This support included clandestine military shipments managed by the **U.S. Central Intelligence Agency (CIA)**, the mining of Managua harbor, and an illegal arms-shipment program managed by Marine Lieutenant Colonel Oliver North.

Discussion Questions

1. Che Guevara is revered by many on the left as a principled revolutionary. He believed that a revolutionary spark was needed to create revolution throughout Latin America. Guevara was killed in Bolivia trying to prove his theory. Was Che Guevara an internationalist freedom fighter?

2. The United States used sabotage to destabilize Cuba's economy and government and plotted to assassinate Fidel Castro. Did the United States engage in state-sponsored terrorism? Compare this to Soviet support of their allies. Is there a difference?

3. The Soviet Union sponsored the Cuban troop presence in Africa during the 1970s. The wars in Angola, Ethiopia/Somalia, and Mozambique were particularly bloody. Did the Soviet Union engage in state-sponsored terrorism? Compare this to U.S. support of their allies. Is there a difference?

4. During the Soviet-United States rivalry in Angola, Jonas Savimbi commanded the pro-Western UNITA army. His U.S. patrons considered him a freedom fighter. Savimbi never overthrew the MPLA government. Promising efforts to share power after an election in 1992 ended in the resumption of the war when Savimbi refused to acknowledge his electoral defeat, and a 1994

cease-fire collapsed. From the U.S. perspective, has Jonas Savimbi's status as a freedom fighter changed? If so, when and how?

5. The Sandinistas overthrew a violent and corrupt government. The Contras were presented by the Reagan administration as an army of freedom fighters fighting against a totalitarian Communist government. Contra atrocities against civilians were documented. Were the Contras freedom fighters? How do their documented atrocities affect your opinion?

Notes

a. At the time, the First World was defined as the developed Western democracies, the Second World was the Soviet bloc, and the Third World was the developing world, composed of newly emerging postcolonial nations.

b. At least one plot allegedly proposed using an exploding cigar.

c. See R. W. Cross, ed., *20th Century* (London: Purnell Reference Books, 1979), 2365; Cross, "The OAU and the New Scramble for Africa," in *20th Century,* 2372–2373.

Key Terms and Concepts

The following topics were discussed in this chapter and can be found in the glossary.

Dirty War

Extremism

"Extremism in defense of liberty is no vice."

Freedom fighter

Hamas (Islamic Resistance Movement)

"It became necessary to destroy the town to save it."

"Kill one man, terrorize a thousand."

Liberation Tigers of Tamil Eelam (LTTE)

Mao Zedong

"One man willing to throw away his life is enough to terrorize a thousand."

"One person's terrorist is another person's freedom fighter."

Provisional Irish Republican Army (Provos)

Soft targets

Sun Tzu

Terrorism

Wu Ch'i

Terrorism on the Web

Log on to the Web-based student study site at **www.sagepub.com/martinessstudy** for additional Web sources and study resources.

Web Exercise

Using this chapter's recommended Web sites, conduct an online investigation of the fundamental characteristics of extremism.

1. What commonalities can you find in the statements of these groups?
2. Is there anything that strikes you as being particularly extremist?
3. Why or why not?

For an online search of different approaches to defining extremism and terrorism, readers should activate the search engine on their Web browser and enter the following keywords:

"Definitions of terrorism"
"Extremism"

Recommended Readings

The following publications provide discussions for defining terrorism and terrorism's underlying extremist motivations.

Hamm, Mark S., ed. *Hate Crime: International Perspectives on Causes and Control.* Highland Heights, KY: Academy of Criminal Justice Sciences and Cincinnati, OH: Anderson, 1994.

Howard, Lawrence, ed. *Terrorism: Roots, Impact, Responses.* New York: Praeger, 1992.

Laqueur, Walter. *The New Terrorism: Fanaticism and the Arms of Mass Destruction.* New York: Oxford University Press, 1999.

Lawrence, Frederick M. *Punishing Hate: Bias Crimes Under American Law.* Cambridge, MA: Harvard University Press, 1999.

Sederberg, Peter C. *Terrorist Myths: Illusion, Rhetoric, and Reality.* Englewood Cliffs, NJ: Prentice Hall, 1989.

2 Historical Perspectives and Ideological Origins

Terrorism has been a dark feature of human behavior since history was first recorded. Great leaders have been assassinated, groups and individuals have committed acts of incredible violence, and entire cities and nations have been put to the sword—all in the name of defending a greater good.

The modern era of terrorism is primarily, though not exclusively, a conflict between adversaries waging, on one side, a self-described war on terrorism and, on the other, a self-described holy war in defense of their religion. It is an active confrontation, evidenced by the fact that the incidence of significant terrorist attacks often spikes to serious levels. Although such trends are disturbing, it is critically important for one to keep these facts in perspective, because the modern terrorist environment is not unique in history.

It will become clear in the following pages that the history of terrorist behavior extends into antiquity and that themes and concepts recur. State terrorism, dissident terrorism, and other types of political violence are common to all periods of civilization. It will also become clear to readers that certain justifications—rooted in basic ideological beliefs—have been used to rationalize terrorist violence throughout history. The following themes are introduced here:

♦ Historical perspectives on terrorism
♦ Ideological origins of terrorism
♦ September 11, 2001 and the new era of terrorism

Historical Perspectives on Terrorism

It is perhaps natural for each generation to view history narrowly, from within its own political context. Contemporary commentators and laypersons tend to interpret modern events as though they have no historical precedent. However, terrorism is by no means a modern phenomenon, and has in fact a long history. Nor does terrorism arise from a political vacuum.

Antiquity

In the ancient world, cases and stories of state repression and political violence were common. Several ancient writers championed **tyrannicide** (the killing of tyrants) as necessary for the greater good of the citizenry and to delight the gods. Some assassins were honored by the public. For example, after Aristogeiton and Harmodius were executed for assassinating the typrant Hipparchus, statues were erected to honor them.[1] Conquerors often set harsh examples by exterminating entire populations or forcing the conquered into exile. An example of this practice is the Babylonian Exile, which followed the conquest of the kingdom of Judea. Babylon's victory resulted in the forced removal of the Judean population to Babylon in 598 and 587 B.C.E. Those in authority also repressed the expression of ideas from individuals they deemed dangerous, sometimes violently. In ancient Greece, Athenian authorities sentenced the great philosopher Socrates to death in 399 B.C.E. for allegedly corrupting the city-state's youth and meddling in religious affairs. He drank hemlock and died among his students and followers.

The Roman Age

During the time of the Roman Empire, the political world was rife with many violent demonstrations of power, which were arguably examples of what we would now term state terrorism. These include the brutal suppression of Spartacus's followers after the Servile War of 73–71 B.C.E., after which the Romans crucified surviving rebels along the Appian Way. **Crucifixion** was a common form of public execution: The condemned were affixed to a cross or other wooden frame, either tied or nailed through the wrist or hand, and later died by suffocation as their bodies sagged.

Warfare was waged in an equally hard manner, such as the final conquest of the North African city-state of Carthage in 146 B.C.E. The city was reportedly allowed to burn for 10 days, the rubble was cursed, and salt was symbolically plowed into the soil to signify that Carthage would forever remain desolate. During another successful campaign in 106 C.E., the Dacian nation (modern Romania) was eliminated, its population was enslaved, and many Dacians perished in gladiatorial games. In other conquered territories, conquest was often accompanied by similar demonstrations of terror, always with the intent to demonstrate that Roman rule would be wielded without mercy against those who did not submit to the authority of the empire.

Regicide (the killing of kings) was also fairly common during the Roman period. Perhaps the best-known was the assassination of Julius Caesar in 44 B.C.E. by rivals in the Senate. Other Roman emperors also met violent fates: Caligula and Galba were killed by the Praetorian Guard in 41 and 68 C.E., respectively; Domitian was stabbed to death in 96 C.E.; a paid gladiator murdered Commodus in 193 C.E.; and Caracalla, Elagabalus, and other emperors either were assassinated or died suspiciously.[2]

The Ancient and Medieval Middle East

Cases exist of movements in the ancient and medieval Middle East that used what modern analysts would consider to be terrorist tactics. For example, in *History of the Jewish War*—a seven-volume account of the first Jewish rebellion against Roman occupation (66–73 C.E.)—the historian Flavius Josephus describes how one faction of the rebels, the **sicarii** (who took their name from their preferred weapon, the *sica*, a short curved dagger), attacked both Romans and members of the Jewish establishment.[3] They were masters of guerrilla warfare and the destruction of symbolic property and belonged to a group known as the **Zealots** (from the Greek *zelos,* meaning ardor or strong spirit), who opposed the Roman occupation of Palestine. The modern term *zealot,* used to describe uncompromising devotion to radical change, is derived from the name of this group. Assassination was a commonly used tactic. Some *sicarii* zealots were present at the siege of Masada, a hilltop fortress that held out against the Romans for 3 years before the defenders committed suicide in 74 C.E. rather than surrender.

The French Revolution: Prelude to Modern Terrorism

During the French Revolution, the word *terrorism* was coined in its modern context by British statesman and philosopher **Edmund Burke**. He used the word to describe the régime de la terreur, commonly known in English as the **Reign of Terror** (June 1793 to July 1794).[4] The Reign of Terror, led by the radical Jacobin-dominated government, is a good example of state terrorism carried out to further the goals of a revolutionary ideology.[5] During the Terror, thousands of opponents to the Jacobin dictatorship—and others merely perceived as enemies of the new revolutionary Republic—were arrested and put on trial before the **Revolutionary Tribunal**. Those found to be enemies of the Republic were beheaded by a new instrument of execution—the guillotine. With the capability to execute victims one after the other in assembly-line fashion, it was regarded by Jacobins and other revolutionaries of the time as an enlightened and civilized tool of revolutionary justice.[6]

The ferocity of the Reign of Terror is reflected in the number of victims. Between 17,000 and 40,000 persons were executed, and perhaps 200,000 political prisoners died in prisons from disease and starvation.[7] Two incidents illustrate the communal nature of this violence. In Lyon, 700 people were massacred by cannon fire in the town square. In Nantes, thousands were drowned in the Loire River when the boats they were detained in were sunk.[8]

The Revolutionary Tribunal, a symbol of revolutionary justice and state terrorism, has its modern counterparts in 20th-century social upheavals. Recent examples include the **struggle meetings** of revolutionary China (public criticism sessions, involving public humiliation and confession) and revolutionary Iran's komitehs (ad hoc people's committees).[9]

CHAPTER PERSPECTIVE 2.1

The Gunpowder Plot of Guy Fawkes

The reign of James I, king of England from 1603 to 1625, took place in the aftermath of a religious upheaval. During the previous century, King Henry VIII (1509–1547) wrested from Parliament the authority to proclaim himself the head of the church in England. Henry had requested permission from Pope Clement VII to annul his marriage to Catherine of Aragon when she failed to give birth to a male heir to the throne. His intention was to then marry Anne Boleyn. When the Pope refused his request, Henry proclaimed the Church of England and separated it from Papal authority. The English crown confiscated Catholic Church property and shut down Catholic monasteries. English Catholics who failed to swear allegiance to the crown as supreme head of the church were repressed by Henry, and later by Queen Elizabeth I (1558–1603).

When James I was proclaimed king, Guy Fawkes and other conspirators plotted to assassinate him. They meticulously smuggled gunpowder into the Palace of Westminster, intending to blow it up along with King James and any other officials in attendance on the opening day of Parliament. Unfortunately for Fawkes, one of his fellow plotters attempted to send a note to warn his brother in law to stay away from Westminster on the appointed day. The note was intercepted, and Fawkes was captured on November 5, 1605 while guarding the store of gunpowder.

Guy Fawkes suffered the English penalty of the day for treason. He was dragged through the streets, hanged until nearly dead, his bowels were drawn from him, and he was cut into quarters—an infamous process known as hanging, drawing, and quartering. Fawkes had known that this would be his fate, and thus when the noose was placed around his neck took a running leap, hoping to break his neck. Unfortunately, the rope broke and the executioner proceeded with the full ordeal.

Nineteenth-Century Europe: Two Examples From the Left

Modern left-wing terrorism is not a product of the 20th century. Its ideological ancestry dates to the 19th century, when anarchist and communist philosophers began to advocate the destruction of capitalist and imperial society—what Karl Marx referred to as the "spectre . . . haunting Europe."[10] Some revolutionaries readily encouraged the use of terrorism in the new cause. One theorist, **Karl Heinzen** in Germany, anticipated the late-20th-century fear that terrorists might obtain weapons of mass destruction when he supported the acquisition of new weapons technologies to destroy the enemies of the people. According to Heinzen, these weapons should include poison gas and new high-yield explosives.[11]

During the 19th century, several terrorist movements championed the rights of the lower classes. These movements were prototypes for 20th-century groups and grew out of social and political environments unique to their countries. To illustrate this point, two examples are drawn from early industrial England and semifeudal Russia of the late 19th century.

The **Luddites** were English workers in the early 1800s who objected to the social and economic transformations of the Industrial Revolution. Their principal objection was that industrialization threatened their jobs, and thus they targeted the machinery of the new textile factories. They attacked, for example, stocking looms that mass-produced stockings at the expense of skilled stocking weavers who made them by hand.

A mythical figure, Ned Ludd, was the supposed founder of the Luddite movement. The movement was active from 1811 to 1816 and was responsible for sabotaging and destroying wool and cotton mills and weaving machinery. The British government eventually suppressed the movement by passing anti-Luddite laws, including establishing the crime of "machine breaking," which was punishable by death. After 17 Luddites were executed in 1813, the movement gradually died out. Modern antitechnology activists and terrorists, such as the Unabomber, **Theodore "Ted" Kaczynski**, in the United States, are sometimes referred to as neo-Luddites.

People's Will (Narodnaya Volya) in Russia was a direct outgrowth of student dissatisfaction with the czarist regime in the late 19th century. Many young Russian university students, some of whom had studied abroad, became imbued with the ideals of anarchism and Marxism. Many became radical reformists who championed the rights of the people, particularly the peasant class. A populist revolutionary society, Land and Liberty (Zemlya Volya), was founded in 1876 with the goal of fomenting a mass peasant uprising by settling radical students among them to raise their class consciousness. After a series of arrests and mass public trials, Land and Liberty split into two factions in 1879. One faction, Black Repartition, kept to the goal of a peasant revolution. The other, People's Will, fashioned itself into a conspiratorial terrorist organization.

People's Will members believed that they understood the underlying problems of Russia better than the uneducated masses of people did, and concluded that they were therefore better able to force government change. This was, in fact, one of the first examples of a revolutionary **vanguard strategy**. They believed that they could both demoralize the czarist government and expose its weaknesses to the peasantry. People's Will quickly embarked on a terrorist campaign against carefully selected targets. Incidents of terror committed by People's Will members—and other revolutionaries who emulated them—included shootings, knifings, and bombings against government officials. In one successful attack, Czar Alexander II was assassinated by a terrorist bomb on March 1, 1881. The immediate outcome of the terrorist campaign was the installation of a repressive police state in Russia that, though not as efficient as later police states would be in the Soviet Union or Nazi Germany, succeeded in harassing and imprisoning most members of People's Will.

The Modern Era

It is useful in developing a critical understanding of modern extremist behavior to understand that the growing threat of the New Terrorism adds a unique dimension to the terrorist environment of the 21st century. This is because "the new terrorism is different in character, aiming not at clearly defined political demands but at the destruction of society and the elimination of large sections of the population."[12]

The new breed of terrorists "would feel no compunction over killing hundreds of thousands if they had the means to do so."[13] In addition, the emerging environment is characterized by a horizontal organizational arrangement wherein independent cells operate autonomously without reporting to a hierarchical (vertical) command structure. Many of these new terrorists are motivated by religious or nationalist precepts that may not fit easily into the classical continuum. The attacks in September 2001 in the United States, March 2004 in Spain, and July 2005 in Great Britain and Egypt are examples of this new environment.

Ideological Origins of Terrorism

Ideologies are systems of belief derived from theories that explain human social and political conditions. Literally scores of belief systems have led to acts of terrorist violence. Because there are so many, it is difficult to classify them with precision. Nevertheless, a **classical ideological continuum** rooted in the politics of the French Revolution has endured to the present time.[14] This is instructive for our discussion of politically motivated violence, because the concepts embodied in the continuum continue and will continue to be relevant.

The Classical Ideological Continuum: The Case of the French Revolution

At the beginning of the French Revolution in 1789, a parliament-like assembly was convened to represent the interests of the French social classes. Although its name changed during the revolution—from Estates-General, to National Constituent Assembly, to Legislative Assembly—the basic ideological divisions were symbolically demonstrated by where representatives sat during assembly sessions. On the left side of the assembly sat those who favored radical change, some advocating a complete reordering of French society and culture. On the right side of the assembly sat those who favored either the old order or slow and deliberate change. In the center of the assembly sat those who favored either moderate change or simply could not make up their minds to commit to either the left or right. These symbolic designations—**left, center**, and **right**—have become part of our modern political culture. Table 2.1 summarizes the progression of these designations from their origin during the French Revolution.

It is readily apparent from the French Revolution that the quality of the classical continuum depended very much on the political environment of each society. For example, within American culture, mainstream values include free enterprise, freedom of speech, and limited government.[15] Depending on where one falls on the continuum, the interpretation can be very different. Thus the continuum summarizes the conventional political environment of the modern era. Many nationalist or religious terrorists, however, do not fit easily. For example,

> to argue that the Algerian terrorists, the Palestinian groups, or the Tamil Tigers are "left" or "right" means affixing a label that simply does not fit. . . . The Third World groups . . . have subscribed to different ideological tenets at different periods.[16]

TABLE 2.1   THE CLASSICAL IDEOLOGICAL CONTINUUM: THE CASE OF THE FRENCH REVOLUTION

After the dissolution of the monarchy, the victorious revolutionaries began to completely restructure French society. Perhaps the most important priority was to create a new elective constituent assembly to represent the interests of the people. The configuration of this new assembly changed repeatedly as the revolution progressed from one ideological phase to the next.

Legislative Body	Political and Ideological Orientation		
	Left	**Center**	**Right**
National Constituent Assembly, 1789–1791	Patriots (republicans)	Moderates (constitutional monarchists)	Blacks (reactionaries)
Legislative Assembly, 1791–1792	Mountain (republicans)	Plain (near-republicans)	Constitutionalists (constitutional monarchists)
National Convention, 1792	Mountain (radicals)	Marsh (uncommitted bourgeois)	Girondins (bourgeois republicans)

Nevertheless, the continuum is still useful for categorizing terrorist behaviors and extremist beliefs. Table 2.2 compares the championed groups, methodologies, and desired outcomes of typical political environments.

An Ideological Analysis: From the Extreme Left to the Extreme Right

Fringe left ideology is usually an extreme interpretation of Marxist ideology, using theories of class warfare or ethno-national liberation to justify political violence. At the leftist fringe, violence is seen as a perfectly legitimate option, because the group considers itself at war with an oppressive system, class, or government. The key justification is the notion of the group as a righteous champion of the poor and downtrodden.

This type of ideological movement frequently concerns itself only with destroying an existing order in the name of the championed class or national group, not with building the new society in the aftermath of the revolution. For example, **Gudrun Ensslin**, a leader of the terrorist Red Army Faction in West Germany, stated: "As for the state of the future, the time after victory, that is not our concern. . . . We build the revolution, not the socialist model."[17]

Far left ideology frequently applies Marxist theory to promote class or ethno-national rights. It is best characterized as a radical worldview, because political declarations often direct public attention against perceived forces of exploitation or repression. Far left groups do not necessarily engage in political violence and often fully participate in democratic processes. In Western Europe, for example, Communist parties and their affiliated Communist labor unions have historically been overt in agitating for reform through democratic processes.[18] It is important to note that this environment of relatively peaceful coexistence occurs only in societies where dissent is tolerated. In countries with weaker democratic traditions, far left dissent has erupted in violence and been met by extreme repression. Latin America has many examples of this kind of environment.

Far right ideology is characterized by strong adherence to social order and traditional values. A chauvinistic racial or ethnic dimension is often present, as is an undercurrent of religion or mysticism (the latter is especially prevalent in the United States). Like the far left, far right groups do not necessarily engage in political violence and have fully participated in democratic processes. Organized political expression is often overt. For example, right-wing political parties in many European countries are a common feature of national politics. Their success has been mixed, and their influence varies in different countries. In Spain, Greece, and Great Britain, they have little popular support.[19] However, those in Austria, Belgium, France, and Italy have enjoyed significant popular support in the recent past.

Not all far right political movements are the same, and a comparison of the American and European contexts is instructive. In Europe, some rightist parties are nostalgic and neofascist, such as the German People's Union. Others are more populist, such as the National Front in France.[20] In the United States, the far right is characterized by activism among local grassroots organizations and has no viable political party. Some American groups have a religious orientation, others are racial, others embody a politically paranoid survivalist lifestyle, and some incorporate all three tendencies.

Fringe right ideology is usually rooted in an uncompromising belief in ethno-national or religious superiority, and terrorist violence is justified as a protection of the purity and superiority of the group. Terrorists on the fringe right picture themselves as champions of an ideal

TABLE 2.2 THE CLASSICAL IDEOLOGICAL CONTINUUM: MODERN POLITICAL ENVIRONMENTS

Activism on the left, right, and center can be distinguished by a number of characteristics. A comparison of these is instructive. The representation here compares their championed groups, methodologies, and desired outcomes.

	Left Fringe	Far Left	Liberalism	Moderate Center	Conservatism	Far Right	Fringe Right
Championed groups	Class/ nationality	Class/ nationality	Demographic groups	General society	General society	Race, ethnicity, nationality, religion	Race, ethnicity, nationality, religion
Methodology/ process	Liberation movement	Political agitation	Partisan democratic processes	Consensus	Partisan democratic processes	Political agitation	Order movement
Desired outcome	Radical change	Radical change	Incremental reform	Status quo slow change	Traditional values	Reactionary change	Reactionary change

order that has been usurped, or attacked, by inferior interests or unwanted religious values. Violence is an acceptable option against those who are not members of the group, because they are considered obstacles to the group's natural assumption of power. Like their counterparts on the fringe left, right-wing terrorists often have only a vague notion of the characteristics of the new order after the revolution. They are concerned only with asserting their value system and, if necessary, destroying the existing order. Significantly, rightist terrorists have been more likely than their leftist counterparts to engage in indiscriminate bombings and other attacks that result in higher numbers of victims.

Table 2.3 applies this discussion to the American context.

Ideology in Practice: From Anarchism to Fascism

Anarchism

Anarchism is a leftist philosophy that was an ideological by-product of the social upheavals of mid-19th-century Europe, a time when civil unrest and class conflict swept the continent, and culminated in the revolutions of 1848. Anarchists were among the first antiestablishment radicals who championed what they considered the downtrodden peasant and working classes. They abhorred central government control and private property. Frenchman **Pierre-Joseph Proudhon**, who published a number of articles and books on the virtues of anarchism, coined an enduring slogan among anarchists—**"Property is theft!"**

The radical undercurrent of anarchist thought began with that proposition. **Mikhail Bakunin** and his philosophical associates **Sergei Nechayev** and **Petr Kropotkin**, all Russians, were the founders of modern anarchism. They supported destruction of the state, radical decentralization of power, atheism, and individualism. They also opposed capitalism and Karl Marx's revolutionary doctrine of building a socialist state.

Early anarchists never offered a concrete plan for replacing centralized state authority because they had no clearly defined vision of postrevolutionary society. They considered the destruction of the state their contribution to the future.

Photo 2.1 A portrait of Pierre-Joseph Proudhon, the ideological father of anarchism. His slogan "Property is theft!" became a rallying cry for anarchists during the 19th and 20th centuries.

They advocated achieving propaganda victories by violently pursuing the revolution, which became known as **propaganda by the deed**. Terrorism was advocated as the principal way to destroy state authority. Interestingly, anarchists argued that terrorists should organize themselves into small groups, or cells, a tactic that modern terrorists have adopted. Anarchists actively practiced propaganda by the deed, as evidenced by the many acts of violence against prominently symbolic targets. In Russia, People's Will conducted a terrorist campaign from 1878 to 1881, and other anarchist terrorist cells operated in Western Europe. Around the turn of the 20th century, anarchists assassinated Russian Czar Alexander II, Austro-Hungarian Empress Elizabeth, Italian King Umberto I, and French President Carnot. An alleged anarchist, Leon Czolgosz, assassinated President William McKinley in the United States.

TABLE 2.3 THE CLASSICAL IDEOLOGICAL CONTINUUM: THE CASE OF THE UNITED STATES

The United States is a good illustration of the classical ideological continuum. Its political environment has produced organizations that represent the ideologies included in the continuum.

	Left Fringe	Far Left	Liberalism	Moderate Center	Conservatism	Far Right	Fringe Right
Economic/class agenda	May 19 Communist Organization	Communist Party, USA	American Federation of State, County and Municipal Employees	American Federation of Labor and Congress of Industrial Organizations	Teamsters Union	Lyndon Larouche groups	Posse Comitatus
Activist/group rights agenda	Black Liberation Army	Black Panther Party	National Association for the Advancement of Colored People	National Bar Association	Heritage Foundation	National Association for the Advancement of White People	Aryan Republican Army
Religious/faith agenda	Liberation theology	Catholic Worker movement	American Friends Service Committee	National Conference of Christians and Jews	Southern Baptist Convention	Moral Majority	Army of God
Legal/constitutional agenda	Individual lawyers	National Lawyers Guild	American Civil Liberties Union	American Bar Association	Thomas More Law Center	American Center for Law and Justice	Freemen

Photo 2.2 Architects of Communism. Russian revolutionary leader Vladimir Ilich Lenin (left) with Leon Trotsky, head of the Red Army and future ideological rival of Joseph Stalin.

Marxism

Radical socialism, like anarchism, is a leftist ideology that began in the turmoil of mid-19th-century Europe and the uprisings of 1848. Socialists championed the emerging industrial working class and argued that the wealth produced by these workers should be more equitably distributed, rather than concentrated in the hands of the wealthy elite.

Karl Marx is regarded as the founder of modern socialism. He and his associate **Friedrich Engels**, both Germans, argued that their approach to socialism was grounded in the scientific discovery that human progress and social evolution is the result of a series of historical conflicts and revolutions. Each era was based on the working group's unequal relationship to the **means of production** (e.g., slaves, feudal farmers, and industrial workers) vis-à-vis the ruling group's enjoyment of the fruits of the working group's labor. In each era, a ruling thesis group maintained the status quo and a laboring antithesis group challenged it (through agitation and revolution), resulting in a socioeconomic synthesis that created new relationships with the means of production. Thus human society evolved into the next era. According to Marx, the most advanced era of social evolution would be the synthesis Communist era, which he argued would be built after the antithesis industrial working class overthrew the thesis capitalist system. Marx theorized that the working class would establish the **dictatorship of the proletariat** in the Communist society and build a just and egalitarian social order.

Marx and Engels collaborated on the ***Manifesto of the Communist Party***, a short work completed in 1847 and published in 1848. It became one of the most widely read documents of the 20th century. In it, Marx and Engels explained the revolutionary environment of the industrial era and how this era was an immediate precursor to the Communist era.

Marxist socialism was pragmatic, revolutionary, and action oriented, and many revolutionary leaders and movements throughout the 20th century adopted it. Terrorism, both state and dissident, was used during the revolutions and the consolidations of power after victory. It is interesting to note that none of these Marxist revolutions was led by the industrial working class; all occurred in preindustrial developing nations, often within the context of anticolonial warfare waged by peasants and farmers.

Fascism

Fascism was a rightist ideological counterpoint to Marxism and anarchism that peaked before World War II. Like Marxism and anarchism, fascism's popular appeal grew out of social turmoil in Europe, this time as a reaction to the 1917 Bolshevik (Communist) Revolution in Russia, the subsequent Bolshevik-inspired political agitation elsewhere in Europe, and the widespread unrest during the Great Depression of the 1930s. It was rooted in a brand of extreme nationalism that championed the alleged superiority

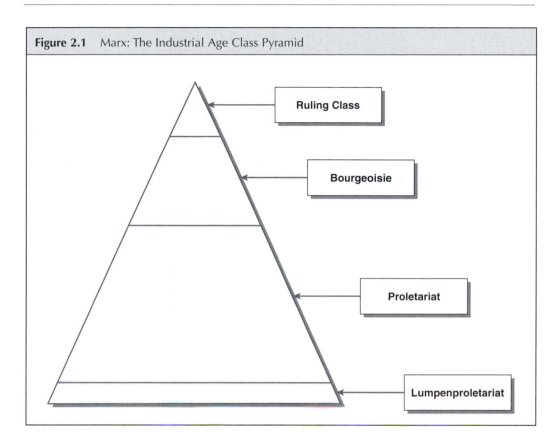

Figure 2.1   Marx: The Industrial Age Class Pyramid

Ruling Class

Bourgeoisie

Proletariat

Lumpenproletariat

of a particular national heritage or ethno-racial group. Fascism was anti-Communist, antimonarchist, antidemocratic, and anti-intellectual (though there were some fascist writers). It demanded extreme obedience to law, order, and the state. Fascism also required cultural conservatism—often looking backward in history to link the ancient past to the modern state. Fascists created their own conceptualizations of traditional values such as military duty, the Christian church, and motherhood. Strong antidemocratic leadership was centralized in the state, usually under the guidance of a single charismatic leader who symbolically embodied the virtues of the state, the people, and the underlying fascist ideology.

Italian dictator **Benito Mussolini** was the first to consolidate power and create a fascist state. Beginning with his March on Rome in 1922, he gradually eliminated all opposition and democratic institutions. He was a mentor to **Adolf Hitler**, who led the fascist National Socialist German Worker's (Nazi) Party to power in Germany in 1933. Both the Italian and German fascist regimes sent troops to fight on the side of right-wing Spanish rebels led by **Francisco Franco** during the Spanish Civil War.[21]

Photo 2.3
Architects of
fascism. German
Führer Adolf Hitler
(front left) stands
beside Italian Duce
Benito Mussolini.
Behind them are
ranking members
of the Nazi and
Fascist regimes.

Although the first fascist movement largely collapsed in 1945, right-wing groups and political parties have continued to promote neofascist ideals. Some terrorist groups in Europe and the United States have been overtly fascist and racist. Dictatorships have also arisen since World War II that adopted many features of prewar fascism. For example, Latin American regimes arose in Chile, Argentina, Uruguay, and El Salvador—to name a few—that fit the fascist pattern.

Chapter Perspective 2.3 discusses the Spanish Falange, which long served as a model for repressive right-wing regimes and movements in Latin America.

CHAPTER PERSPECTIVE 2.3

The Spanish Falange: A Model for Fascism

The Falange party (pronounced fä län' kë in Castilian Spanish) was born from the turmoil of 1930s Spain. A liberal leftist democratic republic had been established when King Alfonso XIII abdicated in 1931. Strong agitation from the far left and the fringe left—anarcho-syndicalists,[a] Communists, and radical socialists—spread disorder in the capital city of Madrid and elsewhere. Fear that a Russian-style Bolshevik state might be established grew and was especially strong among the upper classes, the military, the Catholic Church, and the middle class. In 1936, a rebellion among elite right-wing Spanish native troops and Foreign Legionnaires in Morocco sparked the

Spanish Civil War, which lasted until 1939. This was a classic case of the depiction of the military as national savior. The war ended with the victory of the rightists.

The Spanish military was commanded by a Spanish traditionalist right-wing officer corps, many of whom had served in the recent—and bloody—colonial war in Spanish Morocco against the Moorish leader Abd el Krim.[b] At the beginning of the rebellion, the Falange party was small, traditionalist, and right-wing. It was not overtly fascist in its orientation, but it championed the values of nationalism, the Catholic Church, order, motherhood, antiliberalism, antimonarchism, and antileftism. Francisco Franco, the most prominent rebel general, was eventually accepted as caudillo, or leader, of the rebellion. He successfully revamped the Falange to become his strongest base of support within his broader **Movimiento Nacional**.[c] Falangist fascism was unique, because

> Much of the rebels' propaganda with its appeals to a heroic past, its stress on leadership, and insistence on a Spain untouched by "internationalism" such as Jewry or liberalism or freemasonry was Fascist in tone, but it scarcely went deeper than a justification for the persecution of freemasons and liberals.[d]

Spanish Falangists were strong nationalists, and during World War II some even promoted the fantastic vision of reoccupying Latin America and the Philippines. Falangists formed the core of Francisco Franco's political support apparatus. They were independent nationalists and kept their distance from the German Nazis and Italian fascists during World War II, never entering the conflict other than to send one volunteer division—the Blue—to the Russian Front. This infuriated Adolf Hitler.

Hard-core Falangists could be fanatical, having adopted the slogan **"Long live death!"** during the rebellion against the Spanish Republic in the 1930s. Falangist ideology became popular among elites in Latin America and influenced the political behavior of these elites. Right-wing dictatorships in Argentina, Chile, and elsewhere were influenced by the Falangist example, and the militaries in these societies often adopted the mantle of national savior.

Notes

a. Anarcho-syndicalists were radical prolabor anarchists. They championed the working class and organized labor into anarchist syndicate groups. They believed that these syndicates would be the central means of organized cooperation after the revolution. Government would simply be a representative body of the syndicates.

b. Abd el Krim is a remarkable figure of the 20th century. His war against the Spanish colony in Morocco caused the deaths of thousands of Spanish troops, including 10,000 at Anual in June 1921. After defeating the Spaniards, he then turned against the French in their Moroccan possessions and essentially fought them to a stalemate. He was eventually defeated by combined French and Spanish forces in 1926. Abd el Krim was exiled under honorable terms to the French island of Réunion. He died in 1963.

c. Movimiento Nacional was a coalition of "monarchists, conservatives, business, and the more energetically radical social Fascists of the Falange." Christopher Campbell,"Franco's Spain," in *20th Century*, ed. R. W. Cross (London: Purnell Reference Books, 1979), 2349.

d. Hugh Thomas, "The Spanish Civil War," in *20th Century*, 1602.

Photo 2.4 The Guardia on patrol. Salvadoran troops patrol a village during the brutal civil war in the 1980s. The Guardia and rightist death squads were responsible for thousands of civilian deaths.

The Just War Doctrine

The just war doctrine is an ideal and a moralistic philosophy rather than an ideology. The concept has been used by ideological and religious extremists to justify acts of extreme violence. Throughout history, nations and individuals have gone to war with the belief that their cause was just and their opponents' cause unjust. Similarly, attempts have been made for millennia to write fair and just laws of war and rules of engagement. For example, in the late 19th and early 20th centuries, the Hague Conventions produced at least 21 international agreements on the rules of war.[22]

The just war debate asks who can morally be defined as an enemy and what kinds of targets it is morally acceptable to attack. In this regard,

> there are two separate components to the concept of just war (which philosophers call the just war tradition): the rationale for initiating the war (war's ends) and the method of warfare (war's means). Criteria for whether a war is just are divided into jus ad bellum (justice of war) and jus in bello (justice in war) criteria.[23]

Thus, ***jus in bello*** is correct behavior while waging war, and ***jus ad bellum*** is having the correct conditions for waging war in the first place. These concepts have been debated by philosophers and theologians for centuries. The early Christian philosopher **Augustine** concluded in the 5th century that war is justified to punish injuries inflicted by a nation that has refused to correct wrongs committed by its citizens. Augustine was, of course, referring to warfare between nations and cities, and church doctrine long held that an attack against state authority was an offense against God.[24] Likewise, the Hague Conventions dealt only with rules of conflict between nations and afforded no legal rights to spies or antistate rebels. Neither system referred to rules of engagement for nonstate or antistate conflicts.

In the modern era, both dissidents and states have adapted the just war tradition to their political environments. Antistate conflict and reprisals by states are commonplace. Dissidents always consider their cause just and their methods proportional to the force the agents of their oppressors use. Antiterrorist reprisals launched by states are also justified as appropriate and proportional applications of force—in this case as a means to root out bands of terrorists.

It is important to remember that rules of war and the just war tradition are the result of many motivations. Some rules and justifications are self-serving, others are pragmatic, and others are grounded in ethno-nationalist or religious traditions. Hence, the just war concept can be easily adapted to justify ethnic, racial, national, and religious extremism in the modern era.

Table 2.4 summarizes the ideals and ideologies discussed here.

_____ September 11, 2001 and the New Era of Terrorism

Many saw the attacks of September 11, 2001, as a turning point in the history of political violence. In their aftermath, journalists, scholars, and national leaders repeatedly described the emergence of a new international terrorist environment. It was argued that within this new environment, terrorists were now quite capable of using—and very willing to use—weapons of mass destruction to inflict unprecedented casualties and destruction on enemy targets. These attacks seemed to confirm warnings from experts during the 1990s that a new asymmetrical terrorism[25] would characterize the terrorist environment in the new millennium.

September 11, 2001

The worst incident of modern international terrorism occurred on U.S. soil on the morning of September 11, 2001, carried out by 19 Al Qaeda terrorists on a suicidal martyrdom mission. The goal was to strike at symbols of American (and Western) interests in response to what they perceived to be a continuing process of domination and exploitation of Muslim countries. The 19 hijackers were religious terrorists fighting in the name of a holy cause against perceived evil from the West. Their sentiments were born in the religious, political,

TABLE 2.4	A COMPARISON OF IDEOLOGIES			

Social conflict in the 20th century was deeply rooted in applying ideals and ideologies to practice. The adoption of these social and philosophical systems often motivated both individuals and movements to engage in armed conflict with perceived enemies. This presentation matches proponents, outcomes, and case studies of four ideals and ideologies.

	Ideological Orientation			
	Anarchism	**Marxism**	**Fascism**	**Just War**
Proponents	Proudhon/ Bakunin	Marx/Engels	Mussolini/ Hitler	Augustine
Desired Social Outcome	Stateless society	Dictatorship of the Proletariat	New order	Legitimized conflict
Applications	People's Will (Narodnaya Volya)	Russian Revolution	World War II-era Italy and Germany	State and dissident violence

and ethno-national ferment that has characterized the politics of the Middle East for much of the modern era.

Nearly 3,000 people were killed in the attack. The sequence of events occurred as follows:

7:59 a.m. American Airlines Flight 11, carrying 92 people, leaves Boston's Logan International Airport for Los Angeles.

8:20 a.m. American Airlines Flight 77, carrying 64 people, takes off from Washington's Dulles Airport for Los Angeles.

8:14 a.m. United Airlines Flight 175, carrying 65 people, leaves Boston for Los Angeles.

8:42 a.m. United Airlines Flight 93, carrying 44 people, leaves Newark, New Jersey, International Airport for San Francisco.

8:46 a.m. American Flight 11 crashes into the north tower of the World Trade Center.

9:03 a.m. United Flight 175 crashes into the south tower of the World Trade Center.

9:37 a.m. American Flight 77 crashes into the Pentagon. Trading on Wall Street is called off.

9:59 a.m. Two World Trade Center—the south tower—collapses.

10:03 a.m. United Flight 93 crashes 80 miles southeast of Pittsburgh, Pennsylvania.

10:28 a.m. One World Trade Center—the north tower—collapses.[26]

Photo 2.5 The aftermath of September 11, 2001. Rescue workers amid the smoke and debris of the World Trade Center in New York City.

The United States had previously been the target of international terrorism at home and abroad, but had never suffered a strike on this scale on its territory. The most analogous historical event was the Japanese attack on the naval base at Pearl Harbor, Hawaii, on December 7, 1941. The last time so many people had died from an act of war on American soil was during the Civil War in the mid-19th century.

After the Al Qaeda assault and the subsequent anthrax crisis, American culture shifted away from openness to security. The symbolism of the attack, combined with its sheer scale, drove the United States to war and dramatically changed the American security environment. Counterterrorism in the United States shifted from a predominantly law enforcement mode to a security mode. Measures included unprecedented airport and seaport security, border searches, visa scrutiny, and more intensive immigration procedures. Hundreds of people were administratively detained and questioned during a sweep of persons fitting the profile of the 19 attackers. These detentions set off a debate about the constitutionality of the methods and the fear of many that civil liberties were in jeopardy. In October 2001, the USA PATRIOT Act was passed, granting significant authority to federal law enforcement agencies to engage in surveillance and other investigative work. On November 25, 2002, 17 federal agencies (later increased to 22) were consolidated to form a new Department of Homeland Security.

The symbolism of a damaging attack on homeland targets was momentous because it showed that the American superpower was vulnerable to small groups of determined revolutionaries. The Twin Towers had dominated the New York skyline since 1972. They were a symbol of global trade and prosperity and the pride of the largest city in the United States. The Pentagon, of course, is a unique building that symbolizes American military power, and its location across the river from the nation's capital showed the vulnerability of the seat of government to attack.

On May 30, 2002, a 30-foot-long steel beam, the final piece of debris from the September 11 attack, was ceremoniously removed from the Ground Zero site in New York City.

CHAPTER PERSPECTIVE 2.2

Waging War in the Era of the New Terrorism

A war on terrorism was declared in the aftermath of the September 11, 2001, attacks. This is a new kind of conflict against a new form of enemy. From the outset, policy makers understood that the war would be fought unconventionally, primarily against shadowy terrorist cells and elusive leaders. It is not a war against a nation, but rather against ideas and behavior.

The mobilization of resources in this war necessitated the coordination of law enforcement, intelligence, and military assets in many nations across the globe. Covert operations by special military and intelligence units became the norm rather than the exception. Suspected terrorist cells were identified and dismantled by law enforcement agencies in many countries, and covert operatives worked secretly in other countries. Although many suspects were detained at the U.S. military base in Guantanamo Bay, Cuba, other secret detention facilities were also established.

However, the war has not been fought solely in the shadows. In contrast to the deployment of small law enforcement and covert military or intelligence assets, the U.S.-led invasions of Afghanistan and Iraq involved the commitment of large conventional military forces. In Afghanistan, reasons given for the invasion included the need to eliminate state-sponsored safe havens for Al Qaeda and other international mujahideen (holy warriors). In Iraq, reasons given for the invasion included the need to eliminate alleged stockpiles of weapons of mass destruction and alleged links between the regime of Saddam Hussein and terrorist networks. The U.S.-led operation in Iraq was symbolically named **Operation Iraqi Freedom**.

One significant challenge for waging war against extremist behavior—in this case, against terrorism—is that victory is not an easily definable condition. For example, on May 1, 2003, U.S. President George W. Bush landed on the aircraft carrier *Abraham Lincoln* to deliver a speech in which he officially declared that the military phase of the Iraq invasion had ended, and that the overthrow of the Hussein government was "one victory in a war on terror that began on September 11th, 2001, and still goes on." Unfortunately, the declaration was premature. A widespread insurgency took root in Iraq, with the resistance employing both classic hit-and-run guerrilla tactics and terrorism. Common cause was found between remnants of the Hussein regime and non-Iraqi Islamist fighters. Thousands of Iraqis and occupation troops became casualties during the insurgency. In particular, the insurgents targeted foreign soldiers, government institutions, and so-called collaborators such as soldiers, police officers, election workers, and interpreters. Sectarian violence also spread, with Sunni and Shi'a religious extremists killing many civilians.

Is the war on terrorism being won? How can victory reasonably be measured? Assuming that the New Terrorism will continue for a time, perhaps the best measure for progress in the war is to assess the degree to which terrorist behavior is being successfully managed—in much the same way that progress against crime is assessed. As the global community continues to be challenged by violent extremists during the new era of terrorism, the definition of victory is likely to continue to be refined and redefined by nations and leaders.

The New Terrorism

It is clear from human history that terrorism is deeply woven into the fabric of social and political conflict. This has not changed, and in the modern world, states and targeted populations are challenged by the **New Terrorism**, which is characterized by the following:

♦ Loose, cell-based networks with minimal lines of command and control

♦ Desired acquisition of high-intensity weapons and weapons of mass destruction

♦ Politically vague, religious, or mystical motivations

♦ Asymmetrical methods that maximize casualties

♦ Skillful use of the Internet and manipulation of the media

The New Terrorism should be contrasted with traditional terrorism, which is typically characterized by the following:

♦ Clearly identifiable organizations or movements

♦ Use of conventional weapons, usually small arms and explosives

♦ Explicit grievances championing specific classes or ethno-national groups

♦ Relatively surgical selection of targets

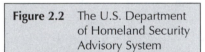

Figure 2.2   The U.S. Department of Homeland Security Advisory System

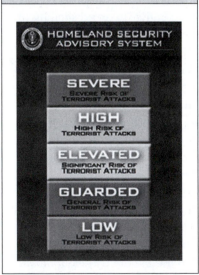

SOURCE: U.S. Department of Homeland Security.

New information technologies and the Internet create unprecedented opportunities for terrorist groups, and violent extremists have become adept at bringing their wars into the homes of literally hundreds of millions of people. Those who specialize in suicide bombings, car bombs, or mass-casualty attacks correctly calculate that carefully selected targets will attract the attention of a global audience. Cycles of violence thus not only disrupt normal routines, but also produce long periods of global awareness. Such cycles can be devastating. For example, during the winter and spring of 2005, Iraqi suicide bombings increased markedly in intensity and frequency, from 69 in April 2005 (a record rate) to 90 in May.[27] These attacks resulted in many casualties, including hundreds of deaths, and greatly outpaced the previous cycle of car bombings by more than two to one.

All of these threats offer new challenges for policy makers about how to respond to the behavior of terrorist states, groups, and individuals. The war on terrorism launched in the aftermath of the September 11 attacks seemed to herald a new resolve to end terrorism. This has proven a difficult task. The war has been fought on many levels, exemplified by the invasions of Afghanistan and Iraq and the disruption of terrorist cells on several continents. Serious terrorist strikes have occurred in Madrid, Spain, Bali, Indonesia, London, England, and Sharm el Sheikh, Egypt. In addition, differences have arisen within the post–September 11 alliance, creating significant strains. It is clear that the war will be a long-term prospect, likely with many unanticipated events.

Chapter Summary

This chapter introduced readers to some of the historical and modern attributes of terrorism, with a central theme that terrorism is deeply rooted in the human experience. The impact of extremist ideas on human behavior should not be underestimated, because certain historical examples of political violence in some ways parallel modern terrorism.

The relationship between extremist ideas and terrorist events was discussed as a nexus, whereby terrorism is the violent manifestation of extremist beliefs. Ideologies are the belief systems at the root of political violence.

Whether terrorist acts are *mala in se* or *mala prohibita* is often relative. Depending on one's perspective, gray areas may challenge us to be objective about the true nature of political violence. Most, if not all, nations promote an ideological doctrine to legitimize the power of the state and to convince the people that their systems of belief are worthy of loyalty, sacrifice, and (when necessary) violent defense. Conversely, when a group of people perceives that an alternative ideology or condition should be promoted, revolutionary violence may occur against the defenders of the established rival order. In neither case would those who commit acts of political violence consider themselves unjustified in their actions or label themselves as terrorists.

DISCUSSION BOX

This chapter's Discussion Box is intended to stimulate critical debate about the legitimacy of using extreme force against civilian populations.

Total War

Total war is "warfare that uses all possible means of attack, military, scientific, and psychological, against both enemy troops and civilians."[a] It was the prevailing military doctrine applied by combatant nations during World War II.

Allied and Axis military planners specifically targeted civilian populations. In the cases of German and Japanese strategists, the war was fought as much against indigenous populations as against opposing armies. The massacres and genocide directed against civilian populations at Auschwitz, Dachau, Warsaw, Lidice, and Nanking and countless other atrocities are a dark legacy of the 20th century.

The estimated number of civilians killed during the war is staggering:[b]

Belgium 90,000	France 391,000	Poland 6,000,000
Britain 70,000	Germany 2,000,000	Soviet Union 7,700,000
China 20,000,000	Greece 391,000	Yugoslavia 1,400,000
Czechoslovakia 319,000	Japan 953,000	

An important doctrine of the air war on all sides was to bomb civilian populations, so that the cities of Rotterdam, Coventry, London, Berlin, Dresden, and Tokyo were deliberately attacked. It is estimated that the atomic bombs dropped on Hiroshima and Nagasaki killed, respectively, 70,000 and 35,000 people.[c]

Discussion Questions

1. Are deliberate attacks against civilians legitimate acts of war?
2. Were deliberate attacks on civilians during World War II acts of terrorism?
3. If these attacks were acts of terrorism, were some attacks justifiable acts of terrorism?
4. Is there such a thing as justifiable terrorism? Is terrorism *malum in se* or *malum prohibitum*?
5. Is the practice of total war by individuals or small and poorly armed groups different from its practice by nations and standing armies? How so or how not?

Notes

a. *Webster's New Twentieth Century Dictionary of the English Language,* unabridged, 2nd ed. (New York: Publishers Guild, 1966).
b. Derrik Mercer, ed., *Chronicle of the Second World War* (Harlow, Essex, UK: Longman Group, 1990), 668.
c. Edward Jablonski, *Flying Fortress* (New York: Doubleday, 1965), 285.

Key Terms and Concepts

The following topics were discussed in this chapter and can be found in the Glossary:

Augustine	Ensslin, Gudrun	Ideologies
Bakunin, Mikhail	Falange	*jus ad bellum*
Burke, Edmund	Far left	*jus in bello*
Classical ideological	Far right	Just War doctrine
continuum	Franco, Francisco	Kaczynski, Theodore "Ted"
Crucifixion	Fringe left	(the Unabomber)
Dictatorship of	Fringe right	Komiteh
the Proletariat	Hague Conventions	Kropotkin, Petr
Engels, Friedrich	Hitler, Adolph	Left, center, right

Luddites
*Manifesto of the
    Communist Party*
Marx, Karl
Mussolini, Benito
Nechayev, Sergei
New Terrorism

People's Will (Narodnaya Volya)
Propaganda by the deed
Property is theft!
Proudhoun, Pierre-Joseph
Regicide
Reign of Terror (régime de la
    terreur)

Revolutionary
    Tribunal
Sicarii
Struggle meetings
Total war
Tyrannicide
Zealots

Terrorism on the Web

Log on to the Web-based student study site at **www.sagepub.com/martinessstudy** for additional Web sources and study resources.

Web Exercise

Using this chapter's recommended Web sites, conduct an online investigation of organizations that monitor extremist sentiment and terrorist behavior. Compare and contrast these organizations.

1. What are the primary agendas of these organizations?
2. How would you describe the differences between research, government, and social activist organizations?
3. In your opinion, are any of these organizations more comprehensive than other organizations? Less comprehensive?

For an online search of research and monitoring organizations, readers should activate the search engine on their Web browser and enter the following keywords:

"Terrorism Research"
"Human Rights Organizations"

Recommended Readings

The following publications provide an introduction to terrorism.

Combs, Cindy C. *Terrorism in the Twenty-First Century,* 3d ed. Englewood Cliffs, NJ: Prentice Hall, 2003.
Griset, Pamala L., and Sue Mahan. *Terrorism in Perspective.* Thousand Oaks, CA: Sage, 2002.
Hoffman, Bruce. *Inside Terrorism.* New York: Columbia University Press, 1998.
National Commission on Terrorist Attacks Upon the United States. *The 9/11 Commission Report.* New York: W. W. Norton, 2004.
Schechterman, Bernard, and Martin Slann, eds. *Violence and Terrorism.* Guilford, CT: Dushkin/ McGraw-Hill, 1999.
Simonsen, Clifford E., and Jeremy R. Spindlove. *Terrorism Today: The Past, the Players, the Future.* Englewood Cliffs, NJ: Prentice Hall, 2000.
White, Jonathan R. *Terrorism: An Introduction,* 3d ed. Belmont, CA: Wadsworth, 2002.

3

Causes of Terrorist Violence

This chapter investigates the causes of terrorism. Many explanations have been given for terrorism, and scholars and other experts have devoted a great deal of effort to explaining terrorist behavior. This has not been a simple task. Explanatory models consider many factors, including political history, government policy, contemporary politics, cultural tensions, ideological trends, economic trends, individual idiosyncrasies, and other variables.

In the following discussion, readers will identify factors that explain why individuals and groups choose to engage in terrorist violence. Readers will also explore and critically assess the sources of ideological belief systems and activism and the reasons why such activism sometimes results in terrorist violence. For example, is the terrorist option somehow forced on people who have no other alternative? Is terrorism simply one choice from a menu of options? Is politically motivated violence perhaps a pathological manifestation of personal or group dysfunction?

It is useful in the beginning of our discussion to identify broad causes of terrorism at the individual and group levels.

At the individual level, some experts have distinguished rational, psychological, and cultural origins:

> Rational terrorists think through their goals and options, making a cost-benefit analysis.... Psychological motivation for resorting to terrorism derives from the terrorist's personal dissatisfaction with his/her life and accomplishments.... A major cultural determinant of terrorism is the perception of "outsiders" and anticipation of their threat to ethnic group survival.[1]

At the group level, terrorism can grow out of an environment of political activism, when a group's goal is to redirect a government's or society's attention toward the grievances of an activist social movement. It can also grow out of dramatic events in the experience of a people or a nation. Although these two sources—social movements and dramatic events—are generalized concepts, it is instructive to briefly review their importance.

Social Movements. Social movements are campaigns that try either to promote change or to preserve something that is perceived as threatened. Movements involve mass action on behalf of a cause, not simply the actions of individuals promoting their personal political beliefs. Examples include the Irish Catholic civil rights movement of the 1960s in Northern Ireland and the African American civil rights movement in the American South during the same decade. Proponents of this type of movement seek the moral high ground as a way to rally sympathy and support for their cause and to bring pressure on those at odds with the cause. In both cases, radicalized sentiment grew out of frustration with the slow pace of change and the violent reaction of at some opponents.

Dramatic Events. Also called traumatic events, these occur when an individual, a nation, or an ethno-national group suffers from an event that has a traumatizing and lasting effect. At the personal level, children of victims of political violence may grow up to oppose perceived oppressors with violence. This is likely to occur in regions of extended conflict, such as the war between Tamils and Sinhalese in Sri Lanka, the Troubles in Northern Ireland, or the Palestinian **intifada**.

Regardless of the specific precipitating cause of a particular terrorist's behavior, the fact that so many individuals, groups, and nations resort to terrorist violence suggests that common motives and explanations can be found. The discussion in this chapter will review the following:

♦ Political violence as the fruit of injustice
♦ Political violence as strategic choice
♦ The morality of political violence

Political Violence as the Fruit of Injustice

Intergroup Conflict and Collective Violence: Sociological Explanations of Terrorism

The sociological approach argues that terrorism is a group-based phenomenon, selected by weaker groups as the only available strategy. From the perspective of an opponent group, "terrorism and other forms of collective violence are often described as 'senseless,' and their participants are often depicted as irrational."[2] However, this is not an entirely complete analysis, because

> if "rational" means goal directed . . . then most collective violence is indeed rational . . . Their violence is indeed directed at achieving certain, social change-oriented goals, regardless of whether we agree with those goals or with the violent means used to attain them. If "rational" further means sound, wise, and logical, then available evidence indicates that collective violence is rational . . . because it sometimes can help achieve their social goals.[3]

In essence, the disadvantaged group asserts its rights by selecting a methodology—in this case, terrorism—that from the group's perspective is its only viable option. The selection process is based on the insurgent group's perceptions and its analysis of those perceptions.

Theoretical Foundations

Two sociological concepts, structural theory and relative deprivation theory, provide useful explanatory analysis for this process.[4]

Structural theory has been used in many policy and academic disciplines to identify social conditions (structures) that affect group access to services, equal rights, civil

Photo 3.1 Photographs that shocked the world. An American soldier pulls an Iraqi detainee with a leash in Abu Ghraib prison in Baghdad in late 2003. Such images served to rally opposition to the U.S.-led occupation.

protections, freedom, or other quality-of-life measures. Examples of social structures include government policies, administrative bureaucracies, spatial (geographic) location of the group, the role of security forces, and access to social institutions. Applying this theory to the context of terrorism,

structural theories of revolution emphasize that weaknesses in state structures encourage the potential for revolution. . . . According to this view, a government beset by problems such as economic and military crises is vulnerable to challenges by insurgent forces. . . . Other governments run into trouble when their . . . policies alienate and even anger elites within the society.[5]

The state is the key actor in structural theories of revolution. Its status is the precipitating factor for popular revolutions. Popular discontent, the alienation of elites, and a pervasive crisis are the central ingredients for bringing a society to the brink of revolution.[6]

Relative deprivation theory essentially holds that "feelings of deprivation and frustration underlie individual decisions to engage in collective action."[7] When a group's rising expectations are met by sustained repression or second-class status, the group's reaction may include political violence. Their motive for engaging in political violence is their observation that they are relatively deprived, in relation to other groups, in an unfair social order. This should be contrasted with **absolute deprivation**, when a group has been deprived of the necessities for survival by a government or social order. This condition can also lead to political violence.

One observation must be made about relative deprivation theory: Although it was, and is, a popular theory among many experts, three shortcomings have been argued:

♦ Psychological research suggests that aggression happens infrequently when the conditions for relative deprivation are met.
♦ The theory is more likely to explain individual behavior than group behavior.
♦ Empirical studies have not found an association between relative deprivation and political violence.[8]

International Cases

Examples of movements that are motivated against a government or social order include ethno-nationalist movements among Irish Catholics in Northern Ireland and Palestinians in Israel. Sociological explanations for these movements are summarized below.

Irish Catholic nationalism in Northern Ireland dates to the 16th century, when English King James I granted Scottish Protestant settlers land in Ireland, thus beginning a long process of relegating Irish Catholics to second-class status in their own country. Protestant

(Scotch-Irish) and English domination was secured in 1690 at the Battle of the Boyne. Catholic independence was finally won in 1919 and 1920, but the island was formally divided between the independent Irish Republic in the south and the British-administered six-county region of Northern Ireland. Since that time, some Irish republicans in the north, especially the Provisional Irish Republican Army, engaged in armed resistance against Protestant and British political domination. They seek union with the southern republic.

Palestinian nationalism dates to the formal creation of the state of Israel on May 14, 1948. The next day, the Arab League—Lebanon, Egypt, Jordan, and Syria—declared war on Israel. Israel was victorious, and in the subsequent consolidation of power, hundreds of thousands of Palestinians either left Israel or were expelled. Since that time, Palestinian nationalists, especially the Palestine Liberation Organization and Hamas, have fought a guerrilla and terrorist war against Israel to establish a Palestinian state.

Table 3.1 summarizes the constituencies and enemies of groups promoting these causes.

Rationality and Terrorist Violence: Psychological Explanations of Terrorism

Psychological approaches to explaining terrorism broadly examine the effects of internal psychological dynamics on individual and group behavior. This kind of analysis incorporates many of the concepts discussed earlier in this chapter, such as moral convictions and simplified definitions of good and evil.

At the outset, it is useful to examine the presumption held by a number of people— experts, policy makers, and laypersons—that terrorism is a manifestation of insanity or

TABLE 3.1	NATIONALISM AND SOCIOLOGICAL EXPLANATIONS OF TERRORISM: CONSTITUENCIES AND ADVERSARIES	
Nationalism is an expression of ethno-national identity. Nationalist activism can range in scale from promoting cultural heritage to armed insurrection. Its goals can range from a desire for equal political rights to complete national separation. Some ethno-national groups have engaged in nationalist activism to preserve their cultural heritage and have opposed what they consider national and cultural repression. Within these groups, violent extremists have engaged in terrorism.		
	Activity Profile	
Group	**Constituency**	**Adversary**
Irish Republican Army factions	Northern Irish Catholics	British and Ulster Protestants
ETA factions	Spanish Basques	Spaniards
Secular and religious Palestinian groups	Palestinians	Israelis
FLQ	French-speaking residents of Quebec (Québécois)	English-speaking Canadians

mental illness or that terrorism is the signature of a lunatic fringe. This presumption suggests that terrorism is a priori (fundamentally) irrational behavior and that only deranged individuals or deranged collections of people would select terrorist violence as a strategy. Most experts agree that this blanket presumption is incorrect. Although individuals and groups do act out of certain idiosyncratic psychological processes, their behavior is neither insane nor necessarily irrational.

Individual-Level Explanations

Some experts argue that the decision to engage in political violence is frequently an outcome of significant events in individual lives that give rise to antisocial feelings. They actively seek improvement in their environment or desire redress and revenge from the perceived cause of their condition.

Research has not found a pattern of psychopathology among terrorists. In comparing nonviolent and violent activists, studies reported "preliminary impressions . . . that the family backgrounds of terrorists do not differ strikingly from the backgrounds of their politically active counterparts."[9] There is evidence of some psychosocial commonalities among violent activists. For example, research on 250 West German terrorists reported "a high incidence of fragmented families," "severe conflict, especially with the parents," conviction in juvenile court, and "a pattern of failure both educationally and vocationally."[10]

Group-Level Explanations

In a number of social and political contexts, political violence is a familiar social phenomenon for some people. When it is combined with "the pronounced need to belong to a group,"[11] individuals can in the end "define their social status by group acceptance." Thus, at the group level,

> another result of psychological motivation is the intensity of group dynamics among terrorists. They tend to demand unanimity and be intolerant of dissent . . . [and] pressure to escalate the frequency and intensity of operations is ever present. . . . Compromise is rejected, and terrorist groups lean towards maximalist positions.[12]

An important outcome of these dynamics is the development of a self-perpetuating cycle of rationalizations of political violence. This occurs because

> the psychodynamics also make the announced group goal nearly impossible to achieve. A group that achieves its stated purpose is no longer needed; thus, success threatens the psychological well-being of its members.[13]

Generalized Psychological Explanations

Psychological explanations are fairly broad approaches. Both individual and group theories attempt to generalize reasons for the decision to initiate political violence and the processes that perpetuate such violence. These explanations may be summarized as follows:

- ◆ Terrorism is simply a choice among violent and less violent alternatives. It is a rational selection of one methodology.
- ◆ Terrorism is a technique for maintaining group cohesion and focus. Group solidarity overcomes individualism.

◆ Terrorism is a necessary process to build the esteem of an oppressed people. Through terrorism, power is established over others, and the weak become strong. Attention itself becomes self-gratifying.

◆ Terrorists consider themselves an elite vanguard. They are not content to debate the issues, because they have found a truth that needs no explanation. Action is superior to debate.

◆ Terrorism provides a means to justify political violence. The targets are depersonalized and symbolic labels are attached to them. Thus, symbolic buildings become legitimate targets even when occupied by people, and individual victims become symbols of an oppressive system.

In essence, then, psychological explanations of terrorist behavior use theories of individual motivations and group dynamics to explain why people decide to adopt strategies of political violence and why groups continue campaigns of violence. Pressures to conform to the group, combined with those to commit acts of violence, form a powerful psychological drive to carry on in the name of the cause even when victory is logically impossible. These pressures become so prevalent that victory becomes secondary to the unity of the group.[14] Having said this, it is inadvisable to generalize about psychological causes of terrorism, because "most terrorists do not demonstrate serious psychopathology" and "there is no single personality type."[15]

Case: The Stockholm Syndrome. In August 1973, three women and one man were taken hostage by two bank robbers in Stockholm, Sweden. The botched robbery led to a hostage crisis that lasted for 6 days. During the crisis, the robbers threatened to kill the four hostages if the authorities tried to rescue them. At the same time, the hostages received treatment from the robbers that they began to think of as kindness and consideration. For example, one hostage was told that he would not be killed, but rather shot in the leg if the police intervened, and that he should play dead. Another hostage, who suffered from claustrophobia, was let out of the bank vault on a rope leash.

During the 6-day episode, all of the hostages began to sympathize with the robbers and gradually came to identify with them. They eventually denounced the authorities' attempts to free them. After the situation was resolved, the hostages remained loyal to their former captors for months. They refused to testify against them and raised money for their legal defense. One of the female former hostages actually became engaged to one of the robbers. This was, to say the least, surprising behavior. The question is whether this was an isolated phenomenon or whether it is possible for it to occur in other hostage crises.

Experts are divided about whether the **Stockholm syndrome** is a prevalent condition. Those who contend that it can occur and has occurred in other situations argue that the syndrome sets in when a prisoner suffers a psychological shift from captive to sympathizer. In theory, the prisoner will try to keep his or her captor happy in order to stay alive whenever he or she is unable to escape, isolated, or threatened with death. This becomes an obsessive identification with what the captor likes and dislikes, and the prisoner eventually begins to sympathize with the captor. The psychological shift theoretically requires 3 or 4 days to set in.

Chapter Perspective 3.1 investigates the subject of gender and terrorism by discussing women as terrorists.

CHAPTER PERSPECTIVE 3.1

Women as Terrorists

Between October 23 and 26, 2003, Chechen terrorists seized 700 hostages in a Moscow theater. The episode ended with the deaths of scores of hostages and all of the terrorists. Russian authorities reported that many of the hostage-takers were women who had suicide explosive vests strapped to their bodies. The presence of female suicide bombers is not uncommon within the Chechen resistance movement. As a result, the Russian media has dubbed the women among Chechen terrorists **Black Widows** because they are allegedly relatives of Chechen men who have died in the ongoing war in Chechnya (see Chapter 5).

How common is terrorism by women? What motivates women to become terrorists? In which environments are female terrorists typically found?[a]

Women have been active in a variety of roles in many violent political movements.[b] Historically, some women held positions of leadership during terrorist campaigns and were well integrated into the command systems and policy decision-making processes in extremist groups. In the modern era, women were central figures in Sri Lanka's Tamil Tigers, Germany's Red Army Faction,[c] Italy's Red Brigades, Spain's Basque ETA, and the Japanese Red Army. During the Palestinian intifada (shaking off, or uprising) against Israel, a number of Palestinian suicide bombers were young women. More commonly, women serve as combatants rather than leaders, or women are recruited to participate as support functionaries, such as finding safe houses and engaging in surveillance.

Regardless of the quality of participation, it is clear that such involvement belies the common presumption that terrorism is an exclusively male preserve. In fact, some of the most committed revolutionaries around the world are women.

The following examples are instructive:

- Before the 1917 Bolshevik revolution, Russian women were leading members of violent extremist groups such as People's Will and the Social Revolutionary Party.
- Female anarchists such as Emma Goldman in the United States demonstrated that women could be leading revolutionary theorists.
- Leila Khaled became a well-respected and prominent member of the Palestinian nationalist movement after her participation in two airline hijacking incidents.
- During the unrest leading up to the Iranian Revolution in the late 1970s, women participated in numerous antigovernment attacks.
- Gudrun Ensslin, **Ulrike Meinhof**, and other women were leaders and comrades-in-arms within Germany's Red Army Faction during the 1970s.
- During the 1970s and 1980s, other West European terrorist groups such as France's **Direct Action**, Italy's Red Brigades, and Belgium's Communist Combat Cells integrated women into their ranks.
- Women were leaders in the nihilistic Japanese Red Army, which was founded by a woman (Fusako Shigenobu), during the 1970s and 1980s.

- During the final quarter of the 20th century, Provisional Irish Republican Army soldiers were mostly men, despite the fact that the IRA was a nationalist and mildly socialist movement.
- Women became renowned leaders among Sri Lanka's Tamil Tigers group during and after the 1990s when many male leaders were killed or captured, and female terrorists known as Freedom Birds engaged in many attacks, including suicide bombings.
- Among Chechen rebels during the early years of the new millennium, young women were recruited, manipulated, or coerced into becoming suicide fighters known among Russians as Black Widows.
- Around the turn of the new millennium, the Al Aqsa Martyr Brigades unit of the Palestine Liberation Organization actively recruited and deployed a few women as suicide bombers.[d]
- Female combatants have been found in the ranks of many insurgent groups, such as Colombia's **FARC** and ELN, India's Naxalites, the Communist Party of Nepal, Peru's **Shining Path**, and Mexico's Zapatistas.

Active participation of women is arguably more common among left-wing and nationalist terrorist movements than in right-wing and religious movements. Rightist and religious movements yield few cases of women as terrorists, and examples of female leaders are equally few. One reason is that, on the one hand, many leftists adopt ideologies of gender equality and many nationalists readily enlist female fighters for the greater good of the group. On the other hand, right-wing and religious movements often adopt ideologies that relegate women to secondary roles within the group. Among religious movements, ideologies of male dominance and female subordination have been common, so that women rarely participate in attacks, let alone in command systems and policy decision-making processes.

Notes

a. For a good discussion of these and other issues, see Rhiannon Talbot, "Myths in the Representation of Women Terrorists," *Eire-Ireland* Fall 2001.
b. For a discussion of the roles of women in terrorist movements, see Rhiannon Talbot, "The Unexpected Face of Terrorism," *This Is the Northeast* January 31, 2002.
c. For a good discussion of Italian women in violent organizations, see Alison Jamieson, "Mafiosi and Terrorists: Italian Women in Violent Organizations," *SAIS Review* Summer/Fall 2000, 51–64.
d. For interviews with female Al Aqsa Martyr Brigades volunteers, see Michael Tierney, "Young, Gifted and Ready to Kill," *The Herald (Glasgow, UK)*, August 3, 2002.

_____ **Political Violence as Strategic Choice**

Making Revolution: Acts of Political Will

An **act of political will** is an effort to force change. It is a choice, a rational decision from the revolutionary's perspective, to adopt specific tactics and methodologies to defeat an adversary. These methodologies are instruments of rational strategic choice, in which terrorism is adopted as an optimal strategy. All that is required for final victory is the political and strategic will to achieve the final goal. Selecting terrorism is a process based on the experiences of each insurgent group, and thus the outcome of an evolutionary political progression.

As a result, terrorism is simply a tool, an option, selected by members of the political fringe to achieve a goal. Terrorism is a deliberate strategy, and success is ensured as long as the group's political and strategic will remains strong.

The evolution of Marxist revolutionary strategy illustrates the essence of political will. Karl Marx argued that history and human social evolution are inexorable forces that will inevitably end in the triumph of the revolutionary working class. He believed that the prediction of the eventual collapse of capitalism was based on scientific law. **Vladimir Ilich Lenin**, however, understood that capitalism's demise would not come about without a push from an organized and disciplined vanguard organization such as the Communist Party. This organization would lead the working class to victory. In other words, the political will of the people can make history if they are properly indoctrinated and led. An important conceptual example will help readers better understand the theory of revolutionary change through acts of political will. It is a strategy known as **people's war**. The context in which it was first developed and applied was the Chinese Revolution.

Mao Zedong led the Communist Red Army to victory during the Chinese Revolution by waging a protracted war—first against Chiang Kai-shek's Nationalists (Kuomintang), then in alliance with the Nationalists against the invading Japanese, and finally driving Chiang's forces from mainland China in 1949. The Red Army prevailed largely because of Mao's military-political doctrine, which emphasized waging an insurgent people's war. His strategy was simple: Indoctrinate the army; win over the people; and hit, run, and fight forever.

People's war was a strategy born of necessity, originating when the Red Army was nearly annihilated by the Nationalists before and during the famous Long March campaign of 1934 and 1935. During the Long March, the Red Army fought a series of rearguard actions against pursuing Nationalist forces, eventually finding refuge in the northern Shensi province after a reputed 6,000-mile march. After the march, while the Red Army rested and was refitted in Shensi, Mao developed his military doctrine. People's war required protracted warfare (war drawn out over time), fought by an army imbued with an iron ideological will to wear down the enemy.[16]

According to Mao, the Red Army should fight a guerrilla war with roving bands that would occasionally unite. The strategy was to fight by consolidating the countryside and then gradually moving into the towns and cities. Red Army units would avoid conventional battle with the Nationalists, giving ground before superior numbers. Space would be traded for time, and battle would be engaged only when the Red Army was tactically superior at a given moment. Thus, an emphasis was placed on avoidance and retreat. In people's war, assassination was perfectly acceptable, and targets included soldiers, government administrators, and civilian collaborators. Government-sponsored programs and events—no matter how beneficial they might be to the people—were to be violently disrupted to show the government's weakness. A successful people's war required the cooperation and participation of the civilian population, so Mao ordered his soldiers to win their loyalty by treating the people appropriately.

Mao's contribution to modern warfare—and to the concept of political will—was that he deliberately linked his military strategy to his political strategy; they were one and the same. Terrorism was a perfectly acceptable option. The combination of ideology, political indoctrination, guerrilla tactics, protracted warfare, and popular support made people's war a very potent strategy and an effective synthesis of political will.

Lighting the Fuse: Adversaries in the War on Terrorism

One other consideration is necessary to fully appreciate the modern causes of terrorism. This theory is rooted in the political environment that gave rise to the new era of terrorism.

The concept of "one person's terrorist is another person's freedom fighter" is pertinent to how the behavior of the West, and particularly the behavior of the United States, is perceived around the world. When the September 11 attacks occurred, many Americans and other Westerners saw them as an attack on Western-style civilization. Reasons given for the subsequent U.S.-led war on terrorism included the argument that war is necessary to defend civilization from a new barbarism. From the official American and allied point of view, the war is a simply counteraction against the enemies of democracy and freedom. However, many Muslims have a wholly different perspective.

Interestingly, many young Muslims are keen to adopt some degree of Western culture, yet remain loyal to the Muslim community. One student commented,

> most of us here like it both ways, we like American fashion, American music, American movies, but in the end, we are Muslims. . . . The Holy Prophet said that all Muslims are like one body, and if one part of the body gets injured, then all parts feel that pain. If one Muslim is injured by non-Muslims in Afghanistan, it is the duty of all Muslims of the world to help him.[17]

The argument, then, is that the cause of anti-American and Western sentiment is the behavior of those nations—that is, their actions rather than their values or their culture. In the opening paragraph of his controversial book *Imperial Hubris*, former ranking CIA official Michael Scheuer presented the central precept of this argument:

> In America's confrontation with Osama bin Laden, Al Qaeda, their allies, and the Islamic world, there lies a startlingly clear example of how loving something intensely can stimulate an equally intense and purposeful hatred of things by which it is threatened. This hatred shapes and informs Muslim reactions to U.S. policies and their execution, and it is impossible to understand the threat America faces until the intensity and pervasiveness of this hatred is recognized.[18]

Can Muslim perceptions and Western behaviors be reconciled? What are the prospects for mitigating this source of terrorism in the modern era? Several events portend a continued disconnect, at least for the immediate future:

◆ The American-led occupation of Iraq and the protracted insurgency that arose
◆ An open-ended presence of Western troops in or near Muslim countries
◆ Broadcast images of civilian casualties and other collateral damage during military operations
◆ Broadcast images and rumors of the mistreatment of prisoners in American-run detention facilities
◆ Cycles of chronic violence between Israelis and Palestinians and the perception that the U.S. and West unfairly favor Israel

In this regard, a 2004 report by the CIA's National Intelligence Council warned that the war in Iraq created a new training ground for professional terrorists, much as the 1979–1989 Soviet war in Afghanistan created an environment that led to the rise of Al Qaeda and other international **mujahideen** (Islamic holy warriors).[19] The report further warned that veterans of the Iraq war will disperse after the end of the conflict, thus constituting a new generation of international mujahideen who will supplant the first Afghanistan-trained generation. This is a plausible scenario, because many foreigners volunteered to fight in Iraq out of a sense of pan-Islamic solidarity.[20]

The Morality of Political Violence

Although not all extremists become terrorists, some do cross the line. For them, terrorism is a morally acceptable strategy, a specifically selected method to further a just cause.

To facilitate critical understanding of the morality of terrorists, the following four motives are reviewed:

1. Moral convictions of terrorists
2. Simplified definitions of good and evil
3. Seeking utopia
4. Codes of self-sacrifice

Moral Convictions of Terrorists

Moral conviction refers to terrorists' unambiguous certainty of the righteousness of their cause. The goals and objectives of their movement are considered principled beyond reproach and their methods absolutely justifiable. This conviction can arise in several environments, including the following two settings:

In the first, a group of people can conclude that they have been morally wronged and that a powerful, immoral, and evil enemy is arrayed against them. This enemy is considered adept at betrayal, exploitation, violence, and repression. These conclusions can have some legitimacy, especially when a history of exploitation has been documented. This historical evidence is identified and interpreted as being the source of the group's modern problems. For example, many leftist insurgents in Latin America characterized the United States as an imperialist enemy because of its long history of military intervention, economic penetration, and support for repressive regimes in the region.[21]

In later generations, native populations who shared this kind of history and who interpreted it as being part of an ongoing

Photo 3.2 Bloody Sunday (January 30, 1972): A British soldier runs down an Irish Catholic demonstrator during protests and rioting in the city of Londonderry in Northern Ireland. The confrontations resulted in elite paratroopers firing on Catholic civilians. The incident was a seminal event that rallied many Catholics to support the Provisional Irish Republican Army.

pattern in contemporary times developed strong resentment against their perceived oppressor—in this case, the United States and the governments it supported. To them, there was no need to question the morality of their cause; it was quite clear.

A second setting in which moral conviction may arise is when a group or a people concludes that it is inherently and morally superior. This can be derived from ideological convictions, ethno-national values, or religious beliefs. From this perspective, the cause is virtually holy; in the case of religious beliefs, it *is* holy. A sense of moral purity becomes the foundation for the simplification of good and evil. In this setting, extremists decide that no compromise is possible and that terrorism is a legitimate option.

For example, a major crisis began in the Yugoslavian territory of **Kosovo** in 1998 when heavy fighting broke out between Serb security forces, the **Kosovo Liberation Army (KLA)**, and the Serb and Albanian communities. The conflict ended when the North Atlantic Treaty Organization (NATO) and Russian troops occupied Kosovo and the Serb security units were withdrawn. The strong Serb bond with Kosovo had originated in 1389 when the Serb hero, Prince Lazar, was defeated by the Ottoman Turks in Kosovo. Kosovo had been the center of the medieval Serb empire, and this defeat ended the Serb nation. Over the next 500 years, as the Turks ruled the province, Albanian Muslims migrated into Kosovo and gradually displaced Serb Christians. Nevertheless, Serbs have always had strong ethno-national ties to Kosovo, considering it a kind of spiritual homeland. It is at the center of their national identity. Thus, despite the fact that 90% of Kosovo's population was Albanian in 1998, Serbs considered their claim to the territoryparamount to anyone else's. From their perspective, the morality of their position was clear.

Chapter Perspective 3.2 illustrates the application of selective and moral terrorism. It is a quotation from a document captured during the war in Vietnam from the Viet Cong by American soldiers of the First Cavalry Division. It is a directive explaining procedures for suppressing counterrevolutionary elements,[22] remarkable for its instructions on how to correctly conduct the campaign.

CHAPTER PERSPECTIVE 3.2

A Viet Cong Directive Ordering Selective Terrorism

The Viet Cong were southern Vietnamese Communists who fought alongside the North Vietnamese Army against American forces and the Republic of South Vietnam. The Communist forces considered this war one phase in an ongoing effort to unify the North and South into a single nation.

During the American phase of their long war, Vietnamese Communists in South Vietnam routinely used terrorism to eliminate enemies. Assassinations and kidnappings were common, and targets regularly included civilians. Thousands fell victims to this policy.

The following quotation is from a passage concerning the suppression of counterrevolutionaries in areas under American or South Vietnamese control. It is interesting because it orders Viet Cong operatives to be very selective in choosing their targets.[a]

Directive

Concerning a number of problems that require thorough understanding in Z's task of repressing counterrevolutionary elements. . . .

In areas temporarily under enemy control:

(1) We are to exterminate dangerous and cruel elements such as security agents, policemen, and other cruel elements in espionage organizations, professional secret agents in organizations to counter the Revolution, henchmen with many blood debts . . . in village administrative machines, in the puppet system, in enemy military and paramilitary organizations, and in [South Vietnamese political parties].

(2) We are to establish files immediately and prepare the ground for later suppression of dangerous henchmen whom we need not eliminate yet or whose elimination is not yet politically advantageous. . . .

While applying the above-mentioned regulations, we must observe the following:

♦ Distinguish . . . the elimination of tyrants and local administrative personnel while fighting from the continuous task of repressing counterrevolutionaries in the liberated areas.

♦ Distinguish the ringleaders, the commissioned officers, from the henchmen.

♦ Distinguish exploiting elements from the basic elements and distinguish persons determined to oppose the Revolution from those who are forced to do so or those who are brought over and have no political understanding.

♦ Distinguish between persons with much political and religious influence and those who have no influence or very little influence.

♦ Distinguish between historical problems and present-day problems.

♦ Distinguish major crimes with many bad effects from minor crimes or innocence.

♦ Distinguish determined and stubborn antirevolutionary attitudes from attitudes of submission to the Revolution and true repentance, and willingness to redeem by contribution to the Revolution.

♦ Distinguish counterrevolutionary elements from backward and dissatisfied persons among the masses.

Note

a. Quoted in Jay Mallin, *Terror and Urban Guerrillas: A Study of Tactics and Documents* (Coral Gables, FL: University of Miami Press, 1971), 33, 39–40.

Simplified Definitions of Good and Evil

Revolutionaries universally conclude that their cause is honorable, their methods are justifiable, and their opponents are representations of implacable evil. They arrive at this conclusion in innumerable ways, often—as in the case of Marxists—after devoting considerable intellectual energy to political analysis. Nevertheless, their final analysis is uncomplicated: Our cause is just, and the enemy's is unjust. Once this line has been clearly drawn between good and evil, the methods used in the course of the struggle are justified by the ennobled goals and objectives of the cause.

A good example of the application in practice of simplified delineations of good and evil is found in the influential ***Mini-Manual of the Urban Guerrilla***, written by Brazilian revolutionary **Carlos Marighella**.[23] In this document, Marighella clearly argues that terrorism is necessary against a ruthless enemy. The *Mini-Manual* was read, and its strategy implemented, by leftist revolutionaries throughout Latin America and Europe. Marighella advocated terrorism as a correct response to the oppression of the Brazilian dictatorship. He wrote:

> The accusation of assault or terrorism no longer has the pejorative meaning it used to have. . . . Today to be an assailant or a terrorist is a quality that ennobles any honorable man because it is an act worthy of a revolutionary engaged in armed struggle against the shameful military dictatorship and its monstrosities.[24]

One fact is clear: There is a moment of decision among those who choose to rise in rebellion against a perceived oppressor. This moment of decision is a turning point in the lives of individuals, people, and nations.

Seeking Utopia

The book *Utopia,* written by the English writer Sir Thomas More in the 16th century, was a fictional work that described an imaginary island with a society having an ideal political and social system. Countless philosophers, including political and religious writers, have since created their own visions of the perfect society.[25] Terrorists likewise envision some form of **utopia,** though for many this can simply mean the destruction of the existing order. For such **nihilist dissidents,** any system is preferable to the existing one, and its destruction alone is a justifiable goal. The question is what kind of utopia terrorists seek. This depends on their belief system. For example, religious terrorists seek to create a God-inspired society on earth that reflects the commandments, morality, and values of their religious faith. Political terrorists define their ideal society according to their ideological perspective. Regardless of which belief system terrorists adopt, they uniformly accept the proposition that the promised good (a utopia) outweighs their present actions, no matter how violent those actions are. The revolution will bring utopia after a period of trial and tribulation, so that **the end justifies the means**. This type of reasoning is particularly common among religious, ethno-nationalist, and ideological terrorists.

Codes of Self-Sacrifice

Terrorists invariably have faith in the justness of their cause and live their lives accordingly. Many consequently adopt **codes of self-sacrifice** that are at the root of their everyday lives. They believe that these codes are superior codes of living and that those who follow the code are superior to those who do not. The code accepts a basic truth and applies it to everyday life. This truth usually has a religious, ethno-national, or ideological foundation. Any actions taken within the accepted parameters of these codes—even terrorist actions—are justified, because the code cleanses the true believer.

A good example of ideological codes of self-sacrifice is found on the fringe left among the first anarchists. Many anarchists did not simply believe in revolution, they lived the revolution.

Photo 3.3 Members of the Waffen SS on parade in Nazi Germany. German members of the Waffen SS were specially selected as pure members of the Aryan race. They lived according to a code that allowed ruthless treatment of enemies and subhumans.

They crafted a lifestyle that was consumed by the cause. Among some, an affinity for death became part of the revolutionary lifestyle. The Russian anarchist Sergei Nechayev wrote in *Revolutionary Catechism:*

> The revolutionary is a man committed. He has neither personal interests nor sentiments, attachments, property, nor even a name. Everything in him is subordinated to a single exclusive interest, a single thought, a single passion: the revolution.[26]

Codes of self-sacrifice explain much terrorist behavior. Those who participate in movements and organizations adopt belief systems that justify their behavior and absolve them of responsibility for normally unacceptable behavior. Such systems cleanse participants and offer them a sense of participating in a higher or superior morality.

Chapter Perspective 3.3 compares and contrasts the motives and behavior of two Palestinian nationalists, Leila Khaled and Abu Nidal.

CHAPTER PERSPECTIVE 3.3

Profiles of Violent Extremists: Leila Khaled and Abu Nidal

How people become political extremists and terrorists is, of course, idiosyncratic. A comparison of two revolutionaries championing the Palestinian cause is useful in illustrating the origins of the motives and ideologies of politically violent individuals.

Leila Khaled: Freedom Fighter or Terrorist?

During the early 1970s, Leila Khaled was famous both because of her exploits as a Palestinian revolutionary and because she was for a time the best-known airline hijacker in the world.

Khaled was born in Haifa in Palestine. After the Israeli war of independence, she and her family became refugees in a camp in the city of Tyre, Lebanon, when she was a child. Khaled has said that she was politicized from a very young age and became a committed revolutionary by the time she was 15. Politically, she was influenced by leftist theory. One of her revolutionary heroes was Ernesto "Che" Guevara, whom she considered a true revolutionary, unlike other Western radicals.

In August 1969, at the age of 23, Leila Khaled hijacked a TWA flight on behalf of the Popular Front for the Liberation of Palestine (PFLP). The goal was to direct the world's attention to the plight of the Palestinians. It was a successful operation, and she reportedly forced the pilots to fly over her ancestral home of Haifa before turning toward Damascus. In Damascus, the passengers were released into the custody of the Syrians and the plane was blown up. Afterward, a then-famous photograph was taken of her.

In preparation for her next operation, and because the photograph had become a political icon, Khaled underwent plastic surgery in Germany to alter her appearance. She participated in a much larger operation on September 6 and 9, 1970, when the PFLP attempted to hijack five airliners. One of the hijackings failed, one airliner was flown to a runway in Cairo and destroyed, and the remaining three were flown to Dawson's Field in Jordan and blown up by the PFLP on September 12. Khaled had been overpowered and captured during one of the failed attempts on September 6—an El Al (the Israeli airline) flight from Amsterdam. She was released on September 28 as part of a brokered deal exchanging Palestinian prisoners for the hostages.

Leila Khaled published her autobiography in 1973, entitled *My People Shall Live: The Autobiography of a Revolutionary*.[a] She eventually settled in Amman, Jordan, and became a member of the Palestinian National Council, the Palestinian parliament. She has never moderated her political beliefs, has always considered herself a freedom fighter, and takes pride in being one of the first to use extreme tactics to bring the Palestinian cause to the world's attention. Khaled considers the progression of Palestinian revolutionary violence—such as the intifada—a legitimate means to regain Palestine.

Abu Nidal: Ruthless Revolutionary

Sabri al-Banna, a Palestinian, adopted the nom de guerre of Abu Nidal, which has since become synonymous with his **Abu Nidal Organization (ANO)**. He was a radical member of the umbrella Palestine Liberation Organization (PLO) from an early point in its history. Within the PLO, Yasir Arafat's nationalist Al Fatah organization was the dominant group. Unlike the Fatah mainstream, Abu Nidal strongly advocated a dissident ideology that was **pan-Arabist**, meaning that national borders in the Arab world are not believed sacrosanct. Abu Nidal long argued that Al Fatah membership should be open to all Arabs, not just Palestinians. In support of the Palestinian cause, he argued that Palestine must be established as an Arab state and that its borders must stretch from the Jordan River in the east to the Mediterranean sea. According to pan-Arabism, however, this is only one cause among many in the Arab world.

After the 1973 Yom Kippur war, when Israel soundly defeated invading Arab armies, many in the mainstream Al Fatah group argued that a political solution with Israel should be an option. In 1974, Abu Nidal split from Al Fatah and began his rejectionist movement to carry on a pan-Arabist armed struggle. He and his followers immediately began engaging in high-profile international terrorist attacks, believing that the war should not be limited to the Middle East. At different periods in his struggle, he successfully solicited sanctuary from Iraq, Libya, and Syria—all of which have practiced pan-Arabist ideologies.

The ANO became one of the most prolific and bloody terrorist organizations in modern history. It carried out attacks in approximately 20 countries and was responsible for killing or injuring about 900 people. The ANO's targets included fellow Arabs, such as the PLO, Arab governments, and moderate Palestinians. Its non-Arab targets included the interests of France, Israel, the United Kingdom, and the United States. Many of these attacks were spectacular, such as an attempted assassination of the Israeli ambassador to Great Britain in June 1982, simultaneous attacks on the Vienna and Rome airports in December 1985, the hijacking of a Pan Am airliner in September 1986, and several assassinations of top PLO officials in several countries. It has been alleged that Abu Nidal collaborated in the 1972 massacre of 11 Israeli athletes by the Black September group at the Munich Olympics.

Abu Nidal remained a dedicated pan-Arabist revolutionary and never renounced his worldwide acts of political violence. His group has several hundred members, a militia in Lebanon, and international resources. The ANO operated under numerous names, including the Al Fatah Revolutionary Council, Arab Revolutionary Council, Arab Revolutionary Brigades, Black September, Black June, and Revolutionary Organization of Socialist Muslims. The group seemingly ended its attacks against Western interests in the late 1980s. The only major attacks attributed to the ANO in the 1990s were the 1991 assassinations of PLO deputy chief Abu Iyad and PLO security chief Abu Hul in Tunis, and the 1994 assassination of the senior Jordanian diplomat Naeb Maaytah in Beirut.

The whereabouts of Abu Nidal were usually speculative, but he relocated to Iraq in December 1998. In August 2002, he was found dead in Iraq of multiple gunshot wounds. The official Iraqi account was that he committed suicide. Other unofficial accounts suggested that he was shot when Iraqi security agents came to arrest him, dying either of self-inflicted wounds or during a shootout.

Notes

a. Leila Khaled, *My People Shall Live: The Autobiography of a Revolutionary* (London: Hodder & Stoughton, 1973).

Chapter Summary

This chapter introduced readers to the theoretical causes of terrorism and presented examples that represent some of the models developed by scholars and other researchers. Individual profiles, group dynamics, political environments, and social processes are at the center of the puzzle of explaining why people and groups adopt fringe beliefs and engage in terrorist behavior. Social movements and dramatic (or traumatic) events have been identified as two sources of terrorism, with the caveat that they are generalized explanations.

Not all extremists become terrorists, but certainly all terrorists are motivated by extremist beliefs. Motives behind terrorist behavior include a range of factors. One is morality, an unambiguous conviction of the righteousness of one's cause. Terrorists believe that the principles of their movement are unquestionably sound. A second is simplified notions of good and evil, with no shades of gray, when terrorists presume that their cause and methods are completely justifiable because their opponents represent inveterate evil. A third is utopian ideals, whereby an idealized end justifies violence. Such ends are often vague concepts, such as Karl Marx's dictatorship of the proletariat. The fourth factor is critical to understanding terrorist behavior. It is codes of self-sacrifice, when an ingrained belief system forms the basis for a terrorist's lifestyle and conduct. Collectively, these factors form a useful theoretical foundation for explaining terrorist motives.

Explanations of terrorism also include a range of factors. The theory of acts of political will is a rational model in which extremists choose to engage in terrorism as an optimal strategy to force change. Sociological explanations of terrorism look at intergroup dynamics, particularly conflict that results in collective violence. Perception is an important factor in the decision. Psychological explanations broadly explain individual motivations and group dynamics. Psychological theories also help explain the cohesion of terrorist organizations and why they perpetuate violent behavior even when victory is logically impossible.

One final point should be considered when evaluating the causes of terrorism. When experts build models and develop explanatory theories for politically motivated violence, their conclusions sometimes "reflect the political and social currents of the times in which the scholars writing the theories live."[27] It is plausible that

> to a large degree, the development of theories . . . reflects changing political and intellectual climates. When intellectuals have opposed the collective behavior of their times . . . they have tended to depict the behavior negatively. . . . When scholars have instead supported the collective behavior of their eras . . . they have painted a more positive portrait of both the behavior and the individuals participating in it.[28]

This is not to say that analysts are not trying to be objective or that they are purposefully disingenuous in their analyses. But it is only logical to presume that new explanatory theories will be affected by factors such as new terrorist environments and new ideologies that encourage political violence. The progression of explanations by the social and behavioral sciences in the future will naturally reflect the sociopolitical environments of the times in which they are developed.

DISCUSSION BOX

This chapter's Discussion Box is intended to stimulate critical debate about seminal incidents in the history of national groups.

Bloody Sunday and Black September

Bloody Sunday

In the late 1960s, Irish Catholic activists calling themselves the Northern Ireland Civil Rights Association attempted to emulate the African American civil rights movement as a strategy to agitate for equality in Northern Ireland. They thought that the same force of moral conviction would sway British policy to improve the plight of the Catholics. Their demands were similar to those of the American civil rights movement: equal opportunity, better employment, access to housing, and access to education. This ended when mostly peaceful demonstrations gradually became more violent, leading to rioting in the summer of 1969, an environment of generalized unrest, and the deployment of British troops. After 1969, the demonstrations continued, but rioting, fire bombings, and gun battles gradually became a regular feature of strife in Northern Ireland.

On January 30, 1972, elite British paratroopers fired on demonstrators in Londonderry. Thirteen demonstrators were killed. After this incident, many Catholics became radicalized and actively worked to drive out the British. The Irish Republican Army received recruits and widespread support from the Catholic community. In July 1972, the Provos launched a massive bombing spree in central Belfast.

Black September

When Leila Khaled and her comrades attempted to hijack five airliners on September 6 and 9, 1970, their plan was to fly all of the planes to an abandoned British Royal Air Force (RAF) airfield in Jordan, hold hostages, broker the release of Palestinian prisoners, release the hostages, blow up the planes, and thereby force the world to focus on the plight of the Palestinian people. On September 12, 255 hostages were released from the three planes that landed at Dawson's Field (the RAF base), and 56 were kept to bargain for the release of seven Palestinian prisoners, including Leila Khaled. The group then blew up the airliners.

Unfortunately for the hijackers, their actions greatly alarmed King Hussein of Jordan. Martial law was declared on September 16, and the incident led to civil war between Palestinian forces and the Jordanian army. Although the Jordanian operation was precipitated by the destruction of the airliners on Jordanian soil, tensions had been building between the army and Palestinian forces for some time. King Hussein and the Jordanian leadership interpreted this operation as confirmation that radical Palestinian groups had become too powerful and were a threat to Jordanian sovereignty.

On September 19, King Hussein asked for diplomatic intervention from Great Britain and the United States when a Syrian column entered Jordan in support of the Palestinians. On September 27, a truce ended the fighting. The outcome of the fighting was a relocation of much of the Palestinian leadership and fighters to its Lebanese bases. The entire incident became known among Palestinians as Black September and was not forgotten by radicals in the Palestinian nationalist movement. One of the most notorious terrorist groups took the name Black September, and the name was also used by Abu Nidal.

Discussion Questions

1. What role do you think these incidents had in precipitating the IRA's and PLO's cycles of violence?

2. Were the IRA's and PLO's tactics and targets justifiable responses to these incidents?

3. What, in your opinion, would have been the outcome in Northern Ireland if the British government had responded to the Irish Catholics' emulation of the American civil rights movement?

4. What, in your opinion, would have been the outcome if the Jordanian government had not responded militarily to the Palestinian presence in Jordan?

5. How should the world community have responded to Bloody Sunday and Black September?

Key Terms and Concepts

The following topics were discussed in this chapter and can be found in the glossary:

Absolute deprivation	The end justifies the means	Nihilist dissidents
Abu Nidal	intifada	People's war
Abu Nidal Organization	Khaled, Leila	Relative deprivation theory
Act of political will	Kosovo Liberation Army	Stockholm
Black September	Marighella, Carlos	syndrome
Black Widows	*Mini-Manual of the Urban*	Structural theory
Bloody Sunday	*Guerilla*	Utopia
Codes of self-sacrifice	Mujahideen	Viet Cong

Terrorism on the Web

Log on to the Web-based student study site at **www.sagepub.com/martinessstudy** for additional Web sources and study resources.

Web Exercise

Using this chapter's recommended Web sites, conduct an online investigation of the causes of extremist agitation and terrorist violence.

1. What issues do these groups consider to have unquestioned merit? What reasons do they give for this quality?
2. What scenarios do you think might cause these groups to engage in direct confrontation or violence?
3. Act as "devil's advocate" and defend one of these causes that you disagree with.

For an online search of factors that are commonly cited as causes for terrorist violence, readers should activate the search engine on their Web browser and enter the following keywords:

"Intifadeh (or Intifada)"
"Just War"

Recommended Readings

The following publications provide discussions about the causes of terrorist behavior.

Djilas, Milovan. *Memoir of a Revolutionary.* New York: Harcourt Brace Jovanovich, 1973.

Huntington, Samuel P. *The Clash of Civilizations and the Remaking of World Order.* New York: Touchstone, 1996.

Khaled, Leila. *My People Shall Live: The Autobiography of a Revolutionary.* London: Hodder & Stoughton, 1973.

Martinez, Thomas, and John Guinther. *Brotherhood of Murder.* New York: Simon & Schuster, 1988.

Perry, Barbara. *In the Name of Hate: Understanding Hate Crimes.* New York: Routledge, 2001.

Reich, Walter, ed. *Origins of Terrorism: Psychologies, Ideologies, States of Mind.* Washington, DC: Woodrow Wilson Center Press, 1998.

PART II

Terrorist Environments

Part II Photo   Asymmetrical warfare. The remains of a vehicular bomb inside the heavily fortified Green Zone of Baghdad. A suicide bomber attacked the U.S.-occupied area during the post-2003 insurgency phase of the war in Iraq.

4 Terrorism by the State

This chapter explores the characteristics of terrorism from above—state terrorism—committed by governments and quasi-governmental agencies and personnel against perceived enemies. State terrorism can be directed externally against foreign adversaries or internally against domestic enemies. Readers will explore the various types of state terrorism and acquire an appreciation for the qualities that characterize each state terrorist environment. A state terrorist paradigm will be discussed and cases will be examined to understand what is meant by terrorism as foreign policy and as domestic policy.

It is important to understand conceptually that political violence by the state is the most organized, and potentially the most far-reaching, application of terrorist violence. Because of the many resources available to the state, its ability to commit acts of violence far exceeds that perpetrated by dissident terrorists in scale and duration. Only communal dissident terrorism (group-against-group violence) potentially approximates the scale and duration of state-sponsored terror.[1] Terrorism by states is characterized by official government support for policies of violence, repression, and intimidation. Such violence and coercion is directed against perceived enemies that the state has determined threaten its interests or security. Although the perpetrators of state terrorist campaigns are frequently government personnel acting in obedience to directives originating from government officials, those who carry out the violence are also quite often unofficial agents the government uses and encourages.

For example, some governments have adopted a strategy of using violent state-sponsored paramilitaries as an instrument of official state repression. The rationale behind supporting these paramilitaries is that they can be deployed to violently enforce state authority, and at the same time permit the state to deny responsibility for their behavior. Such deniability can be useful for propaganda purposes, because the government can officially argue that its paramilitaries represent a spontaneous grassroots reaction against their opponents.

The discussion in this chapter will review the following:

- Perspectives on terrorism by governments
- Domestic terrorism by the state
- Terrorism as foreign policy

Perspectives on Terrorism by Governments

Experts and scholars have designed a number of models to describe state terrorism. These constructs have been developed to identify distinctive patterns of state-sponsored terrorist behavior. Experts agree that several models can be differentiated. For example, one[2] describes state-level participants in a security environment as including the following:

♦ Sponsors of terrorism, meaning those states that actively promote terrorism and have been formally designated as rogue states or state sponsors under U.S. law.[3]

♦ Enablers of terrorism, or those states that operate in an environment in which "being part of the problem means not just failing to cooperate fully in countering terrorism but also doing some things that help enable it to occur."[4]

♦ Cooperators in counterterrorism efforts, including unique security environments in which "cooperation on counterterrorism is often feasible despite significant disagreements on other subjects."[5]

Photo 4.1 A Rewards for Justice poster indicating that the search for Saddam Hussein has ended with his capture. Although Hussein had minimal (if any) ties to Islamist groups such as Al Qaeda, his regime did give safe haven to a number of wanted nationalist terrorists such as Abu Nidal.

Another model describes the scale of violence as including the following:

♦ In warfare, the conventional military forces of a state are marshaled against an enemy. The enemy is either a conventional or guerrilla combatant and may be an internal or external adversary. This is a highly organized and complicated application.

♦ In genocide, the state applies its resources toward the elimination of a scapegoat group. The basic characteristic of state-sponsored genocidal violence is that it does not differentiate between enemy combatants and enemy civilians; all members of the scapegoat group are considered enemies. Like warfare, this is often a highly organized and complicated application.

♦ Assassinations are selective applications of homicidal state violence, in which a person or a specified group is designated for elimination. This is a lower-scale application.

♦ Torture is used by some states as an instrument of intimidation, interrogation, and humiliation. Like assassinations, it is a selective application of state violence directed against a single person or a specified group of people. Although it is often a lower-scale application of state violence, many regimes will make widespread use of torture during states of emergency.[6]

Understanding State-Sponsored Terrorism

Links between regimes and terrorism can range from very clear lines of sponsorship to very murky and indefinable associations. Governments inclined to use terrorism as an instrument of statecraft are often able to control the parameters of their involvement, so that they can sometimes manage how precisely a movement or an incident can be traced back to personnel.

Thus, state sponsorship of terrorism is not always a straightforward process. In fact, it is usually a covert, secret policy that allows states to claim deniability when accused of sponsoring terrorism. Because of these veiled parameters, a distinction must be made between **state patronage** and **state assistance**. These two subtly distinct concepts are summarized in Table 4.1.

State Sponsorship: The Patronage Model

State patronage for terrorism refers to active participation in and encouragement of terrorist behavior. Its basic characteristic is that the state, through its agencies and personnel, actively takes part in repression, violence, and terrorism. Thus, state patrons adopt policies that initiate terrorism and other subversive activities—including directly arming, training, and providing sanctuary for terrorists.

Patronage in Foreign Policy

In the foreign policy domain, state patronage for terrorism occurs when a government champions a politically violent movement or group—a proxy—that is operating beyond its borders. Under this model, the state patron will directly assist the proxy in its cause and continue its support even when the movement or group has become known to commit acts of terrorism or other atrocities. When these revelations occur, patrons typically reply to this information with rationalizations. The patron will do one of four things:

♦ Accept the terrorism as a necessary tactic
♦ Deny that what occurred should be labeled as terrorism
♦ Deny that an incident occurred in the first place
♦ Issue a blanket and moralistic condemnation of all such violence as unfortunate

The 1981 to 1988 U.S.-directed guerrilla war against the Sandinista regime in Nicaragua incorporated elements of the state patronage model.[7] Although it was not a terrorist war

---

TABLE 4.1   STATE SPONSORSHIP OF TERRORISM

State participation in terrorist and extremist behavior can involve either direct or indirect sponsorship and can be conducted in foreign or domestic policy domains. State patronage refers to relatively direct links between a regime and political violence. State assistance refers to relatively indirect links.

Domain	Type of Sponsorship	
	Patronage	**Assistance**
International	International violence conducted on government orders	International violence with government encouragement and support
	Case: Assassination of former Iranian Prime Minister Shahpour Bakhtiar in France by Iranian operatives	*Case:* Pakistani assistance for anti-Indian extremists in Jammu and Kashmir, including the Jammu Kashmir Liberation Front
Domestic	Domestic repression by government personnel	Domestic repression by progovernment extremists
	Case: Argentina's Dirty War conducted by the military	*Case:* Violence by rightist paramilitaries in El Salvador during the 1980s civil war

per se, the U.S. proxy did commit human rights violations. It is therefore a good case study of state patronage for a proxy that was quite capable of engaging in terrorist behavior. The most important component was U.S. support for anti-Sandinista Nicaraguan counterrevolutionaries, known as the contras.

From December 1981 until July 1983, the contras were sustained by U.S. arms and funding. Without this patronage, they would not have been able to operate against the Sandinistas. Unfortunately for the United States, evidence surfaced that implicated the contras in numerous human rights violations. These allegations were officially dismissed or explained away.

Patronage in Domestic Policy

In the domestic policy domain, state patronage of terrorism occurs when a regime engages in direct violent repression against a domestic enemy. Patronage is characterized by the use of state security personnel in an overt policy of political violence. State patrons typically rationalize policies of repression by arguing that they are necessary to

◆ Suppress a clear and present domestic threat to national security
◆ Maintain law and order during times of national crisis
◆ Protect fundamental cultural values that are threatened by subversives
◆ Restore stability to governmental institutions that have been shaken, usurped, or damaged by a domestic enemy

The Syrian government's 1982 suppression of a rebellion by the Muslim Brotherhood is an apt case study. The Muslim Brotherhood is a transnational Sunni Islamic fundamentalist movement very active in several North African and Middle Eastern countries. Beginning in the early 1980s, it initiated a widespread terrorist campaign against the Syrian government. In 1981, the Syrian army and other security units moved in to crush the Muslim Brotherhood in Aleppo and the city of Hama, killing at least 200 people. Syrian president Hafez el-Assad increased security restrictions and made membership in the organization a capital offense. In 1982, another Muslim Brotherhood revolt broke out in Hama. The Syrian regime sent in troops and tanks, backed by artillery, to put down the revolt; they killed approximately 25,000 civilians and destroyed large sections of the city.

State Sponsorship: The Assistance Model

State assistance for terrorism refers to tacit participation in and encouragement of terrorist behavior. Its basic characteristic is that the state, through sympathetic proxies and agents, implicitly takes part in repression, violence, and terrorism. In contrast to state patrons of terrorism, state assisters are less explicit in their sponsorship, and links to state policies and personnel are more ambiguous. State assistance includes policies that help sympathetic extremist proxies engage in terrorist violence, whereby the state will indirectly arm, train, and provide sanctuary for terrorists.

Assistance in Foreign Policy

In the foreign policy domain, state assistance for terrorism occurs when a government champions a politically violent proxy operating beyond its borders. Under this model, the assistance will be indirect, and the state may or may not continue its support if the movement or group becomes known for committing acts of terrorism or other atrocities. When the proxy's terrorism becomes known, state assisters typically weigh political costs and benefits when crafting a reply to the allegations. The ambiguity that the assister has built into its links with the proxy is intended to provide deniability when accused of complicity.

The contra insurgency against the Sandinistas was introduced as a case study of the state patronage model—with the caveat that it was not per se a terrorist war. The later phases of the war are also a good example of the state assistance model.

Several incidents undermined congressional support for the Reagan administration's policy in Nicaragua. In December 1982, Congress passed the **Boland Amendment**, which forbade the expenditure of U.S. funds to overthrow the Sandinista government. In late 1984, a second Boland Amendment forbade all U.S. assistance to the contras. These legislative measures were the catalyst for a highly covert effort to continue supplying the contras. The most effective effort to circumvent the congressional ban was the resupply network set up by Marine Lieutenant Colonel Oliver North, an official of the National Security Council. Lt. Col. North successfully set up a resupply program that shipped large amounts of arms to the contras. This program was intended to wait out congressional opposition to arming the contras and was successful, because in June 1986 Congress approved $100 million in aid for the contras. Congressional support for this disbursement was severely shaken when a covert American cargo plane was shot down inside Nicaragua, an American mercenary was captured, and the press published reports about Lt. Col. North's operations. This combination of factors—known as the **Iran-Contra Scandal**—ended congressional support for the contra program.

Assistance in Domestic Policy

In the domestic policy domain, state assistance for terrorism occurs when a regime engages in indirect violent repression against an enemy. Under this model, the assistance is characterized by the use of sympathetic proxies. This can occur in an environment where the proxy violence coincides with that of state security personnel. Thus, the overall terrorist environment may include both state patronage (direct repression) and state assistance (indirect repression). State assisters typically rationalize policies of indirect repression by adopting official positions that

- Blame an adversary group for the breakdown of order and call on the people to assist the government in restoring order
- Argue that the proxy violence is evidence of popular patriotic sentiment to suppress a threat to national security
- Call on all parties to cease hostilities but focus blame for the violence on an adversary group
- Assure everyone that the government is doing everything in its power to restore law and order but that the regime is unable to immediately end the violence

The Great Proletarian Cultural Revolution in China, which lasted from 1965 to 1969,[8] is a good example of state assistance for an ideologically extremist movement. Launched by national leader Mao Zedong and the Chinese Communist Party Central Committee, it was a mass movement that mobilized the young postrevolution generation. Its goal was to eliminate what it called revisionist tendencies in society and create a newly indoctrinated revolutionary generation. The period was marked by widespread upheaval and disorder. Maoists mobilized millions of young supporters in the Red Guards, who waged an ideological struggle to eliminate what were known as the Four Olds: Old Ideas, Old Culture, Old Customs, and Old Habits.[9] The Red Guards were the principal purveyors of the Cultural Revolution and were strongly encouraged to attack the Four Olds publicly and with great vigor. This led to widespread turmoil. For approximately 18 months, beginning in early 1967, the Red Guards seized control of key government bureaucracies. Because they had no experience in government operations, the government ceased to operate effectively. It was not until violent infighting within the Red Guards began that Mao ordered an end to the Cultural Revolution and deployed the People's Liberation Army to restore order. Chapter Perspective 4.1 explores another example of chaos as liberation in Zimbabwe.

CHAPTER PERSPECTIVE 4.1

Chaos as Liberation: State Repression in Zimbabwe

The southern Africa country of Zimbabwe was designated as the British colony of Southern Rhodesia in 1923. In 1965, it declared itself the independent nation of Rhodesia, an act that was unrecognized and denounced by Great Britain and other nations. Rhodesia's constitution established political domination by the country's white minority, thus creating a system of racial oligarchy.

Because of the inherently racial disparities of the Rhodesian system, the United Nations imposed economic sanctions on the country. These sanctions, coupled with a guerrilla war waged by several rebel factions, led to the first free elections in 1979, and the renaming of Rhodesia to Zimbabwe.

Robert Mugabe, the leader of the Zimbabwe African National Union-Patriotic Front (ZANU-PF), became the country's prime minister, and later president. He initially promised democracy and economic growth, but in fact never permitted significant opposition to his rule. In 2000, Mugabe began a policy of land redistribution, in which pro-Mugabe Zimbabweans (many of them veterans of the guerrilla war) were allowed to seize farmland owned by white farmers. White farmers owned large estates (a condition from the colonial past), and Zimbabwe's agricultural sector regularly produced a surplus for export. Many of the land seizures were violent, and a large number of white farmers emigrated. Because of the seizures and the emigration, the agricultural sector came close to collapse, what had once been a food exporting country become a food importing nation, and the economy was severely damaged. At the same time, all opposition to Mugabe and ZANU-PF—largely from poor Zimbabwean city dwellers—was suppressed.

In 2002, Mugabe held an election that was designed to ensure his reelection over the opposition Movement for Democratic Change (MDU). In 2005, another election was held that was widely condemned as unfair and rigged, and involved intimidating MDU supporters. When he was reelected, Mugabe, who was 81 at the time, pledged to rule until he was 100. He also threatened the opposition by warning that all protests would be suppressed by his security forces, and suppressed violently because the MDU is inherently violent.

Also in 2005, ZANU-PF began a major program of social engineering against the poor urban Zimbabweans, apparently in retaliation for their supporting the MDU. Zimbabwean security personnel and ZANU-PF operatives systematically ordered the demolition of urban poor neighborhoods, with the intention of driving residents into the countryside and thereby creating pro-Mugabe strongholds in the cities. Many residents became homeless and some died.

Domestic Terrorism by the State

State terrorism as domestic policy refers to a state's politically motivated application of force inside its own borders. Military, law enforcement, and other security institutions are used to suppress perceived threats and can be supplemented by unofficial paramilitaries and death squads. The purpose of domestically focused terrorism is to demonstrate the supreme power of the government and to intimidate or eliminate the opposition. In environments where the central government perceives its authority to be seriously threatened, this force can be extreme.

South Africa during the final years of apartheid, the system of racial separation, is a good example. When confronted by a combination of antiapartheid reformist agitation, mass unrest, and terrorist attacks, the South African government began a covert campaign to root out anti-apartheid leaders and supporters. This included government support for the Zulu-based **Inkatha Freedom Party** in its violence against the multiethnic and multiracial **African National Congress (ANC)**. The South African government also assigned security officers to command death squads called **Askaris**, who assassinated ANC members both inside South Africa and in neighboring countries.

Legitimizing State Authority

Every type of regime seeks to legitimize its authority and maintain its conception of social order. How this is done often depends on the political environment at a particular point in a nation's history, and is done with varying degrees of restraint. Stable democracies with strong constitutional traditions are usually characterized by measured restraint. Regimes with weak constitutional traditions, or those in a period of national crisis, often show little or no restraint. Examples of state domestic authority can be summarized as follows:

◆ Democracy is a system of elected government in which authority is theoretically delegated by the people to elected leaders. Under this model, a strong constitution grants authority for elected leaders to govern the people and manage the affairs of government. The power of the state is clearly delimited.

◆ Authoritarianism is a system of government in which authority and power come from the state rather than being delegated by the people to elected leaders. Law, order, and state authority are emphasized. Authoritarian regimes can have elected leaders, but these leaders have authoritarian power and often rule for indefinite periods. Constitutions do not have enough authority to prohibit abuses by the state.

◆ Totalitarianism is a system under which all national authority originates from the government, which enforces its own vision of an ordered society.

◆ Crazy states[10] are those whose behavior is not rational, those in which the people live at the whim of the regime or a dominant group. Some crazy states have little or no central authority and are ravaged by warlords or militias. Others have capricious, impulsive, and violent regimes in power that act out with impunity.

Table 4.2 illustrates these models of domestic state authority by summarizing sources of state authority and giving examples of these environments.

State Domestic Authority

The following discussion is a domestic state terrorist model adapted from one designed by Peter C. Sederberg.[11] It defines and differentiates broad categories of domestic state terrorism that are useful in critically analyzing the motives and behaviors of terrorist regimes. They signify the varied qualities of state-sponsored terrorism directed against perceived domestic enemies:

◆ Unofficial repression: vigilante domestic state terrorism
◆ Repression as policy: overt and covert official state terrorism
◆ Mass repression: genocidal state terrorism

TABLE 4.2 STATE DOMESTIC AUTHORITY			
Several models illustrate the manner in which state authority is imposed and the degree of coercion that is used to enforce governmental authority. Sources of state authority differ depending on which model characterizes each regime.			
Models of State Authority	**Sources of State Authority**		**Examples of Authority Models**
	Legitimization of Authority	**Center of Authority**	
Democracy	Secondary role of security institutions; strong constitution and rule of law	Government with constrained authority	◆ United States ◆ Western Europe ◆ Japan
Authoritarianism	Central role of official security institutions; strong constitution possible	Government with minimally constrained authority	◆ Egypt ◆ Myanmar (Burma) ◆ Syria
Totalitarianism	Central role of official security institutions	Government with unconstrained authority	◆ China ◆ North Korea ◆ Taliban Regime
Crazy states	Central roles of official and unofficial security institutions	Government with unconstrained authority, or unconstrained paramilitaries, or both	◆ Liberia ◆ Somalia ◆ Uganda under Idi Amin

Unofficial Repression: Vigilante Domestic State Terrorism

Vigilante terrorism is political violence perpetrated by nongovernmental groups and individuals. These groups can receive unofficial support from agents of the state.

Why do regimes encourage vigilante violence? What are the benefits of such support? From the perspective of the state, what are the values being safeguarded by the vigilantes? Vigilante violence committed on behalf of a regime is motivated by the perceived need to defend a demographic group or cultural establishment. The overall goal of vigilante state terrorism is to violently preserve the preferred order. In a classic terrorist rationalization process, the end of an orderly society justifies the means of extreme violence.

The vigilante terrorists, sometimes alongside members of the state security establishment, unofficially wage a violent suppression campaign against an adversarial group or movement. Such a campaign occurs when civilians and members of the state's security forces perceive that the state is threatened. This can occur in warlike environments or when an alternative social movement or ideology challenges the established order. Civilians and members of the security establishment who participate in vigilante violence adhere to a code of duty and behavior, and thus believe that their actions are absolutely justifiable. Nongovernmental vigilantes often organize themselves into paramilitaries and operate as death squads. Death squads have committed many documented massacres and atrocities, including assassinations, massacres, disappearances, and random terrorist attacks.

Repression as Policy: Official Domestic State Terrorism

Photo 4.2 The Guardia on patrol. Salvadoran troops patrol a village during the brutal civil war in the 1980s. The Guardia and rightist death squads were responsible for thousands of civilian deaths.

State-sponsored repression and political violence were practiced regularly during the 20th century. Many regimes deliberately adopted domestic terrorism as a matter of official policy, and directives ordering government operatives to engage in violent domestic repression frequently originated with ranking government officials.

Why do regimes resort to official policies of domestic violence? What are the benefits of such programs? From the perspective of the state, who are the people who deserve this kind of violent repression? The goals of **official state terrorism** are to preserve an existing order and to maintain state authority through demonstrations of state power. Regimes that officially selected violent repression as a policy choice rationalized their behavior as a legitimate method of protecting the state from an internal threat. Two manifestations of official state terrorism in the domestic domain must be distinguished: overt and covert.

Overt official state terrorism refers to the visible application of state-sponsored political violence. A policy of unconcealed and explicit repression against a domestic enemy, is common in totalitarian societies, such as Stalinist Russia, Nazi Germany, Khmer Rouge Cambodia, and Taliban Afghanistan.

Covert official state terrorism refers to the secretive application of state-sponsored political violence. A policy of concealed and implicit repression against a domestic enemy, it is common in countries with extensive secret police services, such as President Hafez el-Assad's Syria, Saddam Hussein's Iraq, General Augusto Pinochet's Chile, and Argentina during the Dirty War.

Official state terrorism is not always directed against subversive elements. It is sometimes conducted to cleanse society of an undesirable social group. These groups are perceived as purveyors of a decadent lifestyle or immoral values, or as otherwise unproductive drains on society. Chapter Perspective 4.2 discusses how extremist regimes have solved this problem by engaging in so-called social cleansing and ethnic cleansing.

Photo 4.3   Soviet leader Joseph Stalin in military regalia. Stalin's totalitarian regime purged and killed thousands of ideological rivals and sent millions of members of ethno-national groups into internal exile. Millions of others died during famines and in work camps.

CHAPTER PERSPECTIVE 4.2

Cleansing Society

Among the euphemisms used by propagandists to characterize state-initiated domestic terrorism, perhaps the most commonly applied is cleansing society. Conceptually, an image is constructed that depicts an undesirable group as little more than a virus or bacterium that has poisoned society. The removal of this group is deemed a necessary remedy for the survival of the existing social order.

This imagery has been invoked repeatedly by extremist regimes. An example from Fascist Italy illustrates this point:

"Terror? Never," Mussolini insisted, demurely dismissing such intimidation as "simply . . . social hygiene, taking those individuals out of circulation like a doctor would take out a bacillus."[a]

For society to solve its problems, the bacterium represented by the group must be removed. Acceptance of this characterization makes domestic terrorism palatable to many extremist regimes. The following cleansing programs are recent examples of this imagery.

Social Cleansing

Social cleansing refers to the elimination of undesirable social elements. These undesirable elements are considered to be blights on society and can include street children, prostitutes, drug addicts, criminals, homeless people, transvestites, and homosexuals. In Colombia, undesirable social elements are commonly referred to as disposables.

Social cleansing has occurred in a number of countries. The term was probably coined in Latin America, where social cleansing took on the attributes of vigilante state domestic terrorism in Brazil, Guatemala, Colombia, and elsewhere. Participants in cleansing campaigns have included members of the police and death squads. In societies where social cleansing has occurred, the disposables have been killed, beaten, and violently intimidated.

Ethnic Cleansing

The term *ethnic cleansing* was coined during the war in Bosnia in the former Yugoslavia. It refers to the expulsion of an ethno-national group from a geographic region as a means to create an ethnically pure society. During the war in Bosnia, Serb soldiers and paramilitaries initiated a cycle of ethnic cleansing. They officially and systematically expelled, killed, raped, and otherwise intimidated Bosnian Muslims to create Serb-only districts. The most intensive campaigns occurred in 1992 and 1993. As the war progressed, Croats and Bosnians also engaged in ethnic cleansing, so that there were periods during the war in which all three groups cleansed areas populated by members of the other groups.

Since the war in Bosnia, the term has become widely used to describe present and past campaigns to systematically and violently remove ethno-national groups from geographic regions.

Note

a. Walter Laqueur, *The Age of Terrorism* (Boston: Little, Brown, 1987), 66, quoted in Bruce Hoffman, *Inside Terrorism* (New York: Columbia University Press, 1998), 24.

Mass Repression: Genocidal Domestic State Terrorism

The word **genocide** was first used by Dr. **Raphael Lemkin** in 1943 and first appeared in print in his influential book *Axis Rule in Occupied Europe*, published in 1944.[12] It is derived from the Greek word *genos*, meaning race or tribe, and the Latin-derived suffix *cide*, meaning killing. Genocide is first and generally defined as the elimination of a group as a matter of state policy or communal dissident violence by one group against another. The second refers to campaigns of substate, group-against-group terrorism.

Whether perpetrated at the state or communal level, genocide is considered an unacceptable social policy and an immoral application of force. It has been deemed a crime under

Photo 4.4   The Killing Fields. Skulls are displayed of victims of Cambodia's genocidal Khmer Rouge regime. The Khmer Rouge waged a campaign of domestic terrorism that claimed the lives of at least 1 million Cambodians.

international law since 1946, when the General Assembly of the United Nations adopted Resolution 96(I). In 1948, the General Assembly adopted the Convention on the Prevention and Punishment of the Crime of Genocide. Under Article 2 of the convention, genocide is formally defined as follows:

> Any of the following acts committed with intent to destroy, in whole or in part, a national, ethnic, racial or religious group, such as:
> a. Killing members of the group;
> b. Causing serious bodily or mental harm to members of the group;
> c. Deliberately inflicting on the group conditions of life calculated to bring about its physical destruction in whole or in part;
> d. Imposing measures intended to prevent births within the group;
> e. Forcibly transferring children of the group to another group.[13]

Why do regimes resort to genocidal policies against their fellow countrymen? What are the benefits to a regime of eliminating a particular group? From the perspective of the state, why do some groups deserve to be eliminated? One practical reason for terrorist regimes is that scapegoating a defined enemy is a useful strategy to rally the nation behind the ruling government. The goal is to enhance the authority and legitimacy of the regime by targeting internal enemies for genocidal violence.

States have available to them, and frequently marshal, enormous resources to use against an undesired group. Such resources can include the military, security services, civilian paramilitaries, legal systems, private industry, social institutions, and propaganda resources. When the decision is made to eliminate or culturally destroy a group, state resources can be brought to bear with devastating efficiency. Chapter Perspective 4.3 discusses a case in point, Iraq's genocidal Anfal campaign against its Kurdish population in 1988.

CHAPTER PERSPECTIVE 4.3

The Anfal Campaign: Genocidal State Terrorism in Iraq

During the regime of Saddam Hussein, political power in Iraq was highly centralized in the Ba'ath Party, which Hussein led. The Ba'ath Party, one faction of which governs Syria under the leadership of Bashar el-Assad, is secular, Arab, and nationalist. In Iraq, Ba'athist functionaries were personally loyal to Hussein and commanded key government institutions, the military, and the security services. Hussein and the Ba'ath Party legitimized state authority through force of arms. Dissent was severely repressed and rebellion was dealt with mercilessly. Iraqi armed forces were unleashed to quell rebellions among the so-called Marsh Arabs in the country's southeastern region and among its Kurdish population in the north.

Iraq's Anfal campaign was an instance of genocidal state terrorism waged against the Iraqi Kurdish minority from February to September 1988.

Had history been kinder to the Kurds, they might have established their own sovereign nation of Kurdistan. Unfortunately, Kurdistan is a geographic region politically divided between Iraq, Iran, Turkey, and Syria. The Iraqi Kurds have historically sought political self-governance in their region. Some Iraqi Kurds seeking complete independence resorted to armed insurrection. Iraqi policy during its suppression campaigns against Kurdish rebellions was brutal and included the use of poison gas.

In 1987, Saddam Hussein's cousin Ali Hassan al-Majid was assigned the task of securing Iraqi governmental authority in Kurdistan. His carefully planned campaign was code-named al-Anfal, or *the spoils*, named for passages in the Quran that describe the Prophet Muhammed's revelations after his first great victory and advise the prophet to severely punish unbelievers when necessary. When the Anfal campaign was launched, it had eight phases, during which civilians were uprooted, many killed en masse. Males in particular were singled out for death. During the campaign, Iraq dropped chemical weapons, including mustard gas and nerve agents, on Kurdish villages.[a] One attack was described as follows:

> The planes dropped bombs. They did not produce a big noise. A yellowish cloud was created and there was a smell of rotten parsley or onions. There were no wounds. People would breathe the smoke, then fall down and blood would come from their mouths.[b]

Between 50,000 and 100,000 Kurds may have died.[c] After the campaign, that particular round of the Kurdish rebellion was quelled, and approximately 2.5 million Iraqi Kurds were forced into exile. It is a good case study of how regimes in which state power is centralized in a ruling body may use terrorism and other violent repression to maintain domestic authority.

As a postscript, Ali Hassan al-Majid, nicknamed Chemical Ali and The Butcher of Kurdistan for his actions in the Anfal campaign, was captured by U.S. troops in August 2003 and condemned to death by an Iraqi court in June 2007.

Notes

a. Walter Laqueur, The Age of Terrorism (Boston: Little, Brown, 1987), 66, quoted in Bruce Hoffman, Inside Terrorism (New York: Columbia University Press, 1998), 24.
b. Cited in Jessica Stern, The Ultimate Terrorists (Cambridge, MA: Harvard University Press, 1999), 107.
c. See Human Rights Watch, Iraq's Crime of Genocide: The Anfal Campaign Against the Kurds (New York: Human Rights Watch Publications, 1994).

Genocidal state terrorism occurs, then, when the resources of a nation are mobilized to eliminate a targeted group. The group can be a cultural minority—such as a racial, religious, or ethnic population—or a designated segment of society—such as believers in a banned ideology or a socioeconomically unacceptable group. When ideological or socioeconomic groups are singled out, the resulting environment is one in which members of the same ethnic or religious group commit genocide against fellow members, a practice known as auto-genocide (self-genocide). This occurred during the reign of the Khmer Rouge in Cambodia, which Chapter Perspective 4.4 explores.

CHAPTER PERSPECTIVE 4.4

The Khmer Rouge: Genocidal State Terrorism in Cambodia

Colonial Cambodia was governed as part of French Indochina, which also included what are now the countries of Laos and Vietnam. In 1944, Vietnamese nationalists led by Ho Chi Minh rebelled against the French colonial regime, which had essentially collaborated with Japanese occupiers during World War II. Cambodian intellectuals and activists joined with the Vietnamese during their resistance against the French. One Cambodian activist was Pol Pot, who joined the underground Indochinese Communist Party in about 1946.

When the French withdrew after their defeat in 1954 at Dien Bien Phua in Vietnam, Cambodia became a republican monarchy ruled by Prince Norodom Sihanouk.

Many of the Cambodians who participated in anti-French agitation during World War II were nationalist communists. Some of them gathered abroad after the war and established strong ideological bonds. Pol Pot lived and studied for 4 years in France, where he was an unremarkable student but met with other Indochinese activists and became a dedicated Marxist revolutionary. When he returned to Cambodia in 1953, he became a member of the People's Revolutionary Party of Kampuchea, which was Communist. In 1960, the Cambodian Communist Party was founded. Pol Pot became its leader in 1963 and soon fled with his followers into the jungle after Prince Sihanouk suppressed Communist Party activities.

It was in the jungle in the mid-1960s that the guerrilla movement known as the Khmer Rouge (Red Khmers, given that most Cambodians are ethnic Khmers), led by Pol Pot, took up arms. The Khmer Rouge was an extreme left-wing movement that sought to completely redesign and restructure society. In 1967, it staged an armed rebellion that was put down by Sihanouk's forces. In 1970, a civil war broke out after a right-wing coup—led by Cambodian Prime Minister Lon Nol and Prince Sirik Matak—toppled the Sihanouk government. Prince Sihanouk eventually joined Pol Pot and the Khmer Rouge to fight against the new regime. Khmer Rouge forces were swelled with new recruits when U.S. forces bombed and invaded Cambodia in 1970.b Despite massive U.S. assistance, the Lon Nol government suffered one defeat after another at the hands of the Khmer Rouge.

In January 1975, the Khmer Rouge seized the capital of Phnom Penh, now swollen by 2 million refugees, renamed the country Democratic Kampuchea, set the calendar to Year Zero, and began a 4-year reign of state domestic terror. Their first act was to depopulate Phnom Penh through mass evacuations of the inhabitants to the countryside. There, they toiled in the fields to build the Khmer Rouge's vision of an ideal communist society. The era of the Killing Fields had begun, when hundreds of thousands of Cambodians (possibly as many as 2 million) died in the countryside from executions, starvation, and arduous forced labor.

Notes

a. France lost 15,000 soldiers at Dien Bien Phu. For a good history of the battle, see Bernard Fall, Hell in a Very Small Place: The Siege of Dien Bien Phu (Philadelphia: Lippincott, 1963).

b. See William Shawcross, Sideshow: Kissinger, Nixon, and the Destruction of Cambodia (New York: Simon & Schuster, 1979).

Unlike vigilante and official state terrorism, the scale of violence during campaigns of state-sponsored genocidal terrorism can be virtually unlimited. In some cases, no check whatsoever is placed on what violence can be used against an adversary group, and the targeted group may suffer casualties in the many thousands or millions.

It is important to understand that the elimination of a group does not necessarily entail their physical extermination. The state's goal might also be to destroy a culture. This can be accomplished through forced population removals or prohibitions against practicing religious, linguistic, or other measures of cultural identification.

Most cases of state genocide are not examples of a precipitous policy under which the security services or paramilitaries are suddenly unleashed against a targeted group. More commonly, the methodology and purpose behind genocidal policies require a coordinated series of events, perhaps in phases over months or years. During these phases, cultural or other measures of identification can be suppressed in a number of ways—perhaps with the ultimate goal of physical extermination.

Table 4.3 identifies several examples of state-sponsored genocidal campaigns directed against domestic groups. Table 4.4 identifies several paramilitaries that have operated in Latin America with the support of government security services.

TABLE 4.3 STATE-INITIATED GENOCIDE

Genocidal state terrorism is directed against populations within countries that the state declares to be undesirable. When this occurs, governments and extremist regimes have designed policies of elimination that can include cultural destruction, mass resettlement, violent intimidation, or complete extermination. Historically, state-initiated genocide is not an uncommon policy selection.

Country	Activity Profile		
	Incident	Target Group	Outcome
Rwanda	Rwandan president Habyarimana assassinated	Tutsis and Hutu moderates	Genocidal violence; approximately 500,000 people killed by Rwandan army and Hutu militants
Cambodia	Victory on the battlefield by the Khmer Rouge; imposition of a new regime	City dwellers, educated people, upper class, Buddhists, fellow Khmer Rouge	The Killing Fields; up to 2 million deaths
Bosnia	The breakup of Yugoslavia and Serb resistance to the declarations of independence by Slovenia, Croatia, and Bosnia	Muslims living in territory claimed by Serbs	Ethnic cleansing assisted by Serbia, population removals, massacres, systematic rape, and cultural destruction
Germany	Racially motivated genocide by the Nazi regime	German Jews	The Holocaust; deaths of most of Germany's Jewish population
United States	Conquering the frontier and 19th-century frontier wars	Native Americans	Annihilation of some tribes; resettlement of others on reservations

Terrorism as Foreign Policy

During the 20th century, states used military forces to pursue policies of aggression, conquest, and cultural or ethnic extermination. In the latter half of the century, and especially in the latter quarter, many governments used terrorism as an instrument of foreign policy. As a policy option, state-sponsored terrorism is a logical option, because states cannot always deploy conventional armed forces to achieve strategic objectives. As a practical matter for many governments, it is often not logistically, politically, or militarily feasible to confront an adversary directly. For example, few states can hope to be victorious in a conventional military confrontation with the United States—as Saddam Hussein's well-entrenched Iraqi army learned

TABLE 4.4 VIGILANTE TERRORISM: THE CASE OF THE PARAMILITARIES

Paramilitaries are armed nongovernmental groups or gangs. Progovernment paramilitaries generally consider themselves the defenders of an established order under attack from a dangerously subversive counterorder. Some are well armed and receive direct official support from government personnel. Others are semi-independent vigilante groups. Reactionary right-wing paramilitaries in Latin America have been implicated in numerous atrocities, including massacres and assassinations.

Paramilitary Group	Benefactor	Target
United Self-Defense Groups (AUC)	Colombian security services, Colombian landholders	Marxist FARC rebels and suspected supporters
Civil Patrols (CAP)	Guatemalan security services, Guatemalan landholders	Marxist rebels and suspected supporters
Chiapas Paramilitaries	Possibly Mexican security force members; Mexican landholders	Zapatistas and suspected supporters
Argentine Anti-Communist Alliance (Triple-A)	Argentine security services	Leftists

both in Kuwait during the Gulf War of 1990–1991 and the U.S.-led invasion of March and April 2003. Terrorism thus becomes a relatively acceptable alternative for states pursuing an aggressive foreign policy against a superior adversary.

Governments use terrorism and other confrontational propaganda because, from their point of view, it is an efficient way to achieve strategic objectives. As a practical matter for aggressive regimes, state terrorism in the international domain is advantageous in several respects:

State terrorism is inexpensive. The costs of patronage and assistance for terrorist movements are relatively low. Even poor nations can strike and injure a prosperous adversary in a single spectacular incident.

State terrorism has limited consequences. State assisters that are clever can distance themselves from culpability for a terrorist incident. They can cover their involvement, disclaim responsibility, and thus escape possible reprisals or other penalties.

State terrorism can be successful. Weaker states can raise the stakes beyond what a stronger adversary is willing to bear. Aggressor states that wish to remain anonymous can likewise successfully destabilize an adversary in a proxy movement. They can do so with one or more spectacular incidents or by assisting in a campaign of terror.

To simplify matters for our discussion, we will discuss the following four policy frameworks. They signify the varied qualities of state-sponsored terrorism in the international domain:

1. Moral support: politically sympathetic sponsorship
2. Technical support: logistically supportive sponsorship
3. Selective participation: episode-specific sponsorship
4. Active participation: joint operations

Table 4.5 summarizes these policy frameworks by placing them within the context of state patronage and state assistance for terrorism.

Moral Support: Politically Sympathetic Sponsorship

Politically sympathetic sponsorship occurs when a government openly embraces the main beliefs and principles of a cause. This embrace can range in scope from political agreement with a movement's motives, though not its tactics, to complete support for both. Such support may be delivered either overtly or covertly. Although politically sympathetic governments act as ideological role models for their championed group, such support is often a way for the state to pursue its national agenda.

Iran's support for several violent movements in the Middle East is an example of unambiguous policy of mentorship for groups known to have engaged in acts of terrorism. Iran

---

TABLE 4.5   STATE-SPONSORED TERRORISM: THE FOREIGN POLICY DOMAIN

State participation in terrorism in the international domain can involve several types of backing for championed causes and groups, which can range in quality from relatively passive political sympathy to aggressive joint operations.

Type of State Backing	Type of Sponsorship	
	Patronage	**Assistance**
Politically sympathetic (moral support)	Overt political support and encouragement for a championed group's motives or tactics	Implicit political support and encouragement for a championed group's motives or tactics
	Case: Official Arab governments' political support for the objectives of the PLO	*Case:* Iran's ideological connection with Lebanon's Hezbollah
Logistically supportive (technical support)	Direct state support, such as sanctuary, for a championed cause to the group	The provision of state assistance to a group, such as providing matériel (military hardware
	Case: Jordanian facilitation of PLO fedayeen bases inside Jordan for raids on Israel prior to Black September	*Case:* Syria's provision of sanctuary, resupply, and other facilities for extremist movements
Episode-specific (selective participation)	Direct involvement by government personnel for a specific incident or campaign	The provision of state assistance to a group or movement for a specific goal
	Case: Yugoslavia's deployment of Yugoslav Army units to Bosnia during the Bosnian civil war	*Case:* Iran's attempted delivery of 50 tons of munitions to the PLO in January 2002
Joint operations (active participation)	Operations carried out by government personnel jointly with its proxy	Indirect state support for a proxy using allied personnel
	Case: The American-South Vietnamese Phoenix Program during the war in Vietnam	*Case:* Soviet deployment of Cuban troops to Ethiopia

consistently provided politically sympathetic and logistically supportive sponsorship for several movements, including Lebanon's Hezbollah, **Palestine Islamic Jihad (PIJ)**, and Hamas. All of these organizations adopted religious revolutionary ideologies—including strong anti-Israel goals—which created a sense of revolutionary common cause among religious hardliners in Iran.

Technical Support: Logistically Supportive Sponsorship

Logistically supportive sponsorship occurs when a government provides aid and comfort to a championed cause. This can include directly or indirectly facilitating training, arms resupply, safe houses, or other sanctuary for the movement. These options are relatively passive types of support that allow state sponsors of terrorism to promote an aggressive foreign policy agenda but deny their involvement in terrorist incidents.

An excellent case study is the foreign policy Syria adopted during the rule of Hafez el-Assad. During that time, from February 1971 to June 2000, Syria fought two wars against Israel, strongly backed the Palestinian cause, occupied the Beka'a Valley in Lebanon, and supported the Lebanese militias Amal and Hezbollah. Assad's regime could certainly be aggressive in the international domain, but despite this activism Syria was rarely linked directly to terrorist incidents. "There is no evidence that either Syria or Syrian government officials have been directly involved in the planning or execution of international terrorist attacks since 1986."[14]

Selective Participation: Episode-Specific Sponsorship

Episode-specific sponsorship refers to government support for a single incident or series of incidents. For this type of operation, the government will provide as much patronage or assistance as is needed. Sometimes members of the proxy will carry out the incident, and at other times agents of the state sponsor will.

One example of episode-specific support was the bombing of **Pan Am Flight 103**, which exploded over Lockerbie, Scotland, on December 21, 1988. Two hundred seventy people were killed, all 259 passengers and crew and 11 persons on the ground. In November 1991, the United States and Great Britain named two Libyan nationals as the masterminds of the bombing. The men—**Abdel Basset al-Megrahi** and **Lamen Khalifa Fhima**—were alleged to be agents of Libya's **Jamahiriya Security Organization (JSO)**. This was a significant allegation, because the JSO was repeatedly implicated in numerous acts of terrorism, including killing political rivals abroad, laying mines in the Red Sea, attacking Western interests in Europe, and providing logistical support and training facilities for terrorists from around the world. Libyan leader **Muammar el-Qaddafi** denied any involvement of the Libyan government or its citizens.

Active Participation: Joint Operations

Joint operations are when government personnel carry out campaigns in cooperation with a championed proxy. Close collaboration is typical, with the sponsor providing primary operational support for the campaign. Joint operations are often undertaken during a large-scale and ongoing conflict.

An example is the **Phoenix Program**, a campaign conducted during the Vietnam War to disrupt and eliminate the administrative effectiveness of the **Viet Cong**, the communist guerrilla movement recruited from among southern Vietnamese. It was a 3-year program that focused on the infrastructure of the Viet Cong. Both American and allied South Vietnamese squads were to wage the campaign by pooling intelligence information and making lists of persons to be targeted. The targets were intended to be hard-core communist agents and administrators, and they were supposed to be arrested rather than assassinated.

In practice, although the communists were significantly disrupted, many innocent Vietnamese were swept up in the campaign. Also, "despite the fact that the law provided only for the arrest and detention of the suspects, one-third of the 'neutralized agents' were reported dead."[15] Corruption was rampant among South Vietnamese officials.

Thus, terrorism and sponsorship for subversive movements are methods of statecraft that many types of governments have adopted, from stable democracies to aggressive and revolutionary regimes. It is certainly true that democracies are less likely to engage in this behavior than aggressively authoritarian states. However, democracies have been known to resort to terrorist methods when operating within certain security or political environments.

Chapter Perspective 4.5 discusses the officially defined threat inherent in the authoritarian government of Saddam Hussein that precipitated the U.S.-led invasion of Iraq in 2003.

CHAPTER PERSPECTIVE 4.5

Calculation or Miscalculation? Weapons of Mass Destruction and the Iraq Case

One of the most disturbing scenarios involving state-sponsored terrorism is the delivery of weapons of mass destruction (WMDs) to motivated terrorists by an aggressive authoritarian regime. This scenario was the underlying rationale given for the March 2003 invasion of Iraq by the United States and several allies.

In January 2002, U.S. president George W. Bush identified Iraq, Iran, and North Korea as an **axis of evil**, and promised that the United States "will not permit the world's most dangerous regimes to threaten us with the world's most destructive weapons." In June 2002, President Bush announced during a speech at the U.S. Military Academy at West Point that the United States would engage in preemptive warfare if necessary.

Citing Iraq's known possession of weapons of mass destruction in the recent past, and its alleged ties to international terrorist networks, Bush informed the United Nations in September 2002 that the United States would unilaterally move against Iraq if the UN did not certify that Iraq no longer possessed WMDs. Congress authorized an attack on Iraq in October 2002. UN weapons inspectors returned to Iraq in November 2002. After a 3-month military buildup, Iraq was attacked on March 20, 2003, and Baghdad fell to U.S. troops on April 9, 2003.

The Bush administration had repeatedly argued that Iraq still possessed a significant arsenal of WMDs at the time of the invasion, that Hussein's regime had close ties to terrorist groups, and that a preemptive war was necessary to prevent the delivery of these weapons to Al Qaeda or another network. Although many experts discounted links between Hussein's regime and religious terrorists, it was widely expected that WMDs would be found. Iraq was known to have used chemical weapons against Iranian troops during the Iran-Iraq War of 1980–1988 and against Iraqi Kurds during the Anfal campaign of 1987 (discussed in Chapter Perspective 4.4).

In actuality, UN inspectors identified no weapons of mass destruction before the 2003 invasion, and U.S. officials found none during the occupation of Iraq. Additionally, little evidence was uncovered to substantiate allegations of strong ties between Hussein's Iraq and Al Qaeda or similar networks. The search for WMDs ended in December 2004, and an inspection report submitted to Congress by U.S. weapons-hunter Charles A. Duelfer essentially "contradicted nearly every prewar assertion about Iraq made by Bush administration officials."[a]

Policy makers and experts bear two fundamental questions for critical analysis and debate:

- Did the reasons given for the invasion reflect a plausible threat scenario?
- Was the invasion a well-crafted policy option centered on credible political, military, and intelligence calculations?

Note

a. Dafna Linzer, "Search for Banned Arms in Iraq Ended Last Month," Washington Post, January 12, 2005.

Chapter Summary

This chapter introduced readers to the terror from above that characterizes state-sponsored terrorism and outlined the nature of state terrorism. The purpose of this discussion was to identify and define several state terrorist environments, to differentiate state terrorism in the foreign and domestic policy domains, and to provide cases in point for these concepts.

Readers were introduced to public and private agencies that monitor state terrorism. The U.S. Department of State's list of sponsors of state terrorism is a useful compilation of information about states active in the foreign policy domain. **Human Rights Watch** and **Amnesty International** are private activist organizations with extensive databases on state terrorism in the domestic policy domain.

The state terrorism paradigm identified several approaches experts use to define and describe state terrorism. Included in this discussion was a comparison of the underlying characteristics of the state patronage and state assistance models of terrorism. The patronage model is characterized by situations in which regimes act as active sponsors of, and direct participants in, terrorism. Under the assistance model, regimes tacitly participate in violent extremist behavior and indirectly sponsor terrorism.

The discussion of state terrorism as foreign policy applied a model that categorized terrorism in the foreign domain as politically sympathetic, logistically supportive, episode specific, or joint operations. Each category described different aspects in the scale of support and directness of involvement by state sponsors. Several examples clarified the behavioral distinctions across categories.

In the domestic policy domain, several models of state domestic authority and legitimacy were identified and summarized, and their sources of authority and centers of power contrasted. The models were democracy, authoritarianism, totalitarianism, and crazy states. Because the methodologies of state domestic terrorism differ from case to case, several models provided a useful approach to understanding the characteristics of particular terrorist environments. These models were vigilante, overt official, covert official, and genocidal state domestic terrorism.

DISCUSSION BOX

This chapter's Discussion Box is intended to stimulate critical debate about the application of authoritarian methods by democratic governments and the justifications these governments use for such methods.

Authoritarianism and Democracy

Democracies are constrained by strong constitutions from summarily violating the rights of citizens. Most democracies have due process requirements in place when security services wish to engage in surveillance, search premises, seize evidence, or detain suspects. However, when confronted by serious security challenges, democracies have resorted to authoritarian security measures. Germany, Italy, France, the United Kingdom, and the United States have all adopted aggressive policies to suppress perceived threats to national security.

For example:

In the United States, periodic anticommunist Red Scares occurred when national leaders reacted to the perceived threat of communist subversion. Government officials reacted by adopting authoritarian measures to end the perceived threats. The first Red Scare occurred after the founding of the Communist Party—USA in 1919, and a series of letter bombs were intercepted. President Woodrow Wilson allowed Attorney General R. Mitchell Palmer to conduct a series of raids—the so-called Palmer Raids—against Communist and other leftist radical groups. Offices of these groups were shut down, leaders were arrested and put on trial, and hundreds were deported.

A second Red Scare occurred in the 1930s and led to the creation of the House Un-American Activities Committee and the passage of the Smith Act in 1940, which made advocacy of the violent overthrow of the government a federal crime. In the late 1940s, high-profile investigations, such as that of Alger Hiss, were common and communists were prosecuted.

A third Red Scare occurred in the 1950s, when Senator Joseph McCarthy of Wisconsin held a series of hearings to expose communist infiltration in government, industry, and Hollywood. Hundreds of careers were ruined and many people were blacklisted, that is, barred from employment.

In Northern Ireland, the British government has periodically passed legislation granting British forces authoritarian powers to combat terrorism by the IRA. One such law was the 1973 Northern Ireland Emergency Provisions Act, which permitted the military to temporarily arrest and detain people and to search homes without warrants. Under the act, the army detained hundreds of people and searched more than 250,000 homes. This sweep was actually fairly successful, because thousands of weapons were found and seized.

Discussion Questions

1. Are authoritarian methods morally compatible with democratic principles and institutions?

2. Under what circumstances are authoritarian policies justifiable and necessary, even in democracies with strong constitutional traditions?

3. Many have labeled the postwar Red Scare investigations in the United States as witch hunts. Were these investigations nevertheless justifiable, considering the external threat from the Soviet Union?

4. The British security services detained hundreds of innocent people and searched the homes of many thousands of non-IRA members. Considering the threat from the IRA, were these inconveniences nevertheless justifiable?

5. Assume for a moment that some security environments justify the use of authoritarian measures by democracies. What kind of watchdog checks and balances are needed to ensure that democracies do not move toward permanent authoritarianism?

Key Terms and Concepts

The following topics were discussed in this chapter and can be found in the glossary:

African National Congress (ANC)
al-Megrahi, Abdel Basset
Anfal campaign
Apartheid
Askaris
Assassination
Auto-genocide
Ba'ath Party
Blacklisting
Boland Amendment
Contras
Crazy state
Death squads
Episode-specific sponsorship

Ethnic cleansing
Fhima, Lamen Khalifa
Four Olds, The
Genocidal state terrorism
Genocide
Great Proletarian Cultural Revolution
House Un-American Activities Committee
Hussein, Saddam
Inkatha Freedom Party
Jamahiriya Security Organization (JSO)
Joint operations
Khmer Rouge

Khomeini, Ayatollah Ruhollah
Kurds
Logistically supportive sponsorship
McCarthy, Senator Joseph
Muslim Brotherhood
Official state terrorism
Palmer Raids
Pan Am Flight 103
Paramilitaries
Phoenix Program
Politically sympathetic sponsorship
Pol Pot

Qaddafi, Muammar el-
Red Guards
Red Scares
Sandinistas
Social cleansing

Somoza Debayle, Anastasio
State assistance for
   terrorism
State patronage for
   terrorism

Viet Cong
Vigilante state
   terrorism
Warfare
Year Zero

Terrorism on the Web

Log on to the Web-based student study site at **www.sagepub.com/martinessstudy** for additional web sources and study resources.

Web Exercise

Using this chapter's recommended Web sites, conduct an online investigation of state terrorism.

1. Are there certain governmental or institutional profiles that distinguish repressive regimes from nonrepressive regimes?
2. Read the mission statements of the monitoring organizations. Do they reflect objective and professionally credible approaches for monitoring the behavior of states?
3. In your opinion, how effective are these organizations?

For an online search of state terrorism, readers should activate the search engine on their Web browser and enter the following keywords:

"State Terrorism"
"Terrorist States"

Recommended Readings

The following publications provide discussions on state-sponsored terrorism.

Bullock, Alan. Hitler: A Study in Tyranny. New York: Harper, 1958.
Dror, Yehezkel. Crazy States: A Counterconventional Strategic Problem. New York: Kraus, Milwood, 1980.
Goren, Roberta, and Jillian Becker, eds. The Soviet Union and Terrorism. London: Allen & Unwin, 1984.
Stohl, Michael, and George Lopez, eds. The State as Terrorist: The Dynamics of Governmental Violence and Repression. Westport, CT: Greenwood, 1984.
Tucker, Robert C. Stalin in Power: The Revolution From Above, 1928–1941. New York: Norton, 1990.

5 Terrorism by Dissidents

This chapter discusses the characteristics of terrorism from below—**dissident terrorism**—committed by nonstate movements and groups against governments, ethno-national groups, religious groups, and other perceived enemies. Readers will probe the different types of dissident terrorism and develop an understanding of the qualities that differentiate each dissident terrorist environment.

It is important to understand that political violence by nonstate actors has long been viewed as a necessary evil by those sympathetic to their cause. Revolutionaries, terrorists, and assassins have historically justified their deeds as indispensable tactics to defend a higher cause. The methods can range in intensity from large-scale wars of national liberation—such as the many anticolonial wars of the 20th century—to individual assassins who strike down enemies of their cause. In the United States, for example, when Confederate sympathizer **John Wilkes Booth** assassinated President Abraham Lincoln during a play at Ford's Theater in Washington, DC, he leaped from Lincoln's balcony to the stage after shouting "sic semper tyrannis!" ("Thus always to tyrants!").

The U.S. Department of State publishes an annual report that identifies and describes an official list of foreign terrorist organizations. Table 5.1 reproduces a typical list of these organizations.

Why do people take up arms against governments and social systems? What weapons are available to the weak when they decide to confront the strong? Do the ends of antistate dissident rebels justify their chosen means? State repression and exploitation are frequently cited as grievances to explain why nonstate actors resort to political violence. Such grievances are often ignored by state officials, who refuse to act until they are forced to do so.

An example illustrating this grievance-related concept is the rebellion in Mexico waged by rebels calling themselves the **Zapatista National Liberation Front** (*Ejército Zapatista de Liberacion Nacional*). The Zapatistas were leftists who championed the cause of Indians native to Mexico's Chiapas state, where starvation and disease were endemic and the government had long supported large landowners in exploiting Indian peasants. In January 1994, the Zapatistas began attacking Mexican army troops and police stations in Chiapas. During this initial campaign, approximately 145 people were killed before the rebels retreated into the jungle to continue the conflict. A low-intensity guerrilla insurgency continued, with the government gradually agreeing to address the grievances of all of Mexico's 10 million Indians. By 2001, the Zapatistas had evolved into an aboveground political movement lobbying for the civil rights of Mexico's Indians and peasants. A key reason for the Zapatistas' success was their ability to adopt a Robin Hood image for their movement, and thereby garner support from many Mexicans.

The discussion in this chapter will review the following:

◆ Perspectives on violent dissent
◆ The practice of dissident terrorism
◆ Dissidents and the new terrorism

TABLE 5.1   FOREIGN TERRORIST ORGANIZATIONS, 2004

Title 22 of the U.S. Code, Section 2656f, which requires the Department of State to provide an annual report to Congress on terrorism, requires the report to include, among other things, information on terrorist groups and umbrella groups under which any terrorist group falls, known to be responsible for the kidnapping or death of any U.S. citizen during the preceding five years; groups known to be financed by state sponsors of terrorism about which Congress was notified during the past year in accordance with Section 6(j) of the Export Administration Act; and any other known international terrorist group that the Secretary of State determined should be the subject of the report.

17 November	Lashkar i Jhangvi (LJ)
Abu Nidal Organization (ANO)	Liberation Tigers of Tamil Eelam (LTTE)
Abu Sayyaf Group (ASG)	Libyan Islamic Fighting
Al-Aqsa Martyrs Brigade	Group (LIFG)
Ansar al-Islam (AI)	Mujahedin-e Khalq Organization (MEK)
Armed Islamic Group (GIA)	National Liberation
Asbat al-Ansar	Army (ELN)
Aum Shinrikyo (Aum)	Palestine Liberation Front (PLF)
Basque Fatherland	Palestinian Islamic Jihad (PIJ)
and Liberty (ETA)	Popular Front for the Liberation of Palestine
Communist Party of Philippines/	(PFLP)
New People's Army (CPP/NPA)	Popular Front for the Liberation of Palestine-
Continuity Irish Republican	General Command (PFLP-GC)
Army (CIRA)	Al-Qa'ida
Gama'a al-Islamiyya (IG)	Real IRA (RIRA)
HAMAS	Revolutionary Armed Forces of Colombia
Harakat ul-Mujahidin (HUM)	(FARC)
Hizballah	Revolutionary Nuclei (RN)
Islamic Movement of	Revolutionary People's Liberation Party/Front
Uzbekistan (IMU)	(DHKP/C)
Jaish-e-Mohammed (JEM)	Salafist Group for Call and
Jemaah Islamiya	Combat (GSPC)
Organization (JI)	Shining Path (SL)
Al-Jihad (AJ)	Tanzim Qa'idat al-Jihad fi Bilad al-Rafidayn
Kahane Chai (Kach)	(QJBR)
Kongra-Gel (KGK)	United Self-Defense Forces of Colombia
Lashkar e-Tayyiba (LT)	(AUC)

SOURCE: U.S. Department of State, *Country Reports on Terrorism 2004* (Washington, DC: U.S. Department of State, April 2005), 92.

Photo 5.1   The assassination of American president Abraham Lincoln by Confederate sympathizer John Wilkes Booth.

Perspectives on Violent Dissent

Policy experts and academics have designed a number of models that define dissident terrorism. For example, one model places dissident terrorism into a larger framework of "three generalized categories of political action"[1] that include the following:

◆ Revolutionary terrorism—the threat or use of political violence aimed at effecting complete revolutionary change

◆ Subrevolutionary terrorism—the threat or use of political violence aimed at effecting various changes in a particular political system (but not aimed at abolishing it)

◆ Establishment terrorism—the threat or use of political violence by an established political system against internal or external opposition[2]

Other models develop specific types of dissident terrorism, such as single issue, separatist, and social revolutionary.[3] Likewise, insurgent terrorism has been defined as violence "directed by private groups against public authorities [that] aims at bringing about radical political change."[4]

To simplify our analysis, the discussion here presents a dissident terrorist model adapted from one Peter Sederberg designed.[5] It defines and differentiates broad categories of dissident terrorism useful for critically analyzing terrorist motives and behaviors. Although each category—**revolutionary, nihilist,** and **nationalist dissident terrorism**—is specifically defined for our discussion, one should keep in mind that the same terms are applied by experts in many different contexts.

Revolutionary Dissident Terrorism: A Clear World Vision

The goals of revolutionary dissidents are to destroy an existing order through armed conflict and to build a relatively well-designed new society. This vision can be the result of nationalist aspirations, religious principles, ideological dogma, or some other goals.

Revolutionaries view the existing order as regressive, corrupt, and oppressive; their envisioned new order will be progressive, honest, and just. Revolutionary dissident terrorists are not necessarily trying to create a separate national identity; they are activists seeking to build a new society on the rubble of an existing one. Many Marxist revolutionaries, for example, have a general vision of a Communist Party–led egalitarian classless society with centralized economic planning. Many Islamic revolutionaries also have a grand vision—that of a spiritually pure culture justly based on the application of shari'a, or God's law. An example of the latter is the Hezbollah (**Party of God**) organization in Lebanon, which is actively agitating for its own vision of a spiritually pure Lebanon; to that end, Hezbollah has its own political movement, armed militia, and social services. Various factions of the Muslim Brotherhood also advocate a rather clear program.

As a practical matter, revolutionary dissidents are often outnumbered and outgunned by the established order. Their only hope for victory is to wage an unconventional war to destabilize the central authority. Terrorism thus becomes a pragmatic tactical option to disrupt government administration and symbolically demonstrates the weakness of the existing regime.

Good case studies for terrorism as a legitimate tactic are found in the Marxist revolutionary movements in Latin America during the 1950s to the 1980s. For example, during the Cuban Revolution, which began in 1956, rebels operating in rural areas waged classic hit-and-run guerrilla warfare against the Batista government's security forces. **Fidel Castro** and Ernesto "Che" Guevara led these rural units. In urban areas, however, attacks were commonly carried out by the rebels, who successfully disrupted government administration and thereby undermined public confidence in Batista's ability to govern. This model was repeated throughout Latin America by Marxist revolutionaries, usually unsuccessfully, so that urban terrorism became a widespread phenomenon in many countries during the period.

Nihilist Dissident Terrorism: Revolution for the Sake of Revolution

Nihilism was a 19th-century Russian philosophical movement of young dissenters who believed that only scientific truth could end ignorance. Religion, nationalism, and traditional values (especially family values) were, they believed, at the root of ignorance. Nihilists had no vision for a future society, asserting only that the existing society was intolerable. Nihilism was, at its core, a completely negative and critical philosophy. Modern nihilist dissidents exhibit a similar disdain for the existing social order but, despite a vague goal of justice, offer no clear alternative for the aftermath of its destruction. The goal of modern nihilists is to destroy the existing order through armed conflict with little forethought to the configuration of the new society; victory is defined simply as the destruction of the old society.

Nihilist dissidents have never been able to lead broad-based revolutionary uprisings among the people or to mount sustained guerrilla campaigns against conventional security forces. Thus, the only armed alternative among hard-core nihilists has been to resort to terrorism. Examples of modern nihilist dissident terrorists include the leftist **Red Brigade** in Italy and **Weather Underground Organization** in the United States, each of which had only a vaguely Marxist model for postrevolutionary society. Another example is the Palestinian terrorist Abu Nidal, who had no postrevolution vision. Arguably, Osama bin Laden's Al Qaeda network fits the model, because though Al Qaeda has a generalized goal of defending Islam and fomenting a pan-Islamic revival, it offers no specific model for how the postrevolution world will be shaped, and its long-term goals are not clearly defined.

Nationalist Dissident Terrorism: The Aspirations of a People

Nationalist dissidents champion the national aspirations of groups of people distinguished by their cultural, religious, ethnic, or racial heritage. The championed people generally live in an environment in which their interests are subordinate to the interests of another group or a national regime. The goal of nationalist dissidents is to mobilize a particular demographic group against another group or government. They are motivated by the desire for some degree of national autonomy, such as democratic political integration, regional self-governance, or national independence.

Nationalist sentiment has been commonplace—particularly during the 19th and 20th centuries—and can arise in many social and political environments. For example, the championed group may be a minority living among a majority group, such as the Basques of northern Spain. It may also be a majority national group living in a region politically dominated by the government of another ethnic group, such as Tibetans and the Chinese. The group may be a minority with a separate cultural and linguistic identity, such as the French Canadians in Quebec. Some national groups have a distinct cultural, ethnic, and regional identity that exists within the borders of several countries, such as the Kurds, whose Kurdistan is divided across Iran, Iraq, Turkey, and Syria.

Many nationalist dissidents have used terrorism to achieve their goals. This has often been a practical option, because their opponents have overwhelming military and political superiority and would quickly prove victorious during a guerrilla or conventional conflict. An example is the Provisional Irish Republican Army in Northern Ireland. In other contexts, the armed opposition must operate in urban areas, which always favors the dominant group or regime because of the impossibility of maneuver, the concentration of security forces, and sometimes the lack of mainstream support from the championed group. An example is the **Basque Fatherland and Liberty (*Euskadi Ta Azkatasuna*, or ETA)** organization in the Basque region of northern Spain. These are logical operational policies, because for nationalists, "the basic strategy is to raise the costs to the enemy occupiers until they withdraw."[6]

Chapter Perspective 5.1 explores a troubling practice found among many revolutionary, nihilist, and nationalist paramilitaries and rebel groups. This is the phenomenon of recruiting and training so-called child soldiers.

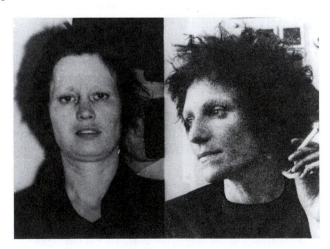

Photo 5.2 Ulrike Meinhof (left) and Gudrun Ensslin, prominent members of West Germany's Red Army Faction, also known as the **Baader-Meinhof Gang**. Both died in prison, apparently by suicide.

CHAPTER PERSPECTIVE 5.1

Child Soldiers

One disturbing—and common—trend among paramilitaries and other armed groups has been conscripting children as fighters. Child soldiers are a serious humanitarian issue, with "children as young as six . . . being used in combat by government and rebel forces in civil wars throughout the world."[a] Around the world, nearly 300,000 children—both boys and girls—have been recruited as fighters in at least 33 armed conflicts.[b] For example,

- In Liberia, rebel leader Charles Taylor formed a unit called the Small Boy Unit during the 1990s.[c] Boys were regularly ordered to commit human rights violations to terrorize civilians.
- In Sierra Leone during the 1990s and early 2000s, the **Revolutionary United Front** abducted thousands of children, and organized those under the age of 15 into Small Boy Units and Small Girl Units.[d]
- In Sri Lanka, the Liberation Tigers of Tamil Eelam have conscripted boys as young as 9 years old. Tamil Tigers have been known to "pull the biggest children out of classes, raid orphanages, and go to homes to demand that families turn over teenage sons and daughters."[e]
- In Mozambique, during its civil war in the 1980s and early 1990s, the Renamo rebel movement used child soldiers extensively.[f]
- In Colombia, both right-wing progovernment paramilitaries and left-wing antigovernment guerrillas conscript children and teenagers.
- In Burundi during the 1990s and early 2000s, the National Council for the Defense of Democracy-Forces for the Defense of Democracy kidnapped hundreds of boys for military service in their movement.[g]

Paramilitaries and rebel movements have assigned child soldiers to heavy combat on the front lines. Some children are drugged before entering into combat and have been known to commit atrocities under orders.

Notes

a. Amnesty International, *Killings by Government* (London: Amnesty International, 1983), quoted in Peter Iadicola and Anson Shupe, *Violence, Inequality, and Human Freedom* (Dix Hills, NY: General Hall, 1998), 255.
b. Human Rights Watch, "Child Soldiers: Facts About Child Soldiers" (New York: Human Rights Watch, 2002), www.hrw.org/campaigns/crp/facts.htm
c. Iadicola and Shupe, *Violence, Inequality, and Human Freedom*.
d. The anarchic war in Sierra Leone is discussed further in Chapter 9.
e. Barbara Crossette, "Tamil Rebels Said to Recruit Child Soldiers," *New York Times*, July 17, 2000.
f. Iadicola and Shupe, *Violence, Inequality, and Human Freedom*.
g. Agence France-Presse, "In Burundi, Rebels Kidnap Children to Fill Ranks," *New York Times*, November 19, 2001.

Revolutionaries, Nihilists, and Nationalists: Freedom Fighters?

Table 5.2 gives examples of how terrorists perceive themselves. Regardless of their ideology, methodology, or goals, there is unanimity in positive self-perception: Terrorists perceive themselves as members of an enlightened fighting elite. The names terrorist organizations adopt reflect this self-perception, but, as indicated in Table 5.2, organizational names often have nothing to do with the reality of the group's composition.

Photo 5.3
Child soldiers in training. Adults drill armed young boys at a stadium in Angola.

The Practice of Dissident Terrorism

Antistate Dissident Terrorism

A good deal of terrorism from below is by definition antistate. It is directed against existing governments and political institutions, and attempts to destabilize the existing order as a precondition to a new society. As discussed earlier, antistate dissidents can have a clear vision of the new society (revolutionary dissidents), a vague vision of the new society (nihilist dissidents), national aspirations (nationalist dissidents), or a profit motive (criminal dissidents). Regardless of which model fits a particular movement, the common goal is to defeat the state and its institutions.

Intensities of Conflict: Antistate Terrorist Environments

With few exceptions, antistate terrorism is directed against specific governments or interests and occurs either within the borders of a particular country or where those interests are found in other countries. Thus, antistate terrorist environments are defined by the idiosyncrasies of each country, each dissident movement, and each terrorist organization. The histories of every nation give rise to specific antistate environments that are unique to their societies. The following examples from North America and Europe illustrate this point.

TABLE 5.2 SELF-PERCEPTION OR SELF-DECEPTION? DISSIDENT TERRORISTS AS FREEDOM FIGHTERS

Dissident terrorists adopt organizational names that characterize themselves as righteous defenders of a group or principle. These are always positive representations that project the higher purpose of the group.

Liberation Fighters	Military Units	Defensive Movements	Retribution Organizations	Inconsequential Alliances
Basque Fatherland and Liberty	Alex Boncayo Brigade	Islamic Resistance Movement	Revolutionary Justice Organization	Aum Shinrikyo (Supreme Truth)
Liberation Tigers of Tamil Eelam	Irish Republican Army	Revolutionary People's Struggle	International Justice Group	Middle Core Faction
Palestine Islamic Jihad	Japanese Red Army	National Council of Resistance	Palestinian Revenge Organization	Al Qaeda ("The Base")
Revolutionary People's Liberation Party	New People's Army	Jewish Defense League	Black September	Orange Volunteers

In the United States, leftist terrorism predominated during the late 1960s through the late 1970s, at the height of the anti–Vietnam War and people's rights movements. Acts of political violence—such as bank robberies, bombings, and property destruction—took place when some black, white, and Puerto Rican radicals engaged in armed protest. This changed in the 1980s, when the leftist remnants either gave up the fight or were arrested. Around this time, right-wing terrorism began to predominate when some racial supremacists, religious extremists, and antigovernment members of the Patriot movement adopted strategies of violence.[7]

In the United Kingdom, the terrorist environment was shaped by the sectarian conflict in Northern Ireland, which was characterized by both antistate and communal violence from 1969 until the peace settlement in 1999. The nationalist Provisional IRA[8] was responsible for most acts of antistate political violence directed against British administration in Northern Ireland. During the same period, Protestant Loyalist terrorism tended to meet the criteria for communal terrorism rather than antistate terrorism, because Loyalist paramilitaries targeted pro-IRA Catholics rather than symbols of governmental authority. The IRA responded in kind, so that more than 3,500 people had been killed on both sides by the 1999 cease-fire.[9]

In West Germany, from the late 1960s through the mid-1980s, the leftist **Red Army Faction (RAF)** engaged in a large number of bank robberies, bombings, assassinations, and other acts of antistate violence aimed at destabilizing the West German government. The RAF also targeted the NATO presence in West Germany, primarily focusing on U.S. military personnel. After the fall of the communist Eastern Bloc in 1989 and the reunification of Germany,

RAF-style leftist terrorism waned. Around this time, rightist neo-Nazi violence increased—much of it directed against non-German *Gastarbeiters,* or guest workers. The perpetrators of this violence were often young skinheads and other neofascist youths. Many of these rightist attacks occurred in the former East Germany.

In Italy, the leftist Red Brigade was responsible for thousands of terrorist incidents from the early 1970s through the mid-1980s. Originating in the student-based activism of the late 1960s and early 1970s, Red Brigade members were young urban terrorists whose campaign is best described as a nihilist attempt to undermine capitalism and democracy in Italy. By the late 1980s, Italian police had eliminated Red Brigade cells and imprisoned their hard-core members. During this period, Italian neofascists also engaged in terrorist violence, eventually outlasting the leftist campaign, and remained active into the 1990s.

In Spain, antistate terrorism has generally been nationalistic or leftist. Without question, the most prominent was the nationalist and vaguely Marxist ETA. ETA was founded in 1959 to promote Basque independence. The Basques are a culturally and linguistically distinct people who live in northern Spain and southwestern France. Although ETA adopted terrorism as a tactic in response to the Franco government's violent repression of Basque nationalism, "of the more than 600 deaths attributable to ETA between 1968 and 1991, 93 per cent occurred after Franco's death."[10] ETA was rife with factional divisions—at least six factions and subfactions were formed—but their terrorist campaign continued, despite the considerable political rights the Spanish government granted them and the loss of popular support among the Basque people. A right-wing terrorist group, Spanish National Action (Accion Nacional Espanila) was formed as a reaction to ETA terrorism.

Sometimes antistate dissident movements, because of their history and political environment, take on elements of both antistate and communal conflict. In Israel, for example, the Palestinian nationalist movement is made up of numerous organizations and movements that have mostly operated under the umbrella of the **Palestine Liberation Organization (PLO)**, founded by **Yasir Arafat** and others in 1959. From its inception, the PLO has sought to establish an independent Palestinian state. Because it claims the same territory as the state of Israel, the PLO and its affiliates have attacked targets inside Israel and abroad. Until recently, Palestinian armed resistance was characterized by a series of dramatic hit-and-run raids, hijackings, bombings, rocket attacks, and other acts of violence. Israeli and Jewish civilians were often targeted.

Since September 28, 2000, Palestinian resistance has taken on the characteristics of a broad-based uprising—and communal terrorism. On that date, Israeli General Ariel Sharon visited the Temple Mount in Jerusalem. The Temple Mount is sacred to both Muslims and Jews. After Sharon's visit, which Palestinians perceived as a deliberate provocation, enraged Palestinians began a second round of massive resistance, or intifada. The new dissident environment included violent demonstrations, street fighting, and suicide bombings. This violence was characterized by bombings, shootings, and other attacks against civilian targets. Thus the Palestinian nationalist movement arguably entered a phase distinguished by the acceptance of communal dissident terrorism as a strategy.

Defeat Is Unthinkable: The Terrorists' Faith in Victory

Why do small groups of individuals violently confront seemingly invincible enemies? Why do they engage powerful foes by force of arms when their envisioned goal is often illogical or unattainable? For antistate dissidents, their armed struggle is never in vain. They believe that their cause is not only likely to end in victory, but that victory is in fact inevitable. To outside observers, terrorists are almost certainly fighting a losing battle yet persist in their war.

Although antistate dissident terrorists avoid direct confrontation out of a pragmatic acceptance of their comparative weakness, they nevertheless believe in the ultimate victory of their cause. They have a utopian vision that not only justifies their means but also guarantees their idealized ends. Violent confrontation in the present—often horrific in scope—is acceptable because of the promised good at the end of the struggle. For example, religious antistate dissidents believe that God will ensure them final victory.

Nonreligious antistate dissidents also hold an enduring faith in final victory. Some have adopted a strategy similar to the urban terrorist (or urban guerrilla) model that Carlos Marighella developed. He maintained that rebels should organize themselves in small cells in major urban areas. He argued that terrorism, when correctly applied against the government, will create sympathy among the population, which in turn forces the government to become more repressive, thus creating an environment conducive to a mass uprising.[11] This model has failed repeatedly, because the people tend not to rise up, and repressive states usually crush the opposition. Nonetheless, it exemplifies the faith antistate dissidents have in their victory scenarios—no matter how far-fetched those scenarios may be.

Chapter Perspective 5.2 summarizes the coalitional features of the Palestinian movement. Attention should be given to the PLO and its role as an umbrella for numerous ideological factions.

CHAPTER PERSPECTIVE 5.2

The Palestinian Movement

Some antistate dissident environments are long-standing and have generated many contending factions. A good example of this phenomenon is the Palestinian movement. Palestinian activism against the state of Israel has as its ultimate goal the creation of an independent Palestinian state. The antistate strategies of most of these groups have been replaced by a broad-based communal dissident environment known as the intifada. The following organizations have been prominent in the Palestinian nationalist movement.[a]

Palestine Liberation Organization

Formed in 1964, the PLO is not a religious movement, but rather a secular nationalist umbrella organization comprising numerous factions. Its central and largest group is al-Fatah, founded by PLO chairman Yasir Arafat in October 1959. The PLO is the main governing body for the Palestinian Authority in Gaza and the West Bank. **Force 17** is an elite unit that was originally formed in the 1970s as a personal security unit for Arafat. It has since been implicated in paramilitary and terrorist attacks. The Al Aqsa Martyr Brigades is a martyrdom society of fighters drawn from al-Fatah and other factions; it includes suicide bombers.

Abu Nidal Organization

Founded by Sabri al-Banna, the ANO was named for al-Banna's nom de guerre. The ANO split from the PLO in 1974 and became an international terrorist organization, launching attacks in 20 countries and killing or wounding 900 people. It fielded several hundred members and a militia in Lebanon and operated under several other names, including Fatah Revolutionary Council, Arab Revolutionary Council, and Black September, from bases in Libya, Lebanon, and Sudan.

Popular Front for the Liberation of Palestine

The PFLP was founded in 1967 by George Habash. It was founded as a Marxist organization advocating a multinational Arab revolution. With about 800 members, the PFLP was most active during the 1970s but continued to commit acts of terrorism. The PFLP has been held responsible for dramatic international terrorist attacks. Its hijacking campaign in 1969 and 1970, its collaboration with West European terrorists, and its mentorship of Carlos the Jackal arguably established the model for modern international terrorism.

Popular Front for the Liberation of Palestine—General Command

Ahmid Jibril formed the Popular Front for the Liberation of Palestine–General Command (PFLP-GC) in 1968 when he split from the PFLP because he considered the PFLP too involved in politics and not committed enough to the armed struggle against Israel. The PFLP-GC fielded several hundred members, was probably directed by Syria, and has been implicated in many cross-border attacks against Israel.

Palestine Liberation Front

The Palestine Liberation Front split from the PFLP-GC in the mid-1970s and further split into pro-PLO, pro-Syrian, and pro-Libyan factions. The pro-PLO faction was led by Abu Abbas, who committed a number of attacks against Israel. It has always been a small organization, with about 50 members.

Democratic Front for the Liberation of Palestine

The Democratic Front for the Liberation of Palestine split from the PFLP in 1969 and further split into two factions in 1991. It was founded as a Marxist organization that advocated ultimate victory through mass revolution. With about 500 members, it committed primarily small bombings and assaults against Israel, including border raids.

Islamic Resistance Movement (Harakt al-Muqaqama al-Islamiya, or Hamas, meaning zeal)

Hamas was founded as an Islamic fundamentalist movement in 1987, with roots in the Palestinian branch of the Muslim Brotherhood. From its inception, Hamas was a comprehensive movement rather than simply a terrorist group, providing social services to Palestinians and committing repeated acts of violence against Israeli interests. Its armed groups were organized as semi-autonomous cells and became known as the **Izzedine al-Qassam Brigade** (named for a famous jihadi in the 1920s and 1930s). Hamas helped to organize and remained at the forefront of the Palestinian intifada.

Palestine Islamic Jihad

The Palestine Islamic Jihad was never a single organization but a loose affiliation of factions. It was founded as an Islamic fundamentalist revolutionary movement that seeks to promote jihad, or holy war, to form a Palestinian state. It has been held responsible for assassinations and suicide bombings. Like Hamas, it helped to organize and actively promote the intifada.

Note

a. Most of these data were found in *Patterns of Global Terrorism 1996* (U.S. Department of State, 1997), 41.

Communal Terrorism

Dissident terrorism is not always directed against a government or national symbols. It is also often directed against entire population groups—people who are perceived to be ethno-national, racial, religious, or ideological enemies. Because the scope of defined enemies is so broad, it is not unusual for this type of terrorism to be characterized by extreme repression and violence on a massive scale. Often deeply rooted in long cultural memories of conflict, communal terrorism sometimes descends into genocidal behavior because

Photo 5.4   Yasir Arafat, chairman of the PLO. Arafat successfully united disparate Palestinian nationalist factions under the PLO umbrella. A number of these factions regularly engaged in terrorism against the state of Israel.

> while the rival combatants often lack the weapons of destruction available to the major powers, they often disregard any recognized rules of warfare, killing and maiming civilians through indiscriminate car bombings, grenade attacks and mass shootings.[12]

Communal terrorism is essentially group against group terrorism, in which subpopulations of society wage internecine (i.e., mutually destructive) violence. As with other types of violence, it occurs in varying degrees of intensity and in many contexts. The scale of violence frequently surprises the world, for these conflicts "often do not command the headlines that rivet world attention on international wars and guerrilla insurgencies, but they frequently prove more vicious and intractable."[13]

There are many sources of communal violence, and it is useful to review a few broad categories and illustrative cases. These categories—ethno-nationalist, religious, and ideological—are explored in the following discussion.

Ethno-Nationalist Communal Terrorism

Ethno-nationalist communal terrorism involves conflict between populations that have distinct histories, customs, ethnic traits, religious traditions, or other cultural idiosyncrasies. Numerous adjectives have been used to describe this type of dissident terrorism, including "separatist, irredentist, . . . nationalist, tribal, racial, indigenous, or minority."[14] It occurs when one group asserts itself against another, many times, to defend its cultural identity. This rationale is not uncommon and has been used in Bosnia, the Caucasus, Sri Lanka, Indonesia, and elsewhere. In these conflicts, all sides believe themselves vulnerable and use this perception to rationalize their violence.

The scale of ethno-nationalist communal violence can vary considerably from region to region, depending on many factors—such as unresolved historical animosities, levels of regional development, and recurrent nationalist aspirations. It can be waged across national borders (as in the Congo-Rwanda-Burundi region of East Africa), inside national borders (as in Afghanistan), within ethnically polarized provinces (as in the Nagorno-Karabakh territory of Azerbaijan), at the tribal level (as in Liberia), and even at the subtribal clan level (as in Somalia).

Religious Communal Terrorism

Sectarian violence refers to conflict between religious groups and is sometimes one element of discord in a broader conflict between ethno-national groups. Many of the world's ethnic populations define their cultural identity partly through their religious beliefs, so that violence committed by and against them has both ethnic and religious qualities. This link is common in regions where ethnic groups with dissimilar religious beliefs have long histories of conflict, conquest, and resistance. In Sri Lanka, for example, the ongoing civil war between the Hindu Tamils and the Buddhist Sinhalese has been exceptionally violent, with massacres and indiscriminate killings a common practice.

The following examples further illustrate this point:

Yugoslavia. Some intraethnic internecine conflict occurs because of combined nationalist aspirations and regional religious beliefs. The breakup of Yugoslavia led to internecine fighting, the worst of which occurred in Bosnia in 1992–1995. During fighting between Orthodox Christian Serbs, Muslim Bosnians, and Roman Catholic Croats, ethnic cleansing—the forcible removal of rival groups from claimed territory—was practiced by all sides. Significantly, all three religious groups are ethnic Slavs.

Israel. Religion is used in Israel by both Jewish and Muslim militants to justify communal violence. This has been encouraged by members of radical organizations such as the late Rabbi Meir Kahane's **Kach** (thus) movement, which has advocated the expulsion of all Arabs from biblical Jewish territories. Settlers generally rationalize their attacks as reprisals for Palestinian attacks and sometimes cite Jewish religious traditions as a basis for their actions.

Intractable religious sentiment exists on both sides of the conflict in Israel and Palestine, then, with Islamic extremists waging a holy war to expel Jews and Jewish settler extremists seeking to reclaim biblical lands and expel Arabs.

Northern Ireland. Communal dissident terrorism between Catholic nationalists (Republicans) and Protestant unionists (Loyalists) became a regular occurrence in Northern Ireland during the unrest that began in 1969. Targets included civilian leaders, opposition

Photo 5.5 Tamil Tigers on parade. Members of the Liberation Tigers of Tamil Eelam stand for review.

sympathizers, and random victims. From 1969 to 1989, of the 2,774 recorded deaths, 1,905 were civilians; of the civilian deaths, an estimated 440 were Catholic or Protestant terrorists.[15] Between 1969 and 1993, 3,284 people died. During this period, Loyalist paramilitaries killed 871 people, Republican paramilitaries killed 829, and British forces killed 203.[16]

Sudan. In Sudan, long-term animosity exists between the mostly Arabized[17] Muslim north and mostly black Christian and animist (traditional religions) south. Civil war has been a feature of Sudanese political life since its independence in 1956, generally between progovernment Muslim groups and antigovernment Christian and animist groups. The war has been fought by conventional troops, guerrilla forces, undisciplined militias, and vigilantes. In addition, the Sudanese government began arming and encouraging Arabized militants in the Darfur region to attack Black Muslims. Tens of thousands died in this conflict, which approached genocide in scale.

Lebanon. In Lebanon, bloody religious communal fighting killed more than 125,000 people during the 16-year Lebanese civil war that began in 1975. Militias were formed along religious affiliations, so that Maronite Christians, Sunni Muslims, Shi'a Muslims, and Druze all contended violently for political power. Palestinian fighters, Syrian troops, and Iranian revolutionaries were also part of this environment, which led to the breakdown of central government authority.

Ideological Communal Terrorism

Ideological communal terrorism in the post–World War II era reflected the global rivalry between the United States and Soviet Union. The capitalist democratic West competed with the authoritarian communist East for influence in the postcolonial developing world and in countries ravaged by invading armies during the war. A common pattern was for civil wars to break out after European colonial powers or Axis armies[18] were driven out of a country. These civil wars were fought by indigenous armed factions drawn from among the formerly occupied population. In China, Yugoslavia, Malaysia, and elsewhere, communist insurgents vied with traditional monarchists, nationalists, and democrats for power. Civilian casualties were high in all of these conflicts.

Examples of ideological communal conflict have occurred in the following countries and regions:[19]

Greece. The 5-year civil war in Greece from 1944 to 1949 was a complicated and brutal affair that in the end took at least 50,000 to 65,000 lives. It involved fighting among conventional troops, guerrilla groups, gendarmerie (armed police), and armed bands. The Greek Communist Party, which had led a resistance group during World War II, fought against the Greek government in several phases after liberation in 1944. The Greek Communist Party eventually lost, in the only attempted Communist takeover in postwar Europe to be defeated by force of arms.

Angola. Former anti-Portuguese allies in Angola fought a long conflict after independence in 1975. The ruling Movement for the Liberation of Angola (MPLA) is a Marxist-Leninist movement whose ideology promotes a multicultural and nationalistic (rather than ethnic or regional) agenda. Its principal adversary is National Union for the Total Independence of Angola (UNITA), mostly made up of the Ovimbundu tribal group. Because the MPLA leadership identified with the international ideological left, the Soviet Union, Cuba, the United States, and South Africa supported either the MPLA or UNITA. This is a rare example of conflict between a multicultural ideological movement and a regional ethnic movement.

Indonesia. The Indonesian Communist Party (PKI) was implicated in an October 1965 abortive coup attempt. The army rounded up PKI members and sympathizers, and many Indonesians took to the streets to purge the Communist presence. During a wave of anti-Communist communal violence, much of it done by gangs supported by the government, roughly 500,000 Communists, suspected Communists, and political opponents to the government were killed.

Ideology has been used repeatedly in the 20th century to bind together nations or distinctive groups. It has become, in many conflicts, a means to discipline and motivate members of a movement. When applied to rationalize behavior in communal conflicts, the effect can be devastatingly brutal.

Dissidents and the New Terrorism

The dissident terrorist paradigm is a good model for analyzing the environments, motives, and behaviors of modern terrorism. Categorizing the goals and strategies of dissident terrorists as revolutionary, nihilistic, or nationalistic is a useful way to understand dissident violence. However, one must remember that terrorism is an evolutionary phenomenon and that terrorist environments are never static. Methodologies and organizational configurations continue to evolve.

Toward the end of the 20th century, two important developments came to characterize the terrorist environment, moving it into a new phase: a new morality and decentralization.

The New Dissident Terrorist Morality

The morality of dissident terrorism in the latter decades of the 20th century differed from 19th-century anarchist terrorism and other violent movements. The new generation did not share the same moralistic scruples of its predecessor. Terrorism in late 19th- and early

20th-century Russia, for example, was surgical in the sense that it targeted specific individuals to assassinate, specific banks to rob, and specific hostages to kidnap. In fact, not only did the **Social Revolutionary Party** in Russia (founded in 1900) engage in an extensive terrorist campaign in the early 20th century, but its tactics actually became somewhat popular because its victims were often hated government officials.

In contrast, during the postwar era, the definitions of who an enemy was, what a legitimate target could be, and which weapons to use became much broader. This led to a new kind of political violence. Late 20th-century dissident terrorism was new in the sense that it was "indiscriminate in its effects, killing and maiming members of the general public . . . , arbitrary and unpredictable . . . , refus[ing] to recognize any of the rules or conventions of war . . . [and] not distinguish[ing] between combatants and non-combatants."[20] Operationally, the new terrorist morality can be spontaneous and gruesome. For example, in March 2004, four American private contractors were killed in an ambush in the Iraqi city of Fallujah. Their corpses were burned, dragged through the streets, and then displayed hanging from a bridge.

When terrorists combine this new morality with the ever-increasing lethality of modern weapons, the potential for high casualty rates and terror on an unprecedented scale is very real. As noted in previous chapters, this combination was put into practice in September 2001 in the United States, in March 2004 in Spain, and in July 2005 in Great Britain and Egypt. It was especially put into practice during the long-term terrorist suicide campaign in Israel during the Palestinians' intifada. Should terrorists obtain high-yield weapons—such as chemical, biological, nuclear, or radiological weapons—the new morality would provide an ethical foundation for their use.

Chapter Perspective 5.3 explores the new morality within the context of Chechen terrorism in Russia. The Chechen Republic lies in the Caucasus region of the Russian Federation. Also known as Chechnya, it has a long history of opposition to Russian rule that dates at least to the 18th century. In the modern era, the region has been at war since 1994.

CHAPTER PERSPECTIVE 5.3

Chechen Terrorism in Russia

During the collapse of the Soviet Union, a group of Chechens perceived an opportunity for independence, and in 1991 declared the new Chechen Republic of Ichkeria to be independent from Russia. Their rationale was that they were no different from the Central Asian, Eastern European, and Baltic states that had also declared their independence. The Russian Federation refused to recognize Chechnya's independence and in 1994 invaded with 40,000 troops. The Chechens resisted fiercely, inflicting severe casualties on Russian forces, and in 1996 Russia agreed to withdraw its troops after approximately 80,000 Russians and Chechens had died.

Tensions mounted again in 1999 as Russian troops prepared to reenter Chechnya. In September 1999, several blocks of apartments were destroyed by terrorist explosions in Dagestan and Moscow; hundreds were killed. The Russian army invaded Chechnya, thus beginning a protracted guerrilla war that has also witnessed repeated Chechen terrorist attacks in Russia. Although guerrillas inside Chechnya were mostly suppressed, approximately 100,000 Russians and Chechens died during the second invasion.

Because Chechnya is a Muslim region, Russian authorities have tried to link their conflict with the global war on terrorism. At the same time, some Chechen fighters have become Islamists and sought support from the Muslim world. Russian president Vladimir Putin repeatedly voiced a strong and aggressive tone against Chechen terrorists, stating on one occasion that "Russia doesn't conduct negotiations with terrorists—it destroys them."[a]

During the Russian occupation, Chechen separatists waged an ongoing terrorist campaign on Russian soil. Their attacks have been dramatic and deadly. Examples of the quality of their attacks include the following incidents:

◆ Between October 23 and 26, 2002, approximately 50 Chechen terrorists seized about 750 hostages during the performance of a musical in a Moscow theater. During the 57-hour crisis, the Chechens wired the theater with explosives and threatened to destroy the building with everyone inside. Several of the female terrorists also wired themselves with explosives. Russian commandos eventually pumped an aerosol anesthetic (possibly manufactured with opiates) into the theater. One hundred and twenty-nine hostages died, most of them from the effects of the gas, which proved to be more lethal than expected in a confined area. All of the Chechens were killed as the commandos swept through the theater during the rescue operation.

◆ On February 6, 2004, a bomb in a Moscow subway car killed 39 people and wounded more than 100.

◆ On August 24, 2004, two Russian airliners crashed, almost simultaneously. Investigators found the same explosive residue at both sites. Chechen suicide bombers were suspected. A group calling itself the Islambouli Brigades of Al Qaeda claimed responsibility.

◆ On August 31, 2004, a woman detonated a bomb near a Moscow subway station, killing herself and nine other people, and wounding 100. The Islambouli Brigades of Al Qaeda claimed responsibility.

◆ On September 1, 2004, Chechens seized a school in Beslan, taking 1,200 hostages. On September 3, as explosives were detonated and special forces retook the school, more than 330 people were killed, about half of them schoolchildren. Russian authorities displayed the bodies of 26 Chechens.

Note

a. Judith Ingram, "Rush Hour Blast Hits Moscow Metro," *Washington Post,* February 6, 2004.

Terrorist Cells and Lone Wolves: New Models for a New War

A newly predominant organizational profile—the **cell**—also emerged as the 20th century drew to a close. Terrorist organizations had traditionally been rather clearly structured, many with hierarchical command and organizational configurations. They commonly had above-ground political organizations and covert military wings.

During the heyday of group-initiated New Left and Middle Eastern terrorism from the 1960s to the 1980s, it was not unusual for dissident groups to issue formal communiqués. These would officially claim credit for terrorist incidents committed on behalf of championed causes. Formal press conferences were also held on occasion.

The vertical organizational models began to be superseded by less-structured horizontal models during the 1990s. Such cell-based movements have indistinct command and organizational configurations. Modern terrorist networks are often composed of a hub that may guide the direction of a movement but has little direct command and control over operational units. These units are typically autonomous or semiautonomous cells that act on their own, often after lying dormant for long periods as sleepers in a foreign country. The benefit of this configuration is that if one cell is eliminated or its members are captured, they can do little damage to other independent cells. It also permits aboveground supporters to have deniability over the tactics and targets of the cells.

A good example of how a cell can be as small as a single person—the **lone wolf model**—is the case of Richard C. Reid, a British resident who converted to Islam. Reid was detected by an alert flight attendant and overpowered by passengers on December 22, 2001, when he attempted to ignite plastic explosives in his shoe on a Boeing 767 carrying 198 passengers and crew from Paris to Miami.[13] Reid was apparently linked to Al Qaeda and had been trained by the organization in Afghanistan. He was sentenced to life imprisonment after pleading guilty before a federal court in Boston.

Chapter Summary

This chapter provided readers with an understanding of the nature of dissident terrorism. The purpose of this discussion was to identify and define several categories of dissident behavior, to classify antistate dissident terrorism, to describe types of communal dissident terrorism, and to provide examples of these concepts.

The dissident terrorist paradigm identified several categories of dissident terrorism—revolutionary, nihilist, and nationalist. These environments were defined and discussed with the underlying recognition that they are ideal categorizations, and that some terrorists will exhibit characteristics of several categories. It should also be understood that new models became more common as the 20th century drew to a close—the cell organizational structures and lone wolf attacks are now integral to the modern terrorist environment.

Antistate dissident terrorism was defined as terrorism directed against existing governments and political institutions to destabilize the existing environment as a precondition to building a new society. Several antistate terrorist environments were presented as examples. The cases included the United States, several European societies, and the nexus of antistate and communal violence in Israel. The seemingly irrational faith in ultimate victory despite overwhelming odds was examined; this faith in the inevitability of success is at the center of antistate dissident campaigns.

Communal terrorism was defined as group against group terrorism, in which subpopulations of society are involved in internecine violence. Several environments were discussed to illustrate differences in motivations, manifestations of violence, and environments conducive to communal conflict. The evaluated categories were ethno-nationalist, religious, and ideological communal terrorism. Cases were identified to illustrate each concept.

DISCUSSION BOX

This Chapter's Discussion Box is intended to stimulate critical debate about the legitimacy of using guerrilla and terrorist tactics by dissident movements.

The Tamil Tigers

The Democratic Socialist Republic of Sri Lanka is an island nation in the Indian Ocean off the southeast coast of India. Its population is about 74% Sinhalese and 18% Tamil; the rest of the population is a mixture of other ethnic groups.[a]

In April 1987, more than 100 commuters were killed when terrorists—most likely **Liberation Tigers of Tamil Eelam (Tamil Tigers)**—exploded a bomb in a bus station in the capital city of Colombo. This type of attack was typical in the Tigers' long war of independence against the Sri Lankan government. The organization was founded in 1976 and champions the Tamil people of Sri Lanka—Hindus, who make up 18% of the population—against the majority Buddhist Sinhalese.

The goal of the movement is to carve out an independent state from Sri Lanka, geographically in the north and east of the island. To accomplish this, the Tamil Tigers have used conventional, guerrilla, and terrorist tactics to attack government, military, and civilian targets. A unit known as the Black Tigers specializes in terrorist attacks, often committing suicide in the process. Sinhalese forces and irregular gangs have often used extreme violence to repress the Tamil uprising.

About half the members of the Tiger movement have been teenagers. Indoctrination of potential Tigers includes spiritual purity, nationalist militancy, a higher morality, and a glorification of death. At the conclusion of training and indoctrination, young Tiger initiates are given a vial of cyanide, which is worn around the neck to be taken if capture is inevitable. Songs, poetry, and rituals glorify the Tamil people and nation. The Tamil Tigers have been very shrewd with public relations, making extensive use of the media, video, and the Internet; they have also established a foreign service presence in numerous countries. They also apparently became adept at transnational organized crime, raising revenue for the cause by trading in arms and drugs.

Estimates of membership numbers range between 6,000 and 15,000 fighters. They are well organized and disciplined. Women, called Freedom Birds, have taken on important leadership positions over time as Tamil male leaders have died. About one third of the movement are women.

Some Tamil Tiger attacks have been spectacular. In May 1991, a Tamil girl detonated a bomb, killing herself and Indian Prime Minister Rajiv Gandhi. In 1996, Tigers surrounded and annihilated a government base, killing all 1,200 troops. Also in 1996, a Tiger bomb at Colombo's Central Bank killed scores and injured 1,400 others. In 1997, the new Colombo Trade Center was bombed, causing 18 deaths and more than 100 injuries. The Tamil Tigers operate a small naval unit of speedboats (the Sea Tigers) that intercept Sri Lankan shipping. Fighting has centered repeatedly on the Jaffna peninsula in the north, with both sides capturing and losing bases.

By 1997, the war had claimed at least 58,000 military and civilian lives, including 10,000 Tigers. By 2002, the combatants had fought to a stalemate. In early 2002, both sides agreed to Norwegian mediation to negotiate terms for a lasting peace settlement. Several hundred thousand Tamils have fled the island, with more than 100,000 now living in India and about 200,000 now in the West.

Discussion Questions

1. Is terrorism a legitimate tactic in a war for national independence? Does the quest for national freedom justify the use of terrorist tactics?
2. When a cause is considered just, is it acceptable to use propaganda to depict the enemy as uncompromisingly corrupt, decadent, and ruthless, regardless of the truth of these allegations?
3. Is suicidal resistance merely fanatical and irrational, or is it a higher form of commitment to one's struggle for freedom? Is this type of indoctrination and myth building necessary to sustain this level of commitment to a just cause?
4. When a cause is just, are arms smuggling and drug trafficking acceptable options for raising funds?
5. Are the Tamil Tigers terrorists or freedom fighters?

Note

a. Data are derived from Central Intelligence Agency, *The World Fact Book 2001* (Washington, DC: U.S. Central Intelligence Agency, 2001).

Key Terms and Concepts

The following topics were discussed in this chapter and can be found in the glossary:

Abbas, Abu	Dissident terrorism	Izzedine al-Qassam Brigade
Al Aqsa Martyr Brigades	Ethno-nationalist	Kach
Arafat, Yasir	communal terrorism	Lone wolf model
Cells	Fatah	Nationalist dissident
Communal terrorism	Freedom Birds	terrorism
Democratic Front for	Hamas	Nihilism
the Liberation of	Ideological communal	Nihilist dissident terrorism
Palestine	terrorism	Palestine Islamic Jihad

Palestine Liberation
    Front
Palestine Liberation
    Organization
Popular Front for
    the Liberation of
    Palestine

Popular Front for the
    Liberation of
    Palestine–General
    Command
Red Army Faction
Red Brigades
Religious communal
    terrorism

Revolutionary dissident
    terrorism
Sectarian violence
Social Revolutionary Party
Tamil Tigers
Weather Underground
    Organization
Zapatista National
    Liberation Front

Terrorism on the Web

Log on to the Web-based student study site at **www.sagepub.com/martinessstudy** for additional Web sources and study resources.

Web Exercise

Using this chapter's recommended Web sites, conduct an online investigation of dissident terrorism.

1. How would you describe the self-images presented by dissident movements?
2. Based on the information given by the monitoring organizations, are some dissident movements seemingly more threatening than others? Less threatening? Why?
3. Compare the dissident Web sites to the monitoring agency sites. Are any of the dissident groups unfairly reported by the monitoring agencies?

For an online search of dissident terrorism, readers should activate the search engine on their Web browser and enter the following keywords:

"Terrorist Organizations (or Groups)"
"Revolutionary Movements"
The names of specific dissident organizations

Recommended Readings

The following publications provide discussions on dissident activism, protest movements, and violence.

Barkan, Steven E., and Lynne L. Snowden. *Collective Violence*. Boston: Allyn & Bacon, 2001.

Bell, J. Boywer. *The IRA 1968–2000: Analysis of a Secret Army*. London: Frank Cass, 2000.

Jaber, Hala. *Hezbollah: Born With a Vengeance*. New York: Columbia University Press, 1997.

Mallin, Jay, ed. *Terror and Urban Guerrillas: A Study of Tactics and Documents*. Coral Gables, FL: University of Miami Press, 1971.

Wickham-Crowley, Timothy P. *Guerrillas and Revolution in Latin America: A Comparative Study of Insurgents and Regimes Since 1956*. Princeton, NJ: Princeton University Press, 1993.

6

Religious Terrorism

Terrorism in the name of religion has become the predominant model for political violence in the modern world. This is not to suggest that it is the only model, because nationalism and ideology remain as potent catalysts for extremist behavior. However, religious extremism has become a central issue for the global community.

In the modern era, religious terrorism has increased in its frequency, scale of violence, and global reach. At the same time, a relative decline has occurred in secular terrorism. The old ideologies of class conflict, anticolonial liberation, and secular nationalism have been challenged by a new and vigorous infusion of sectarian ideologies. Grassroots extremist support for religious violence has been most widespread among populations living in repressive societies that do not permit demands for reform or other expressions of dissent.

What is religious terrorism? What are its fundamental attributes? Religious terrorism is a type of political violence motivated by an absolute belief that an otherworldly power has sanctioned—and commanded—terrorist violence for the greater glory of the faith. Acts committed in the name of the faith will be forgiven by the otherworldly power and perhaps rewarded in an afterlife. In essence, one's religious faith legitimizes violence as long as such violence is an expression of the will of one's deity.

Table 6.1 presents a model that compares the fundamental characteristics of religious and secular terrorism. The discussion in this chapter will review the following:

◆ Historical perspectives on religious violence
◆ The practice of religious terrorism
◆ Trends and projections

_____Historical Perspectives on Religious Violence

Terrorism carried out in the name of the faith has long been a feature of human affairs. The histories of people, civilizations, nations, and empires are replete with examples of extremist true believers who engage in violence to promote their belief system. Some religious terrorists are inspired by defensive motives, others seek to ensure the predominance of their faith, and others are motivated by an aggressive amalgam of these tendencies.

Religious terrorism can be communal, genocidal, nihilistic, or revolutionary. It can be committed by lone wolves, clandestine cells, large dissident movements, or governments. And, depending on one's perspective, there is often debate about whether the perpetrators should be classified as terrorists or religious freedom fighters. The following cases are historical

TABLE 6.1   CASE COMPARISON: RELIGIOUS AND SECULAR TERRORISM

Environment	Activity Profile			
	Quality of Violence[a]	**Scope of Violence**	**Constituency Profile**	**Relationship to Existing System**
Religious	Unconstrained scale of terrorist violence	Expansive target definition	Narrow, insular, and isolated	Alienated "true believers"
	Result: Unconstrained choice of weapons and tactics	*Result:* Indiscriminate use of violence	*Result:* No appeals to a broader audience	*Result:* Completely reconfigured social order
Secular	Constrained scale of terrorist violence	Focused target definition	Inclusive, for the championed group	Liberators
	Result: Relative constraint in choice of weapons and tactics	*Result:* Relative discrimination in use of violence	*Result:* Appeals to actual or potential supporters	*Result:* Restructured or rebuilt society

SOURCE: Bruce Hoffman, *Inside Terrorism* (New York: Columbia University Press, 1998), 94–95.

a. Communal terrorism is rarely constrained and is an example of convergence in the quality of violence used by religious and secular terrorism.

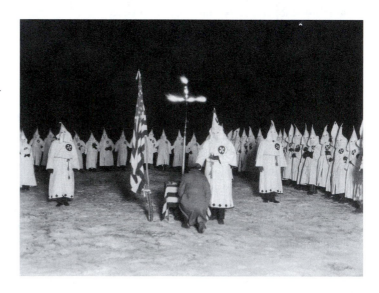

Photo 6.1   A ritualistic Ku Klux Klan induction ceremony in the United States. The KKK is a long-standing racist movement that lives according a code of racial supremacy. Its ceremonies invoke mystical symbols such as hooded gowns and the burning cross, as well as bizarre titles such as Imperial Wizard and Exalted Cyclops.

examples of religious violence. This is a selective survey (by no means exhaustive) that will demonstrate how some examples of faith-based violence are clearly examples of terrorism, how others are not so clear, and how each example must be considered within its historical and cultural context.

Judeo-Christian Antiquity

Within the Judeo-Christian belief system, references in the Bible are not only to assassinations and conquest but also to the complete annihilation of enemy nations in the name of the faith. One such campaign is described in the Book of Joshua.

The story of Joshua's conquest of Canaan is the story of the culmination of the ancient Hebrews' return to Canaan. To Joshua and his followers, this was the Promised Land of the covenant between God and the chosen people. According to the Bible, the Canaanite cities were destroyed and the Canaanites attacked until "there was no one left who breathed."[1] Assuming that Joshua and his army put to the sword all the inhabitants of the 31 cities mentioned in the Bible, and assuming that each city averaged 10,000 people, his conquest cost 310,000 lives.[2]

To the ancient Hebrews, the Promised Land had been occupied by enemy trespassers. To fulfill God's covenant, it was rational and necessary from their perspective to drive them from the land, exterminating them when necessary. Chapter Perspective 6.1 presents the passage that describes the conquest.

CHAPTER PERSPECTIVE 6.1

The Conquest of Canaan

When King Jabin of Hazor heard of this, he sent to [other kings in the region for assistance to defeat the Hebrews]. . . . They came out, with all their troops, a great army, in number like the sand on the seashore, with very many horses and chariots. All these kings joined their forces, and came and camped together at the waters of Merom, to fight with Israel.

And the Lord said to Joshua, "Do not be afraid of them, for tomorrow at this time I will hand over all of them, slain, to Israel. . . ." So Joshua came suddenly upon them with all his fighting force, by the waters of Merom, and fell upon them. And the Lord handed them over to Israel. . . . They struck them down, until they had left no one remaining.

Joshua turned back at that time, and took Hazor, and struck its king down with the sword. . . . And they put to the sword all who were in it, utterly destroying them; there was no one left who breathed, and he burned Hazor with fire. And all the towns of those kings, and all their kings, Joshua took, and struck them with the edge of the sword, utterly destroying them . . . All the spoil of these towns, and the livestock, the Israelites took for their booty; but all the people they struck down with the edge of the sword, until they had destroyed them, and they did not leave any who breathed.

SOURCE: Joshua 11:1, 4-8, 10-14, in *The Holy Bible, New Revised Standard Version*.

Other passages in the Bible are arguably examples of religious communal violence or terrorism, such as the following story from the book of Numbers:

> While Israel was staying at Shittim, the people began to have sexual relations with the women of Moab. . . . Just then one of the Israelites came and brought a Midianite woman into his family. . . . When Phineas . . . saw it, he got up and left the Congregation. Taking a spear in his hand, he went after the Israelite man into the temple, and pierced the two of them, the Israelite and the woman, through the belly. [3]

Christian Crusades

During the Middle Ages, the Western Christian (i.e., Roman Catholic) church launched at least nine invasions of the Islamic east, the first one in 1095. These invasions were called the **Crusades** because they were conducted in the name of the Cross. The purpose of the Crusades was to capture the holy lands from the disunited Muslims, to whom they referred collectively as Saracens.

Christian knights and soldiers answered the call for many reasons. The promise of land, booty, and glory were certainly central. Another important reason was the spiritual promise, made by **Pope Urban II**, that fighting and dying in the name of the Cross would ensure **martyrdom** and thereby guarantee a place in heaven. Liberation of the holy lands would bring eternal salvation. Thus, "knights who with pious intent took the Cross would earn a remission from temporal penalties for all his sins; if he died in battle he would earn remission of his sins."[4] This religious ideology was reflected in the war cry of the early Crusades: *Deus lo volt!* (God wills it!).

During the First Crusade, Western knights—primarily Frankish soldiers—captured a broad swath of biblical lands, including Jerusalem and Bethlehem. When cities and towns were captured, most of the Muslim and Jewish inhabitants were killed outright, a practice common in medieval warfare. When Jerusalem was captured in July 1099, Frankish knights massacred thousands of Muslim, Jewish, and Orthodox Christian residents. An embellished Crusader letter sent to Pope Urban II in Rome boasted that the blood of the Saracens reached the bridles of the Crusaders' horses.

Not all Christian Crusades were fought in Muslim lands. The Western Church also purged its territories of Jews and divergent religious beliefs that were denounced as heresies. The zealousness and violence of these purges became legendary. During the brutal **Albigensian Crusade** in southern France during the 13th century, the story was told that concerns were raised about loyal and innocent Catholics who were being killed along with targeted members of the enemy Cathar sect. The pope's representative, Arnaud Amaury, allegedly replied, "Kill them all, God will know his own!"

The Church-sanctioned invasions and atrocities were deemed to be in accordance with God's wishes and therefore perfectly acceptable. An extreme and unquestioning faith in the cause led to a series of campaigns of terror against the non-Christian (and sometimes the Orthodox Christian) residents of conquered cities and territories. In a typical and tragic irony of the time, the Greek Orthodox city of Constantinople, center of the Byzantine Empire and one of the great cities of the world, was captured and sacked by Western Crusaders in 1204 during the Fourth Crusade. The Crusaders looted the city and created a short-lived Latin Empire, which lasted until 1261.

Photo 6.2 The conquest of Bethlehem. A romanticized depiction of victorious Christian Crusaders, who seized Bethlehem in June 1099 during the First Crusade. The Crusaders subsequently killed virtually all of the town's inhabitants.

The Assassins

The **Order of Assassins**,[5] sometimes referred to as the Brotherhood of Assassins, was founded by **Hasan ibn al-Sabbah** (d. 1124) in 11th-century Persia. Al-Sabbah was a caliph (religious head) of the Ismaili sect of Islam. He espoused a radical version of Ismaili Islam and founded the Order of Assassins to defend this interpretation of the faith. Beginning in 1090, he and his followers seized a string of fortresses in the mountains of northern Persia, the first of which was the strong fortress of Alamut near Qazvin. Because of these origins, al-Sabbah was called The Old Man of the Mountain.

The word *assassin* was allegedly derived from the drug hashish, which some commentators believe al-Sabbah's followers ate before committing acts of violence in the name of the faith.[6] They referred to themselves as hashashins or hashishis, reputedly meaning hashish eaters. During the early years of the movement, Assassin followers spread out of the mountains to the cities of Persia, modern Iraq, Syria, and the Christian Crusader–occupied areas of Palestine. The Assassins killed many people, including fellow Muslims who were Sunnis, and Christians. Suicide missions were common, and some Crusader leaders went so far as to pay tribute to the Assassins so that the Assassins would leave them alone.

The Assassins were very adept at disguise, stealth, and surprise killings, and thus the word *assassination* was coined. A key component of the Assassins' beliefs was the righteousness of their cause and methodology. To kill or be killed was a good thing, because it was done in the name of the faith and ensured a place in paradise after death. This belief is practiced by many of today's religious terrorists.

Although their political impact was negligible and the organization was eliminated in 1256, the Assassins left a profound psychological mark on their era, and in many ways on ours.

A Secret Cult of Murder

In India during the 13th through the 19th centuries, the **Thuggee** cult existed among worshippers of the Hindu goddess Kali, the destroyer. Members were called by various names, including Phansigars (noose operators), Dacoits (members of a gang of robbers), and Thuggees (from which the English word *thug* is derived). They would strangle sacrificial victims—usually travelers—with a noose called a **phansi** in the name of Kali, and then rob and ritually mutilate and bury them. Offerings would then be made to Kali.

The British eventually destroyed the movement during the 19th century, but the death toll of Thuggee victims was staggering: "This secretive cult is believed to have murdered 20,000 victims a year . . . perhaps dispatching as many as several million victims altogether before it was broken up by British officials."[7] There are few debatable counterpoints about this cult—the Thuggees waged a campaign of religious terror for centuries.

Modern Arab Islamist Extremism

The Arab world passed through several important political phases during the 20th century. Overlordship by the Ottoman Empire ended in 1918 after World War I. It was followed by European domination, which ended in the aftermath of World War II. New Arab and North African states were initially ruled primarily by monarchs or civilians who were always authoritarian and frequently despotic. A series of military coups and other political upheavals led to the modern era of governance. These phases had a significant influence on activism among Arab nationalists and intellectuals, culminating in the late 1940s, when the chief symbol of Western encroachment became the state of Israel. Postwar activism in the Arab Muslim world likewise progressed through several intellectual phases, most of them secular expressions of nationalism and socialism. The secular phases included the following:

◆ Anticolonial nationalism, during which Arab nationalists resisted the presence of European administrators and armed forces

◆ Pan-Arab nationalism (Nasserism), led by Egyptian president Gamel Abdel-Nasser, which advocated the creation of a single dynamic United Arab Republic

◆ Secular leftist radicalism, which activists often adopted to promote Marxist or other socialist principles of governance, sometimes in opposition to their own governments

Many activists and intellectuals became disenchanted with these movements when they failed to deliver the political reforms, economic prosperity, and desired degree of respect from the international community. In particular, several humiliating military defeats at the hands of the Israelis—and the seemingly intractable plight of the Palestinians—diminished the esteem and deference the secular movements had once enjoyed. Arab nationalists—both secular and sectarian—had struggled since the end of World War II to resist what they perceived as Western domination and exploitation, and some tradition-oriented nationalists began to interpret Western culture and values as alien to Muslim morality and values.

As a result, new movements promoting Islamist extremism began to overshadow the ideologies of the previous generation. This has placed many Islamists at odds with existing Arab governments, many of which are administered under the principles of the older ideologies.

In the post–Cold War political environment, adopting Islam as a vehicle for liberation is a logical progression. When radical secular ideologies and movements made little progress in resisting the West and Israel, and when secular Arab governments repressed any expressions of domestic dissent, many activists and intellectuals turned to radical interpretations of Islam.

There is a sense of collegiality and comradeship among many Islamists, but there are also differences within the ideologies of many leaders, as well as between the Sunni and Shi'a traditions. The Islamist movement, however, has transcended most ethnic and cultural differences and is a global phenomenon.

Cult Case: Mysticism and Rebellion in Uganda[8]

Phase 1: The Holy Spirit Mobile Force. Uganda in 1987 was a hotbed of rebellion, with several rebel groups opposing the new government of President Yoweri Museveni. One such group was the **Holy Spirit Mobile Force**, inspired and led by the mystical **Alice Auma Lakwena**. Lakwena claimed to be possessed by a spirit called Lakwena, and preached that her movement would defeat Musevani's forces and purge Uganda of witchcraft and superstition. Because her followers championed the Acholi tribe, the Holy Spirit Mobile Force attracted thousands of followers, many of whom were former soldiers from previous Ugandan government armies. In late 1987, she led thousands of her followers against Museveni's army. To protect themselves from death, Holy Spirit Mobile Force fighters anointed themselves with holy oil, which they believed would ward off bullets. When they met Museveni's forces, thousands of Lakwena's followers were slaughtered in the face of automatic weapons and artillery fire. Alice Lakwena fled the country to Kenya, where she lived until her death in January 2007.

Phase 2: The Lord's Resistance Army. Josef Kony, either a cousin or nephew of Alice Lakwena, reorganized the Holy Spirit Mobile Force into the **Lord's Resistance Army**. Kony blended Christianity, Islam, and witchcraft into a bizarre mystical foundation for his movement. Kony proclaimed to his followers that he would overthrow the government, purify the Acholi people, and seize power and reign in accordance with the principles of the biblical Ten Commandments.

From its inception, the Lord's Resistance Army was exceptionally brutal and waged near-genocidal terrorist campaigns—largely against the Acholi people it claimed to champion. The movement destroyed villages and towns, killed thousands of people, drove hundreds of thousands more from the land, abducted thousands of children, and routinely committed acts of mass rape and banditry. With bases in southern Sudan, the Lord's Resistance Army proved extremely difficult for the Ugandan government to defeat in the field.

An estimated 30,000 children became kidnap victims, and 1.6 million Ugandans were displaced into refugee camps. These camps became regular targets of the Lord's Resistance Army, which raided them for supplies, to terrorize the refugees, and to kidnap children. Among the kidnapped children, boys were forced to become soldiers and girls became sex slaves known as bush wives. There has been some hope of ending the conflict. In 2005, a top Lord's Resistance Army commander surrendered, the government claimed a temporary cease-fire, and Sudan began to stabilize its border with Uganda after its own southern civil war ended.

Like the Thuggees, the Lord's Resistance Army is unquestionably an example of a cultic movement that waged a campaign of religious terrorism.

Religious Scapegoating Case: The Protocols of the Learned Elders of Zion

Extremist ideologies have historically scapegoated undesirable groups. Many conspiracy theories have been invented to denigrate these groups and to implicate them in nefarious plans to destroy an existing order. Some of these conspiracy theories possess quasi-religious elements that in effect classify the scapegoated group as being in opposition to a natural and sacred order.

Among right-wing nationalists and racists, a convergence is often seen between scapegoating and mysticism. Just as it is common for rightists to assert their natural and sacred superiority, it is also common for them to demonize a scapegoated group, essentially declaring that it is inherently evil. One quasi-religious conspiracy theory is the promulgation of a document titled *The Protocols of the Learned Elders of Zion.*[9]

The *Protocols* originated in czarist Russia and were allegedly the true proceedings of a meeting of a mysterious committee of the Jewish faith, during which a plot to rule the world was hatched—in league with the Freemasons. They are a detailed record of this alleged conspiracy for world domination but were, in fact, a forgery written by the secret police (**Okhrana**) of Czar Nicholas II around 1895 and later published by professor Sergyei Nilus. Many anti-Semitic groups have used this document to justify the repression of European Jews, and it was an ideological foundation for the outbreak of anti-Jewish violence in Europe, including massacres and **pogroms** (violent anti-Jewish campaigns in eastern Europe).

The National Socialist (Nazi) movement and Adolf Hitler used the *Protocols* extensively. Modern Eurocentric neo-Nazis and Middle Eastern extremists, both secular and religious, continue to publish and circulate the *Protocols* as anti-Semitic propaganda. In this regard, neo-Nazis and Middle Eastern extremists have found common cause in quasi-religious anti-Semitism. In 1993, a Russian court formally ruled that the *Protocols* are a forgery.[10] Excerpts are sampled in Chapter Perspective 6.2.

CHAPTER PERSPECTIVE 6.2

The Protocols of the Learned Elders of Zion: A Conspiracy of the Extreme Right

The Protocols of the Learned Elders of Zion was a forgery written by Czar Nicholas II's secret police (the Okhrana) around 1895. It was reproduced privately in Russia in 1897 and printed publicly in 1905 by Professor Sergyei Nilus. The document was used by the Okhrana to incite anti-Jewish pogroms. It was widely used by other reactionaries in Russia, including the violently nationalist and anti-Semitic **Black Hundreds** group and later by the White Army counterrevolutionaries during their civil war against the communist Bolsheviks, or Reds. It lists a series of 24 protocols from a fictional meeting of a mysterious group called the Elders of Zion, who allegedly conspired to rule the world in league with ancient freemasonry.

The following excerpts are from the *Protocols*.[a]

From Protocol 1: The political has nothing in common with the moral. The ruler who is governed by the moral is not a skilled politician, and is therefore unstable on his throne. He who wishes to rule must have recourse [*sic*] both to cunning and to make-believe. Great national qualities, like frankness and honesty, are vices in politics, for they bring down rulers from their thrones more effectively and more certainly than the most powerful enemy. Such qualities must be the attributes of the kingdoms of the goyim, but we must in no wise be guided by them.

From Protocol 2: In the hands of the States of to-day there is a great force that creates the movement of thought in the people, and that is the Press. The part played by the Press is to keep pointing out requirements supposed to be indispensable, to give voice to the complaints of the people, to express and create discontent. It is in the Press that the triumph of freedom of speech finds its incarnation. But the goyim States have not known how to make use of this force; and it has fallen into our hands.

From Protocol 3: The people under our guidance have annihilated the aristocracy, who were their one and only defence and foster-mother for the sake of their own advantage which is inseparably bound up with the well-being of the people. Nowadays, with the destruction of the aristocracy, the people have fallen into the grips of merciless money-grinding scoundrels who have laid a pitiless and cruel yoke upon the necks of the workers . . . We appear on the scene as alleged saviours of the worker from this oppression when we propose to him to enter the ranks of our fighting forces—Socialists, Anarchists, Communists—to whom we always give support in accordance with an alleged brotherly rule (of the solidarity of all humanity).

From Protocol 7: The intensification of armaments, the increase of police forces—are all essential for the completion of the aforementioned plans. What we have to get at is that there should be in all the States of the world, besides ourselves, only the masses of the proletariat, a few millionaires devoted to our interests, police and soldiers.

From Protocol 14: When we come into our kingdom it will be undesirable for us that there should exist any other religion than ours of the One God with whom our destiny is bound up by our position as the Chosen People and through whom our same destiny is united with the destinies of the world. We must therefore sweep away all other forms of belief.

The *Protocols* are organized as follows:

Protocol I: The Basic Doctrine
Protocol II: Economic Wars
Protocol III: Methods of Conquest
Protocol IV: Materialism Replace Religion
Protocol V: Despotism and Modern Progress
Protocol VI: Take-Over Technique
Protocol VII: World-Wide Wars
Protocol VIII: Provisional Government
Protocol IX: Re-education
Protocol X: Preparing for Power
Protocol XI: The Totalitarian State

Protocol XII: Control of the Press
Protocol XIII: Distractions
Protocol XIV: Assault on Religion
Protocol XV: Ruthless Suppression
Protocol XVI: Brainwashing
Protocol XVII: Abuse of Authority
Protocol XVIII: Arrest of Opponents
Protocol XIX: Rulers and People
Protocol XX: Financial Programme
Protocol XXI: Loans and Credit
Protocol XXII: Power of Gold
Protocol XXIII: Instilling Obedience
Protocol XXIV: Qualities of the Ruler

a. From the 1906 English translation by Victor E. Marsden, a reporter for the *Morning Post*.

The Practice of Religious Terrorism

Understanding Jihad as a Primary Religious Motive: An Observation and Caveat

Keeping the idiosyncratic quality of religious terrorism in mind, it is arguably necessary to make a sensitive observation—and caveat—about the study of religious terrorism in the modern era. The observation is that today, the incidence of religious terrorism is disproportionately committed by radical Islamists:

Popular Western perception equates radical Islam with terrorism. . . . There is, of course, no Muslim or Arab monopoly in the field of religious fanaticism; it exists and leads to acts of violence in the United States, India, Israel, and many other countries. But the frequency of Muslim- and Arab-inspired terrorism is still striking. . . . A discussion of religion-inspired terrorism cannot possibly confine itself to radical Islam, but it has to take into account the Muslim countries' preeminent position in this field.[8]

The caveat is the degree of misunderstanding in the West about the historical and cultural origins of the growth of radical interpretations of Islam. One is the common belief that the concept of holy war is an underlying principle of the Islamic faith. Another is that Muslims are united in supporting **jihad**. This is simplistic and fundamentally incorrect. Although the term jihad is widely presumed in the West to refer exclusively to waging war against nonbelievers, an Islamic jihad is not the equivalent of a Christian Crusade.

Chapter Perspective 6.3 provides some clarification of the concept of jihad.

CHAPTER PERSPECTIVE 6.3

Jihad: Struggling in the Way of God

The concept of jihad is a central tenet in Islam. Contrary to misinterpretations common in the West, the term literally means a sacred struggle or effort rather than an armed conflict or fanatical holy war.[a] Although a jihad can certainly be manifested as a holy war, it more correctly refers to the duty of Muslims to personally strive "in the way of God."[b]

This is the primary meaning of the term as used in the Quran, which refers to an internal effort to reform bad habits in the Islamic community or within the individual Muslim. The term is also used more specifically to denote a war waged in the service of religion.[c]

Regarding how one should wage jihad,

The **greater jihad** refers to the struggle each person has within him or herself to do what is right. Because of human pride, selfishness, and sinfulness, people of faith must constantly wrestle with themselves and strive to do what is right and good. The **lesser jihad** involves the outward defense of Islam. Muslims should be prepared to defend Islam, including military defense, when the community of faith is under attack.[d] [emphasis added]

Thus, waging an Islamic jihad is not the same as waging a Christian crusade—it has a broader and more intricate meaning. Nevertheless, it is permissible—and even a duty—to wage war to defend the faith against aggressors. Under it, warfare is conceptually defensive; by contrast, the Crusades were conceptually offensive. Those who engage in armed jihad are known as mujahideen, or holy warriors. Mujahideen who receive **martyrdom** by being killed in the name of the faith will find that

> awaiting them in paradise are rivers of milk and honey, and beautiful young women. Those entering paradise are eventually reunited with their families and as martyrs stand in front of God as innocent as a newborn baby.[e]

The precipitating causes for the modern resurgence of the armed and radical jihadi movement are twofold: the revolutionary ideals and ideology of the 1979 Iranian Revolution and the practical application of jihad against the Soviet Union's occupation of Afghanistan.

Some radical Muslim clerics and scholars have concluded that the Afghan jihad brought God's judgment against the Soviet Union, leading to the collapse of its empire. As a consequence, radical jihadis fervently believe that they are fighting in the name of an inexorable force that will end in total victory and guarantee them a place in paradise.

Notes

a. Karen Armstrong, *Islam: A Short History*. (New York: Modern Library, 2000), 201.
b. Josh Burke and James Norton. "Q&A: Islamic Fundamentalism: A World-Renowned Scholar Explains Key Points of Islam," *Christian Science Monitor*, October 4, 2001.
c. Armstrong, *Islam*, 201.
d. Burek and Norton, "Q&A: Islamic Fundamentalism."
e. Walter Laqueur, *The New Terrorism. Fanaticism and the Arms of Mass Destruction* (New York: Oxford University Press, 1999), 100.

State-Sponsored Religious Terrorism

Government sponsorship of terrorism is not limited to providing support for ideological or ethno-national movements. It also incorporates state sponsorship of religious revolutionary movements.

National Case: Iran

Iran became a preeminent state sponsor of religious terrorism after the overthrow of the monarchy of **Shah Muhammed Reza Pahlavi** in 1979, and the creation of the theocratic Islamic Republic of Iran soon thereafter.

Iran has been implicated in the sponsorship of a number of groups known to have engaged in terrorist violence, making it a perennial entry on the Department of State's list of state sponsors of terrorism. The Iranian **Revolutionary Guards Corps** has a unit—the **Qods (Jerusalem) Force**—that promotes Islamic revolution abroad and the liberation of Jerusalem from non-Muslims. Members of the Revolutionary Guards have appeared in Lebanon and Sudan.

Case: Iranian Support for Lebanon's Hezbollah

An important example of Iranian support for politically sympathetic groups is the patronage and assistance Iran gives to Lebanon's **Hezbollah** movement. The Iran-Hezbollah relationship is important because of the central role Hezbollah has played in the region's political environment.

Hezbollah (Party of God) is a Shi'a movement in Lebanon that arose to champion the country's Shi'a population. The organization emerged during the Lebanese civil war and Israel's 1982 invasion as a strongly symbolic champion for Lebanese independence and justice for Lebanese Shi'a. It was responsible for hundreds of incidents of political violence during the 1980s and 1990s, which included kidnappings of Westerners in Beirut, suicide bombings, attacks against Israeli interests in South Lebanon, and attacks against Israel proper. Hezbollah operated under various names, such as Islamic Jihad and **Revolutionary Justice Organization**.

Hezbollah has for some time been closely linked to Iran. Hezbollah's leadership, though sometimes guarded about their identification with Iran, have said publicly that they support the ideals of the Iranian Revolution. Their ultimate goal is to create an Islamic republic in Lebanon, and they consider Israel an enemy of all Muslims. Hezbollah tends to consider Iran a "big brother" for its movement. At their root, the ideological bonds between the movement and the Iranian Revolution are strong. These bonds allowed Iran's support to extend beyond ideological identification toward overt sponsorship. Beginning in the 1980s, Iran deployed members of its Revolutionary Guards Corps into Lebanon's **Beka'a Valley**—then under Syrian occupation—to organize Hezbollah into an effective fighting force. Iran provided training, funding, and other logistical support. This was done with the acquiescence of Syria, so that Hezbollah is also a pro-Syrian movement.

Case: Iranian Support for Palestinian Islamists

Iran has also promoted movements that directly confront the Israelis in Gaza, the West Bank, and inside Israel's borders. Since the early days of the Iranian Revolution, the Iranian regime has never been guarded about its goal to liberate Jerusalem. To achieve it, Iran has likewise never been guarded about its support for Palestinian organizations that reject dialogue and negotiations with the Israelis. Iran has, in fact, helped the Palestinian cause significantly by promoting the operations of religious movements. For example, two militant Islamic organizations—Palestine Islamic Jihad (PIJ) and Hamas (Islamic Resistance Movement)—are Palestinian extremist groups that have received important support from Iran. Both groups perpetrated many acts of terrorism, including suicide attacks, bombings, shootings, and other violent assaults.

PIJ is not a single organization but instead a loose affiliation of factions, an Islamic revolutionary movement that advocates violent jihad to form a Palestinian state. Iranian support to PIJ includes military instruction and logistical support. PIJ members have appeared in Hezbollah camps in Lebanon's Beka'a Valley and in Iran, and planning for terrorist attacks has apparently taken place in these locations. Members who received this training were infiltrated back into Gaza and the West Bank to wage jihad against Israel.

Hamas's roots lie in the Palestinian branch of the Muslim Brotherhood. It operates as both a social service organization and an armed resistance group that promotes jihad. Because of its social service component, Iran's **Fund for the Martyrs** has disbursed millions of dollars to Hamas. Hamas posted a representative to Iran who held a number of meetings with top Iranian officials. Iran has also provided Hamas with the same type of support that it provides PIJ, including military instruction, logistical support, training in Hezbollah's Beka'a Valley camps (before the Syrian withdrawal), and training in Iran. Hamas operatives returned from these facilities to Gaza and the West Bank.

Regional Case: Pakistan and India

India and Pakistan are seemingly implacable rivals. Much of this rivalry is grounded in religious animosity between the Hindu and Muslim communities of the subcontinent, but sponsorship of terrorist proxies has kept the region in a state of nearly constant tension. Hindus and Muslims in Southwest Asia have engaged in sectarian violence since 1947, when British colonial rule ended. During and after the British withdrawal, communal fighting and terrorism between Hindus and Muslims led to the partition of British India into mostly Muslim East Pakistan and West Pakistan (now Bangladesh) and mostly Hindu India. Since independence, conflict has been ongoing between Pakistan and India over many issues, including Indian support for Bangladesh's war of independence from Pakistan, disputed borders, support for religious nationalist terrorist organizations, the development of nuclear arsenals, and the disputed northern region of Jammu and Kashmir.

Pakistan, through its intelligence agency, the **Directorate for Inter-Services Intelligence (ISI)**, has a long history of supporting insurgent groups fighting against Indian interests. Religious terrorist groups in the Indian state of Punjab and in Jammu and Kashmir have received Pakistani aid in what has become a high-stakes conflict between two nuclear powers that can also field large conventional armies. The Pakistan-India conflict is arguably as volatile as the Arab-Israeli rivalry but with many times the manpower and firepower. This is especially noteworthy because both countries have nuclear arsenals.

Dissident Religious Terrorism in the Modern Era

Dissident religious terrorism is political violence conducted by groups of fervent religious true believers with faith in the sacred righteousness of their cause. Any behavior in the defense of this cause is considered not only justifiable but also blessed. As discussed, most major religions—in particular, Christianity, Islam, Judaism, and Hinduism—have extremist adherents, some of whom have engaged in terrorist violence. Smaller religions and cults have similar adherents. Among the ubiquitous principles are the convictions that they are defending their faith from attack by nonbelievers, or that their faith as indisputable and universal guiding principle must be advanced for the salvation of the faithful. These principles are manifested in various ways and to varying degrees by religious extremists, but are usually at the core of the belief system.

From the perspective of religious radicals in the Middle East, violence done in the name of God is perfectly rational behavior because God is on their side. Many of the holy sites in the region are sacred to more than one faith, such as Jerusalem, where a convergence of claims exists among Muslims, Jews, and Christians. When these convergences occur, some extremists believe that the claims of other faiths are inherently blasphemous. Because of this sort of indisputable truth, some extremists believe that God wishes for nonbelievers to be driven from sacred sites, or otherwise barred from legitimizing their claims.

Case: The Grand Mosque Incident

The framework for Muslim life is based on the Five Pillars of Islam. The Five Pillars are faith, prayer, charity (*zakat*), fasting during the month of Ramadan, and the hajj (pilgrimage) to the holy city of Mecca, Saudi Arabia. In November 1979, during their hajj, 300 radicals occupied the Grand Mosque in Mecca. Their objective was to foment a popular Islamic uprising against the ruling Saud royal family. After nearly two weeks of fighting, the Grand Mosque was reoccupied by the Saudi army, but not before the Saudis called in French counterterrorist commandos to complete the operation. More than 100 radicals were killed and scores were later executed by the Saudi government. During the fighting, Iranian radio accused the United States and Israel of plotting the takeover, and a Pakistani mob attacked the U.S. embassy, killing two Americans.

Case: The Hebron Mosque Massacre

On February 25, 1994, a New York–born physician, **Baruch Goldstein**, fired on worshippers inside the Ibrahim Mosque at the Cave of the Patriarchs holy site in the city of Hebron, Israel. As worshippers performed their morning prayer ritual, Goldstein methodically shot them with an Israel Defense Forces Galil assault rifle. He fired approximately 108 rounds in about 10 minutes before a crowd of worshippers rushed him and killed him. According to official government estimates, he killed 29 people and wounded another 125;[9] according to unofficial estimates, approximately 50 people died.[10] In reprisal for the Hebron massacre, the Palestinian Islamic fundamentalist movement Hamas launched a bombing campaign that included the first wave of human suicide bombers.

Case: The Rabin Assassination

On November 4, 1995, Israeli Prime Minister Yitzhak Rabin was assassinated by **Yigal Amir**, who had stalked Rabin for about a year,[11] and shot Rabin in the back with hollow-point bullets in full view of Israeli security officers. Amir was a Jewish extremist who said that he acted fully within the requirements of Halacha, the Jewish code.[12]

Case: Sectarian Violence in Iraq

Iraq is a multicultural nation with significant but varying strong ethno-national, tribal, and religious identities. The demography consists of the following subpopulations:[13]

Arab: 75–80%
Kurdish: 15–20%
Turkoman, Assyrian, or other: 5%
Shi'a Muslim: 60–65%
Sunni Muslim: 32–37%
Christian or other: 3%

Tensions that had simmered during the Hussein years led to difficulty in fully integrating all groups into accepting a single national identity. For example, many Arabs who had moved into northern Kurdish regions after native Kurds were forced out became pariahs when Kurds returned to reclaim their homes and land. Some violence was directed against the Arab migrants. More ominously, the Sunni minority—which had dominated the country under Hussein—found itself recast as a political minority when the country began to move toward democracy when an interim government was established in June 2004. Sunnis expressed their dissatisfaction when large numbers refused to participate in elections to form a Transitional National Assembly in January 2005.

Also in Iraq, religious extremists—it is unclear whether they are Sunnis or Shi'a—conducted a series of attacks on non-Muslim cultural institutions. These included liquor stores (often owned by Christians) and barber shops (that offered Western-style haircuts).[14]

Readers should be familiar with the essential distinctions between the Sunni and Shi'a Islamic traditions, which Table 6.2 summarizes.

These examples confirm that religious terrorism in the Middle East occurs between, and within, local religious groups. Radical true believers of many faiths attack not only those of other religions but also "fallen" members of their own. Attacks against proclaimed apostasies can be quite violent.

Movement Case: The International Mujahideen (Holy Warriors for the Faith)

The mujahideen are Islamic fighters who have sworn a vow to take up arms to defend the faith. They tend to be believers in fundamentalist interpretations of Islam who have defined their jihad, or personal struggle, to be one of fighting and dying on behalf of the faith.

The modern conceptualization of the mujahideen began during the Soviet war in Afghanistan, which dated from the Soviet invasion of the country in December 1979 to its withdrawal in February 1989. Although several Afghan rebel groups (mostly ethnically based)

TABLE 6.2   TWO TRADITIONS, ONE RELIGION

Sunni and Shi'a Muslims represent the two predominant traditions in Islam. Demographically, Sunni Islam represents about 85–90% of all Muslims, and Shi'a Islam represents about 10–15%. They are distinct practices that arise from, and worship within, a core system of belief.

Unlike Christian denominations, which can diverge markedly, Sunnis and Shi'a differ less on divergent interpretations of religious faith and more on historical sources of religious authority. They originate in the death of the prophet Muhammed and the question on who among his successors represented true authority within the faith.

Sunni Muslims	Shi'a Muslims
Historically accept all four caliphs as successors to Muhammed, including the caliph Ali, Muhammed's son-in-law and cousin.	Historically reject the first three caliphs before Ali as being illegitimate successors to Muhammed.
Only the prophet Muhammed and the holy Quran are authorities on questions of religion. The Shi'a succession of imams is rejected.	As the first legitimate caliph, Ali was also the first in a historical line of imams, or leaders within Muslim communities.
Historically, leaders within the Islamic world have been political leaders and heads of governments rather than religious leaders.	Imams serve as both political and religious leaders.
There is no strictly organized clergy. For example, no single religious leader can claim ultimate authority, and non-clergy may lead prayers.	Imams have strict authority, and their pronouncements must be obeyed. Imams are without sin, and appoint their successors.

Photo 6.3 Holy war against Communist invaders. Afghan mujahideen during their jihad against occupying Soviet troops.

fought the Soviets, they collectively referred to themselves as mujahideen. To them, their war of resistance was a holy jihad. Significantly, Muslim volunteers from around the world served alongside them. These **Afghan Arabs** played an important role in spreading the modern jihadi ideology throughout the Muslim world.

Reasons for taking up arms as a jihadi[15] vary. Some recruits answer calls for holy war from religious scholars who might declare, for example, that Islam is being repressed by the West. Others respond to clear and identifiable threats to their people or country, such as the Soviet invasion of Afghanistan, the U.S.-led occupation of Iraq, or the Israeli occupation of Gaza and the West Bank. Others may join on behalf of the cause of other Muslims, such as the wars fought by Bosnian Muslims or Algerian rebels. Regardless of the precipitating event, mujahideen are characterized by their faith in several basic values.

The ideology of the modern mujahideen requires selfless sacrifice in defense of the faith. Accepting the title of mujahideen means that one must live, fight, and die in accordance with religious teachings. Mujahideen believe in the inevitability of victory, because the cause is being waged on behalf of the faith and in the name of God; both the faith and God will prevail. In this defense of the faith, trials and ordeals should be endured without complaint, because the pain suffered in this world will be rewarded after death in paradise. If one lives a righteous and holy life, for example, by obeying the moral proscriptions of the Qur'an, one can enjoy these proscribed pleasures in the afterlife.

As applied by the mujahideen, the defensive ideology of jihad holds that when one defends the faith against the unfaithful, death is martyrdom, and through death paradise is ensured.[16]

Organization Case: Al Qaeda's Religious Foundation

The most prominent pan-Islamic revolutionary organization of this era is Saudi national **Osama bin Laden**'s cell-based Al Qaeda (the Base), which seeks to unite Muslims throughout the world in a holy war. Al Qaeda is not a traditional hierarchical revolutionary organization, nor does it call for its followers to do much more than engage in terrorist violence in the name of the faith. It is best described as a loose network of like-minded Islamic revolutionaries. Al Qaeda is different because it

◆ Holds no territory

◆ Does not champion the aspirations of an ethno-national group

◆ Has no top-down organizational structure

◆ Has virtually no state sponsorship

◆ Promulgates vague political demands

◆ Has a completely religious worldview

Al Qaeda's religious orientation is a reflection of Osama bin Laden's sectarian ideological point of view. Bin Laden's worldview was created by his exposure to Islam-motivated armed resistance. As a boy, he inherited between $20 million and $80 million, with estimates ranging as high as $300 million, from his father. When the Soviets invaded Afghanistan in 1979, bin Laden eventually joined with thousands of other non-Afghan Muslims who traveled to Peshawar, Pakistan, to prepare to wage jihad. However, his main contribution to the holy war was to solicit financial and matériel (military hardware) contributions from wealthy Arab sources. He apparently excelled at this. The final leg on his journey toward international Islamic terrorism occurred when he and thousands of other Afghan veterans—the Afghan Arabs—returned to their countries to carry on their struggle in the name of Islam. Beginning in 1986, bin Laden organized a training camp that grew in 1988 into the Al Qaeda group. While in his home country of Saudi Arabia, bin Laden "became enraged when King Fahd let American forces, with their rock music and Christian and Jewish troops, wage the Persian Gulf war from Saudi soil in 1991."[17]

Photo 6.4 A Rewards for Justice poster of Abu Musab al-Zarqawi, Jordanian-born leader of Islamist resistance in Iraq during the U.S.-led occupation. Al-Zarqawi maintained close ties with Osama bin Laden before he was killed in June 2006.

After the Gulf War, bin Laden and a reinvigorated Al Qaeda moved to its new home in Sudan for five years. It was there that the Al Qaeda network began to grow into a self-sustaining financial and training base for promulgating jihad. Bin Laden and his followers configured the Al Qaeda network with one underlying purpose: "launching and leading a holy war against the Western infidels he could now see camped out in his homeland, near the holiest shrines in the Muslim world."[18]

Al Qaeda has inspired Islamic fundamentalist revolutionaries and terrorists in a number of countries. It became a significant source of financing and training for thousands of jihadis. The network is essentially a nonstate catalyst for transnational religious radicalism and violence.

When Al Qaeda moved to Afghanistan, its reputation as a financial and training center attracted many new recruits and led to the creation of a loose network of cells and sleepers in dozens of countries. Significantly, aboveground radical Islamic groups with links to Al Qaeda—such as Abu Sayyaf in the Philippines and Laskar Jihad in Indonesia—took root in some nations and overtly challenged authority through acts of terrorism.

Cult Case: Aum Shinrikyō (Supreme Truth)

Aum Shinrikyō is a Japan-based cult founded by **Shoko Asahara** in 1987. Its goal under Asahara's leadership was to seize control of Japan and then the world. The core belief is that Armageddon is imminent. One component is that the United States will wage World War III against Japan.[19] As one top member of the cult explained, "This evil [of the modern age] will be shed in a 'catastrophic discharge' . . . [and only those who] repent their evil deeds . . . [will survive]."[20]

At its peak membership, Aum Shinrikyō had perhaps 9,000 members in Japan and 40,000 members around the world, thousands of them in Russia.[21] Asahara claimed to be the reincarnation of Jesus Christ and Buddha and urged his followers to arm themselves if they were to survive Armageddon. This apocalyptic creed led to the stockpiling of chemical and biological weapons, including nerve gas, anthrax, and Q-fever.[22] One report indicated that Aum Shinrikyō members had traveled to Africa to acquire the deadly Ebola virus. Several mysterious biochemical incidents occurred in Japan, including one in June 1994 in the city of Matsumoto, where seven people died and 264 were injured from a release of gas into an apartment building.[23]

The Tokyo Subway Nerve Gas Attack. On March 20, 1995, members of Aum Shinrikyō positioned several packages containing sarin nerve gas on five trains in the Tokyo subway system scheduled to travel through Tokyo's Kasumigaseki train station. The containers were simultaneously punctured with umbrellas, thus releasing the gas into the subway system. Twelve people were killed and an estimated 5,000 to 6,000 were injured.[24] Tokyo's emergency medical system was unable to adequately respond to the attack, so that only about 500 victims were evacuated and the remaining victims had to make their own ways to local hospitals.

The police seized tons of chemicals the cult had stockpiled. Asahara was arrested and charged with 17 counts of murder and attempted murder, kidnapping, and drug trafficking. A new leader, Fumihiro Joyu, assumed control of Aum Shinrikyō in 2000 and renamed the group Aleph (the first letter in the Hebrew and Arabic alphabets). He has publicly renounced violence, and the cult's membership has enjoyed new growth in membership. A Japanese court sentenced Shoko Asahara to death by hanging on February 27, 2004.

Aum Shinrikyō is an example of the potential terrorist threat from apocalyptic cults and sects that are completely insular, segregated from mainstream society. Some cults are content to simply prepare for the End of Days, but others—like Aum Shinrikyō—are not averse to giving the apocalypse a violent push. The threat from chemical and other weapons of mass destruction will be explored further in Chapter 10.

Table 6.3 summarizes the activity profiles of several of the terrorist groups and movements discussed in this chapter.

TABLE 6.3   RELIGIOUS TERRORISM

Although religious terrorist groups and movements share the general profile of religious identity and often are rooted in similar belief systems, they arise out of unique historical, political, and cultural environments that are peculiar to their respective countries. With few exceptions, most religious movements are grounded in these idiosyncratic influences.

Group	Activity Profile	
	Constituency	**Adversary**
Aum Shinrikyō	Fellow believers	The existing world order
Lord's Resistance Army	Fellow believers and members of the Acholi tribe	Ugandan government and "nonpurified" Acholis
Palestine Islamic Jihad	Palestinian Muslims	Israel
Hamas	Palestinian Muslims	Israel
Al Qaeda	Faithful Muslims, as defined by Al Qaeda	Secular governments, nonbelievers, the West
Abu Sayyaf	Filipino Muslims	Filipino government, Western influence
Laskar Jihad	Moluccan Muslims	Moluccan Christians
Jammu-Kashmir groups	Jammu-Kashmir Muslims	India
Sikh groups	Punjabi Sikhs	India
Algerian/North African cells	Algerian Muslims and Muslims worldwide	Secular Algerian government, the West

Trends and Projections

Religion is a central feature of the New Terrorism, which is characterized by asymmetrical tactics, cell-based networks, indiscriminate attacks against soft targets, and the threatened use of high-yield weapons technologies. Al Qaeda and its Islamist allies pioneered this strategy, and it serves as a model for similarly motivated individuals and groups. Religious extremists understand that if they take on these characteristics, their agendas and grievances will receive extensive attention and their adversaries will be sorely challenged to defeat them. It is therefore reasonable to presume that religious terrorists will practice this strategy for the near future.

Having made this observation, it is important to critically assess the following questions: What trends are likely to challenge the global community in the immediate future? Who will enlist as new cadres in extremist religious movements? Who will articulate the principles of their guiding ideologies? The following patterns, trends, and events are offered for critical consideration:

◆ *Extremist religious propaganda cannot be prevented.* All religious extremists—Christian, Islamic, Jewish, and others—have discovered the utility of the Internet and the global media. They readily communicate with each other through the Internet, and their Web sites have become forums for propaganda and information.[25] Cable television and other members of the globalized media frequently broadcast interviews and communiqués.

◆ *A new generation of Islamist extremists has been primed.* In a study reported in January 2005, the Central Intelligence Agency's National Intelligence Council concluded that the war in Iraq created a new training and recruitment ground for potential terrorists, replacing Afghanistan in this respect. As one official suggested, "There is even, under the best scenario . . . the likelihood that some of the jihadists [will go home], and will therefore disperse to various other countries."[26]

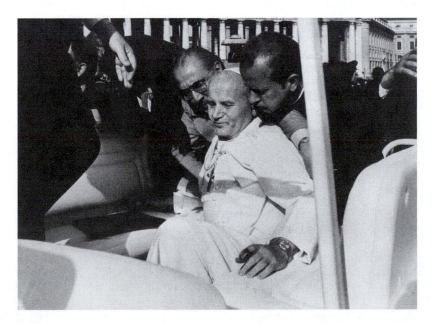

Photo 6.5 Pope John Paul II lies wounded after being shot by Mehmet Ali Agca. Agca was a member of the **Grey Wolves**, a fringe right-wing Turkish nationalist group implicated in numerous acts of terrorism.

◆ *Al Qaeda has become more than an organization—it evolved to become a symbol and ideology.* Osama bin Laden, founder and leader of Al Qaeda, presented himself in a series of communiqués as an elder statesman and intellectual of Islam. He recast himself as a symbolic mentor for the next generation of fighters.[27]

◆ *The jihadi movement has become a globalized phenomenon.* The dissemination of information and images through the media and the Internet created a global sense of solidarity among Islamists. Potential recruits easily access information, and many new volunteers are young people who live in the West, often in Europe.[28]

◆ Christian extremists continue to promote a religious motivation for the war on terrorism. Postings on some Christian Web sites and comments from some Christian leaders, usually in the United States, suggest that the Islamic faith is wrong or evil or both, and the war on terrorism is part of a divine plan pitting the true faith against Islam.29

Chapter Summary

Religious movements are motivated by a belief that an otherworldly power sanctions and commands their behavior. Some terrorists are motivated primarily by faith, whereas others use religion secondarily. The latter movements are motivated by nationalism or other ideology as a primary inspiration, but are united by an underlying religious identity. The goals of both primary and secondary religious terrorism are to construct a new society based on a religious or ethno-national identity. The terrorist behavior of both tendencies is active and public.

State-sponsored religious terrorism arises in governments that pursue international agendas by mentoring and encouraging religious proxies. The case of Iranian support for religious dissident terrorists is an example of a theocracy promoting its own revolutionary agenda. Syria is an example of a secular government that supports religious movements from a sense of common cause against a mutual enemy. Dissident religious terrorism involves attacks by self-proclaimed true believers against members of other faiths and perceived apostasies within their own faith. Some dissident groups espouse mystical or cult-like doctrines outside the belief systems of major religions.

DISCUSSION BOX

This chapter's Discussion Box is intended to stimulate critical debate about faith-motivated terrorism within major religions.

The One True Faith

Most religious traditions have produced extremist movements whose members believe that their faith and value system is superior. This concept of the one true faith has been used by many fundamentalists to justify violent religious intolerance. Religious terrorists are modern manifestations of historical traditions of extremism within the world's major faiths. For example,

- Within Christianity, the medieval Crusades were a series of exceptionally violent military campaigns against Muslims, Jews, and heretical Christian sects. Later, during the 16th and 17th centuries, Catholic and Protestant Christians waged relentless brutal wars against each other. In the modern era, Christian terrorists and extremists have participated in communal fighting in numerous countries and, in the United States, have bombed abortion clinics and committed other acts of violence.
- Within Judaism, the Old Testament is replete with references to the ancient Hebrews' faith-based mandate to wage war against non-Jewish occupiers of the Promised Land. In the modern era, the late **Rabbi Meir Kahane's Kach (Kahane Chai)** movement in Israel has likewise advocated the expulsion of all Arabs from Israel. Two members of the **Jewish Defense League** were arrested in the United States in December 2001 on charges of conspiring to bomb Muslim mosques and the offices of a U.S. congressman in Los Angeles.
- Within Islam, the relative tolerance of the 15th and 16th centuries is counterbalanced by intolerance today among movements such as Afghanistan's Taliban, Palestine's Hamas, and Lebanon's Hezbollah. Examples of political and communal violence waged in the name of Islam are numerous. Overt official repression has also been imposed in the name of the Islamic faith, as in Saudi Arabia's policy of relegating women to second-class status.

Modern religious extremism is arguably rooted in faith-based natural law. Natural law is a so-called philosophical higher law theoretically discoverable through human reason and references to moral traditions and religious texts. In fact, most religious texts have passages that can be selectively interpreted to encourage extremist intolerance. To religious extremists, it is God's law that has been revealed to—and properly interpreted by—the extremist movement.

Discussion Questions

1. Is faith-motivated activism a constructive force for change?
2. At what point does the character of faith-motivated activism become extremist and terrorist?
3. Does faith-based natural law justify acts of violence?
4. Why do religious traditions that supposedly promote peace, justice, and rewards for spiritual devotion have so many followers who piously engage in violence, repression, and intolerance?
5. What is the future of faith-based terrorism?

Key Terms and Concepts

The following topics were discussed in this chapter and can be found in the glossary:

Afghan Arabs
Amir, Yigal
Asahara, Shoko
Aum Shinrikyō
 (Supreme Truth)

Bin Laden, Osama
Crusades
Directorate for
 Inter-Services
 Intelligence (ISI)

Fund for the Martyrs
Goldstein, Baruch
Greater jihad
Grey Wolves
Hezbollah

Holy Spirit Mobile Force
Jihad
Lesser jihad
Lord's Resistance Army
Martyrdom
Okhrana

Order of Assassins
Phansi
Pogroms
Protocols of the Learned
    Elders of Zion

Qods Force
Revolutionary
    Guards Corps
Thuggees

Web

Log on to the Web-based student study site at **www.sagepub.com/martinessstudy** for additional Web sources and study resources.

Web Exercise

Using this chapter's recommended Web sites, conduct an online investigation of religious extremism.

1. What commonalities can you find among the religious Web sites? What basic values are similar? In what ways do they differ?
2. Are the religious sites effective propaganda? How would you advise the site designers to appeal to different constituencies?

For an online search of historical and cultural issues pertaining to religious extremism, readers should activate the search engine on their Web browser and enter the following keywords:

"Christian Crusades"
"Jihad"

Recommended Readings

The following publications discuss the motives, goals, and characteristics of religious extremism.

Bader, Eleanor J., and Patricia Baird-Windle. *Targets of Hatred: Anti-Abortion Terrorism.* New York: Palgrave, 2001.

Halevi, Yossi Klein. *Memoirs of a Jewish Extremist: An American Story.* New York: Little, Brown, 1995.

Huband, Mark. *Warriors of the Prophet: The Struggle for Islam.* Boulder, CO: Westview, 1999.

Lifton, Robert Jay. *Destroying the World to Save It: Aum Shinrikyō, Apocalyptic Violence, and the New Global Terrorism.* New York: Henry Holt, 2000.

Rashid, Ahmed. *Taliban: Militant Islam, Oil and Fundamentalism in Central Asia.* New Haven, CT: Yale Nota Bene, 2001.

Stern, Jessica. *Terror in the Name of God: Why Religious Militants Kill.* New York: Ecco/HarperCollins, 2003.

7

International Terrorism

This chapter investigates the dimensions of international terrorism. International terrorism is terrorism that "spills over" onto the world's stage. Targets are selected because of their value as symbols of international interests and the impact that attacks them will have on a global audience. Terrorism in the international arena has been a common feature of political violence since the late 1960s, when political extremists began to appreciate the value of allowing their revolutionary struggles to be fought in a global arena, primarily for its propaganda benefits. In adopting terrorism on a broader scale, revolutionary movements have been very successful in moving their underlying grievances to the forefront of the international agenda. It is not an exaggeration to conclude that "international terrorism represents one of the defining elements of politics on the world's stage today."[1]

International terrorism is one of the best examples of **asymmetrical warfare**—that is, unconventional, unexpected, and nearly unpredictable acts of political violence. Although it is an old practice, asymmetrical warfare has become a core feature of the New Terrorism. Practicing it, terrorists can theoretically acquire and wield new high-yield arsenals, strike at unanticipated targets, cause mass casualties, and apply unique and idiosyncratic tactics. The dilemma for victims and for counterterrorism policy makers is that by using these tactics, the terrorists can win the initiative and redefine the international security environment by overriding traditional protections and deterrent policies. In an era of immediate media attention, small and relatively weak movements have concluded that worldwide exposure is possible and worthwhile. They have discovered that politically motivated hijackings, bombings, assassinations, kidnappings, extortion, and other criminal acts conducted under an international spotlight guarantee some degree of attention and afford greater opportunities.

Chapter Perspective 7.1 summarizes the changing environments of international terrorism, focusing on the predominant dissident profiles during the latter part of the 20th century.

CHAPTER PERSPECTIVE 7.1

Cooperation Between Terrorists: The European Connection

Terrorists have cooperated with each other for a variety of reasons. Some considered their revolutionary struggle to be part of a global war against a particular nation, interest, or ideology. For example, France's Direct Action directed its violence

(Continued)

against American interests and in solidarity with the Palestinian cause. Many groups have in fact cooperated as a matter of revolutionary solidarity. Some, like the Japanese Red Army (JRA), allowed themselves to be retained by groups such as the PFLP. Others, such as the Red Army Faction or the 2nd of June Movement, regularly collaborated on missions with other movements.

The PFLP established strong links with European terrorist groups. Beginning in the late 1960s, the PFLP provided secure training and networking facilities in Jordan. Its working relationships with Germany's Red Army Faction and 2nd of June Movement were very strong, and members of these groups were welcomed and trained in the PFLP's camps, as were members from Spain's Basque Fatherland and Liberty. The PFLP served as a mentor for Ilich Ramirez Sanchez, also known as Carlos the Jackal, who cooperated extensively with a variety of terrorist groups.

The December 1975 kidnapping of OPEC ministers in Vienna was a dramatic example of cooperation. Led by the Venezuelan terrorist Carlos the Jackal, German and Palestinian terrorists carried out a well-planned and coordinated assault and then escaped safely.

From 1985 to 1987, European terrorists attempted to create an anti-imperialist umbrella group similar to the Palestine Liberation Organization. They hoped to replicate the PLO's success by targeting NATO and European capitalism. The RAF, Direct Action, Italy's Red Brigades, and Belgium's Communist Combat Cells joined forces for approximately 2 years in a temporary united front. However, arrests and political changes in Europe kept the new movement from producing anything like a PLO-style environment.

In the aftermath of the decline and dissolution of the leftist terrorist environment, a new cell-based network of cooperation emerged. It was inspired by Al Qaeda, which has clearly become the most extensive—albeit loosely structured—international network to date. The example of the Al Qaeda network and tactics was adopted by radical Islamists in Europe, and put into action with the March 2004 terrorist attack in Madrid and the July 2005 attack in London.

International terrorism has two important qualities. It is both a specifically selected tactical and strategic instrument of political violence and a type of terrorism. The discussion in this chapter will review the following:

◆ Understanding international terrorism
◆ International terrorist networks
◆ The international dimension of the new terrorism

Understanding International Terrorism

International terrorism occurs when the target is a global symbol and the political-psychological effects go beyond a purely domestic agenda. It will be recalled that state terrorism as foreign policy is often characterized by state sponsorship of dissident movements. In addition, many dissident terrorist groups and extremist movements have regularly acted in solidarity with international interests such as class struggle or national liberation. It is also not uncommon for domestic groups and movements to travel abroad to attack targets symbolizing their domestic conflict or some broader global issue. Especially in the post–World War II world, terrorist groups have selected targets that only tangentially symbolize the sources of their perceived oppression. Links between seemingly domestic-oriented dissident terrorist groups and international terrorism are therefore not uncommon. Table 7.1 illustrates this point by reviewing the activity profiles of several dissident terrorist groups implicated in international incidents.

The Spillover Effect

Those who engage in political violence on an international scale do so with the expectation that it will have a positive effect on their cause at home—thus reasoning that international exposure will bring about compensation for perceived injustices. Using this logic, terrorists will either go abroad to strike at targets or remain at home to strike internationally symbolic targets. The following characteristics distinguish international terrorism as a specific type of terrorism:

◆ Domestic attacks against victims with an international profile
◆ Operations in a foreign country
◆ Unambiguous international implications

TABLE 7.1   DISSIDENT TERRORISM ON THE WORLD'S STAGE

During the latter quarter of the 20th century, dissident terrorist groups attacked symbols of international interests many times. Some groups traveled abroad to strike at targets, whereas others attacked domestic symbols of international interests.

| | Activity Profile | | | |
Dissident Group	Home Country	Incident	Target	International Effect
Abu Sayyaf	Philippines	April 2000 kidnapping	20 Asian and European tourists in Malaysia	Increased profile; $20 million ransom
Provos	Northern Ireland	December 1983 bombing	Harrods department store in London	National security crisis in United Kingdom
Shining Path	Peru	March 2002 bombing	Near U.S. embassy	Renewed profile; domestic spillover
Red Brigades	Italy	December 1981 kidnapping	U.S. Brigadier General James Dozier	Increased profile; NATO security crisis
Black September	Palestinian Diaspora	Summer 1972 kidnapping	Israeli athletes at Summer Olympics	Increased profile; international crisis

Domestic Attacks Against Victims With an International Profile

Most experts would agree that international terrorism by definition does not require terrorists to leave their home countries; they can strike domestically, depending on what the target represents. Examples of domestic targets with symbolic international links include diplomats, businessmen and women, military personnel, and tourists. An important and disturbing result of these attacks has been a repeated trend in which terrorists recognize no qualitative difference between their victim group and the enemy interest it represents. Thus, innocent business travelers, civilians, academics, and military personnel are considered legitimate targets and fair game because they symbolize an enemy interest.

During the long Lebanese civil war in the 1970s and 1980s, militant Islamic groups targeted symbols of international interests with great drama. The Shi'a group Hezbollah was responsible for most of these incidents, examples of which include the bombing of the U.S. embassy in Beirut and attacks against French and American peacekeeping troops in Beirut. In addition, the kidnapping of foreign nationals by Islamic extremists became a prominent characteristic of the violence.

Operations in a Foreign Country

The history of terrorism is replete with examples of extremists who deliberately travel abroad to strike at an enemy. Their targets often include nationals that the terrorists symbolically associate with the policies of their perceived enemy. Some are representatives of these policies, such as diplomats or political officials, whereas others, such as business travelers and tourists, are only minimally linked. For example, the Algerian Armed Islamic Group launched a terrorist campaign in metropolitan France from July to October 1995. Eight people were killed and 180 injured during bombings of trains, schools, cafes, and markets. Terrorist plots today have taken on a decidedly transnational dimension, with cells linked across several countries. Extremist groups have deliberately positioned terrorist operatives and autonomous cells in foreign countries. Their purpose is to attack enemy interests with a presence in those countries. For example, during the 1990s, Latin American and U.S. security officials identified an apparent threat in South America, which indicated that Middle Eastern cells became active in the 1990s along the triangular border region where Brazil, Argentina, and Paraguay meet.

Unambiguous International Implications

Regardless of whether political violence is directed against domestic targets that are international symbols or take place in third countries, the fundamental characteristic of international terrorism is that international consequences are clear. For example, the February 1993 attack on the World Trade Center in New York City had clear international consequences, as did the terrorist attacks of September 11, 2001. In both instances, foreign terrorist cells and the Al Qaeda network had used transnational resources to position terrorist operatives in the United States, their targets were prominent symbols of international trade and power, and their victims were citizens from many countries.

Case: Hijackings as International Spillovers

A good example of terrorist spillovers is the selection of international passenger carriers—that is, ships and aircraft—as targets by terrorists. They are relatively soft targets that attract international media attention when attacked. The passengers are considered legitimate symbolic targets, so that terrorizing or killing them is justifiable in the minds of the terrorists.

Beginning in the late 1960s, terrorists began attacking international ports of call used by travelers. International passenger carriers—primarily airliners—became favorite targets. In the beginning, hijackings were often the acts of extremists, criminals, or otherwise desperate people trying to escape their home countries to find asylum in a friendly country. This profile changed, however, when the Popular Front for the Liberation of Palestine (PFLP) staged a series of aircraft hijackings as a way to **publicize the cause** of the Palestinians before the world community. The first successful high-profile PFLP hijacking was Leila Khaled's attack in August 1969. The PFLP struck again in September 1970, when it attempted to hijack five airliners, succeeding in four of the attempts. These incidents certainly directed the world's attention to the Palestinian cause, but they also precipitated the Jordanian army's Black September assault against the Palestinians. Nevertheless, passenger carriers have since been frequent targets of international terrorists.

Table 7.2 summarizes the interplay between international terrorism and terrorist environments.

TABLE 7.2 INTERNATIONAL TERRORISM AND TERRORIST ENVIRONMENTS				
Although cases of international terrorism exist for most terrorist environments, right-wing and traditional criminal terrorists tend to refrain from violence in the international arena.				
	Activity Profile			
Environment	**Incident**	**Perpetrator**	**Target**	**International Effect**
State	Car bomb in Washington, DC	Chile Security Service, DINA (Chilean secret service)	Orlando Letelier	Minimal effect; domestic spillover
Dissident	Bomb in London	Irish Republican Army	Canary Wharf	Significant effect; ended IRA's cease-fire
Left-wing	Hostages taken in Lima, Peru	Peru's Tupac Amaru Revolutionary Movement	Japanese ambassador's residence	Minimal effect; domestic spillover
Right-wing	Arson attacks in former East Germany	Neo-Nazis and supporters	Foreign "guest workers"	Minimal effect
Religious	Attacks of September 11, 2001	Al Qaeda cells	World Trade Center, Pentagon	Significant effect; international crisis and war
Criminal	Kidnapping, murder	Mexican narcotraficantes	U.S. DEA Special Agent Enrique Camarena	Significant effect; enhanced war on drugs

Reasons for International Terrorism

Extremist groups and movements resort to international terrorism for a number of reasons. Some act in cooperation with others, waging an international campaign against a perceived global enemy. France's Direct Action, Germany's Red Army Faction, and Italy's Red Brigade frequently attacked symbolic targets as expressions of solidarity with international causes, for example. Other groups are motivated by idiosyncrasies unique to the group. For example, the **Armenian Secret Army for the Liberation of Armenia (ASALA)**, the Justice Commandos of the Armenian Genocide, and the Armenian Revolutionary Army waged a campaign of international terrorism against Turkish interests in Europe and the United States during the 1970s and 1980s.

The following discussion summarizes several underlying reasons extremists select international terrorism as a strategy:

◆ Ideological reasons: modern "isms" and international revolutionary solidarity
◆ Practical reasons: perceived efficiency
◆ Tactical reasons: adaptations of revolutionary theory to international operations
◆ Historical reasons: perceptions of international terrorism

Ideological Reasons: Modern "isms" and International Revolutionary Solidarity

Conflict in the postwar era was largely a by-product of resistance in many countries against former colonial powers. These indigenous wars of national liberation were global in the sense that insurgents established a common cause against what they perceived as domination by repressive and exploitative imperial powers and their local allies. Marxism and generalized leftist sentiment provided a common bond among radicals in the developing world and among Western dissidents.

Reasons for this resistance included, of course, the overt presence of foreign troops, administrators, and business interests in the newly emerging countries. From a broader global perspective, other reasons included the policies and ideologies of the Western powers. These policies and ideologies were given negative labels that insurgents affixed to the Western presence in the developing world. If one were to apply Marxist interpretations of class struggle and national liberation, these indigenous wars could easily be interpreted as representing an internationalized struggle against global exploitation by the West. The conceptual labels commonly used by violent extremists—both secular and religious—include the following:

Imperialism. Postwar dissidents fought against what they considered to be the vestiges of European colonial empires. Western powers had traditionally deemed empire building (imperialism) a legitimate manifestation of national prestige and power, an ideology that existed for centuries and was not finally ended until the latter decades of the 20th century. For example, the European scramble for Africa during the late 19th century was considered completely justifiable. During the wars of national liberation, foreign interests and civilians were attacked because they were seen as representatives of imperial powers.

Neocolonialism. Neocolonialism referred to exploitation by Western interests, usually symbolized by multinational corporations. Even when Western armies or administrators were not present and when indigenous governments existed, insurgents argued that economic exploitation still relegated the developing world to a subordinate status. From their perspective, the wealth of the developing world was being drained by Western economic interests. Domestic governments thus became targets because they were labeled as dupes of Western interests. Multinational corporations and other symbols of neocolonialism became targets around the world, so that facilities and employees were attacked internationally.

Zionism. Historically an intellectual movement that sought to establish the proper means, conditions, and timing to resettle Jews in Palestine, Zionism was officially sanctioned by the Balfour Declaration of November 2, 1917; this was a statement by the British government that favored establishing a Jewish nation in Palestine as long as the rights of non-Jewish residents were guaranteed. The concept has become a lightning rod for resistance against Israel and its supporters. One significant difference in the international character of anti-Zionist terrorism— vis-à-vis resistance against imperialism and neocolonialism—is that international terrorists have attacked Jewish civilian targets around the world as symbols of Zionism. This adds a religious and ethno-national dimension to anti-Zionist terrorism that does not necessarily exist in the other concepts discussed.

Chapter Perspective 7.2 explores cooperation between European and Palestinian terrorists, who conducted joint operations out of a sense of global solidarity for their respective agendas.

CHAPTER PERSPECTIVE 7.2

The Changing Environment of International Terrorism

International terrorism is in many ways a reflection of global politics, so that as an environment it is dominated at different times by different groups and cycles of behavior. In this regard, its profile has progressed through several phases in the postwar era, which can be roughly summarized as follows:[a]

◆ From the 1960s through the early 1980s, left-wing terrorists figured prominently in international incidents. For example, West European groups frequently attacked international symbols in solidarity with defined oppressed groups. Only a few leftist groups remain, and most of them (such as Colombia's FARC) do not often practice international terrorism.

◆ From the beginning, that is, since the late 1960s, Palestinian nationalists were the leading practitioners of international terrorism. Participating in their struggle were Western European and Middle Eastern extremists who struck targets in solidarity with the Palestinian cause. By the late 1990s, with the creation of the governing authority on the West Bank and Gaza, Palestinian-initiated terrorism focused primarily on targets inside Israel and the occupied territories. Their radical Western comrades ceased their violent support by the late 1980s, but many Middle Eastern extremists continued to reference the Palestinian cause as a reason for their activism.

◆ Throughout the postwar era, ethno-national terrorism occupied an important presence in the international arena. Its incidence has ebbed and flowed in scale and frequency, but it has never disappeared. By the late 1990s, these groups operated primarily inside their home countries but continued to attack international symbols.

(Continued)

◆ By the end of the 20th century, the most prominent practitioners of international terrorism were religious extremists. Although Islamist movements such as Al Qaeda and other groups were the most prolific international religious terrorists, extremists from every major religion have operated on the international stage.

Note

a. See Paul R. Pillar, *Terrorism and U.S. Foreign Policy* (Washington, DC, 2001), 44–45.

Practical Reasons: Perceived Efficiency

One basic (and admittedly simplified) reason for the high incidence of international terrorism is that it is perceived by many extremists to be "a highly efficient (if repugnant) instrument for achieving the aims of terrorist movements."[2] Using rationale of perceived efficiency as a core motivation, practical reasons for international terrorism include several factors:

The Potential for Maximum Publicity. Terrorists practice propaganda by the deed and understand that international deeds get maximum media exposure. The media has always moved dramatic events to the top of their reporting agenda, thus international terrorism is always depicted as important news. In November 1979, the American embassy was attacked in Tehran, Iran, and 53 Americans were held hostage until January 1981, a total of 444 days. During that time, Iran was at the center of media attention and the revolutionaries were able to publicize their revolutionary agenda. They also embarrassed the most powerful nation in the world.

Inflicting the Potential for Maximum Psychological Anxiety. Terrorism against enemy interests on a global scale theoretically creates security problems and psychological anxiety everywhere that the target commands a political or economic presence. When travelers, airliners, and diplomats are attacked internationally, an enemy will give attention to the grievances of the movement that is the source of the violence.

Pragmatism. Demanding concessions from adversaries who become the focus of worldwide attention has occasionally been successful. Terrorists and extremists have sometimes secured ransoms for hostages, prisoner releases, and other concessions.

Thus, international terrorism is a functional operational decision that offers—from the perspective of the terrorists—greater efficiency in promoting the goals of the cause.

Tactical Reasons: Adapting Theory to Operations

One may observe that international extremists apply modern adaptations of their ideas on a global scale. It is therefore instructive to consider their theories from an international perspective.

1. *Fish Swimming in the Sea of the People.* With the inexorable trend toward globalization, many national, information, and communication barriers have been dramatically altered. Members of dissident movements are now able to live in multiple countries yet remain in regular contact through the Internet and other communications technologies. Adapting Mao Zedong's maxim to the international arena, terrorists can become "fish swimming in the sea of the global community." It is much more difficult to root out a movement that uses cells prepositioned in several countries, as became apparent during the post–September 11, 2001, war against terrorism.

2. *Enraging the Beast.* Relatively weak groups continue to attempt to provoke governments into reacting in ways detrimental to government interests. In doing so, they almost guarantee that new supporters will arise to join the cause. Adapting Carlos Marighella's theory of provoking a powerful adversary into overreacting—and thus creating a heightened revolutionary consciousness among the people—international terrorists can now provoke a nation on a global scale. These provocations provide effective propaganda if the offending nation can be depicted as an international bully. This is not difficult to do, because modern information technology allows for dramatic images and messages to be disseminated worldwide. During the 2001–2002 coalition campaign against the Taliban in Afghanistan and the 2003 invasion of Iraq, images of wounded and killed civilians were regularly and graphically broadcast throughout the Muslim world.

Historical Reasons: Perceptions of International Terrorism

Governments in developing countries do not always share the West's interpretations when political violence is used in the international arena. Western democracies regularly abhor and denounce international terrorism, whereas many regimes and leaders elsewhere have been either weak in their denunciations or have on occasion expressed their approval of international extremist violence.

Three factors illustrate why terrorism is unacceptable to Western governments:

1. First, Western governments have adopted an ideology of democratic justice as a norm. A norm is an accepted standard for the way societies ought to behave. Within the context of these norms, terrorism is perceived as inherently criminal behavior.
2. Second, the West is often a target of terrorism. This is a practical consideration.
3. Third, the West recognizes accepted methods of warfare. These include rules that define which modes of conflict constitute a legal manner to wage war, and that only just wars should be fought. If possible, so-called collateral damage (unnecessary casualties and destruction) is to be avoided.

In the developing world, wars waged to gain independence or to suppress political rivals were commonly fought by using irregular tactics. Combatants often used guerrilla or terrorist tactics against colonial opponents or against indigenous adversaries during civil wars. In fact, many of the statesmen who rose to prominence after the formation of the new nations were former insurgents known by their adversaries as terrorists. From their perspective, terrorism and other violent methods were necessary weapons for waging poor man's warfare against enemies that were sometimes many times stronger than themselves.

Three factors demonstrate why terrorism is often acceptable to governments in the developing world:

1. Many anticolonial extremists, freedom fighters and heroes in the eyes of their people, became national leaders. A large number of revolutionaries became national leaders in Asia, Africa, and the Middle East.
2. Terrorism was used as a matter of practical choice during insurgencies. It provided **armed propaganda**, sowed disorder, and demoralized their adversaries. At some point in many postwar conflicts, the cost of war simply became unacceptable to the colonial powers.

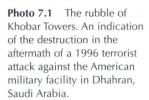

Photo 7.1 The rubble of Khobar Towers. An indication of the destruction in the aftermath of a 1996 terrorist attack against the American military facility in Dhahran, Saudi Arabia.

3. Many developing world insurgents crafted an effective fusion of ideology and warfare. Terrorism was a justifiable and legitimate method of warfare as long as the cause was just and the fighters were cadres who understood why they were fighting.

International Terrorist Networks

Historically, many revolutionary movements have proclaimed their solidarity with other movements. In some cases, these movements engaged in coordinated attacks or other behaviors that intimated the existence of terrorist networks. The concept of an international terrorist network naturally presumes one or more of the following environments:

◆ Terrorist groups talk to each other   ◆ The international terrorist environment
◆ Terrorist groups support each other   is basically conspiratorial
◆ Governments sponsor terrorists

Chapter Perspective 7.3 discusses a noteworthy example of cooperation between Japanese Red Army terrorists and the Popular Front for the Liberation of Palestine.

CHAPTER PERSPECTIVE 7.3

A Remarkable Example of International Terrorism: The JRA and the Lod (Lydda) Airport Massacre

The Japanese Red Army was a nearly fanatical international terrorist organization founded in about 1970. It participated in a series of terrorist incidents, including assassinations, kidnappings, and airline hijackings. On one occasion, it tried to occupy the U.S. embassy in Kuala Lumpur, Malaysia. On another, the group bombed a United Service Organizations club in Naples, Italy. Members have been arrested in a number of countries, including one who was caught with explosives on the New Jersey Turnpike. Others who were not arrested are thought to have joined Latin American revolutionaries in Colombia and Peru.

> The JRA committed a number of terrorist acts in cooperation with other terrorist groups. On May 30, 1972, three members of the organization fired assault rifles into a group of religious pilgrims and other travelers at Israel's Lod (Lydda) Airport. The death toll was high—26 people were killed and about 80 were wounded. Most of the injured were travelers from Puerto Rico on a religious pilgrimage. During the firefight that ensued, two of the terrorists were killed by security guards and one was captured. The JRA had been retained by the Popular Front for the Liberation of Palestine for the attack, and the PFLP had sent the three operatives on their mission on behalf of the Palestinian cause.
>
> The attack is a remarkable example of international terrorism in its purest form: Japanese terrorists killed Christian pilgrims from Puerto Rico arriving on a U.S. airline at an Israeli airport on behalf of the nationalist Popular Front for the Liberation of Palestine.
>
> As a postscript, the survivor of the Lod attack—Kozo Okamoto—was tried and sentenced to life imprisonment in Israel but was later released in a 1985 prisoner exchange with other Palestinian prisoners. He lived in Lebanon's Beka'a Valley until 1997 but was arrested by the Lebanese with four other members of the JRA. After serving 3 years in prison, all five members were freed in 2000.

International Terrorist Environments

Several terrorist environments are theoretically possible at different times and in different regional contexts. These are not, of course, exclusive descriptions of every aspect of international terrorism. They are, however, useful models for framing a generalized interpretation of international terrorism. The following discussion summarizes four environments that range in structure and cohesion from tightly knit single-sourced threats to loosely linked multiple-source threats.[3] The four environments are the following:

- ◆ Monolithic terrorist environments
- ◆ Strong multipolar terrorist environments
- ◆ Weak multipolar terrorist environments
- ◆ Cell-based terrorist environments

Monolithic Terrorist Environment

The emphasis in the **monolithic terrorist environment** model is on state-centered behavior. The old model of a global Soviet-sponsored terrorist threat described an environment in which terrorism was a single threat from a single source. This presupposed that Soviet assistance and Marxist ideology were binding motivations for international terrorism. As a matter of counterterrorism policy, if this had been correct, combating the threat would have been uncomplicated, because in theory the source (i.e., the Soviets) might have been induced or co-opted to reduce or end their sponsorship.

Strong Multipolar Terrorist Environment

In the **strong multipolar terrorist environment**, the emphasis is also on state-centered behavior. This model presumes that state sponsorship guides terrorist behavior but that several governments support their favored groups. It also presumes that there are few truly autonomous international terrorist movements; they all have a link to a state sponsor. Unlike the theory of a monolithic Soviet-sponsored threat, this scenario suggests that regional powers use terrorism to support their agendas and that many proxies exist. To counter this type of environment, the several sources would theoretically each have to be co-opted independent of the others.

Weak Multipolar Terrorist Environment

The model of the **weak multipolar terrorist environment** shifts the emphasis away from the state toward the terrorist movements. It presupposes both state sponsorship and more autonomous terrorist groups. Under this scenario, several governments support their favored groups, but many of the groups are relatively independent. The terrorist groups would be content with any state sponsor as long as they received enough assistance to continue their revolutionary struggle. Countering this environment is more difficult than countering the monolithic or strong multipolar models, because the dissident movements are more flexible and adaptable. Cutting off one source of assistance would theoretically have little more effect than a temporary disruption in operations.

Cell-Based Terrorist Environment

In the model of the **cell-based terrorist environment**, the emphasis is centered on the terrorist movement. This model presumes that state sponsorship may exist to some extent but that the revolutionary movement is independent of governmental constraints, a kind of free floating revolution that maintains its autonomy through its own resources. In this kind of environment, the movement is based on the viability of small cells loosely affiliated with one other. Thus, the movement will prevail even when some of its cells are destroyed or otherwise compromised. Because of its fluid organizational structure, this environment is the most difficult to counter.

Table 7.3 summarizes the activity profiles for these terrorist environments.

TABLE 7.3   UNDERSTANDING INTERNATIONAL TERRORIST ENVIRONMENTS

International terrorism occurs within the context of international social and political environments. These environments are not static and can vary from time to time and region to region.

	Activity Profile			
Environment	Foremost Participant	State Control	Group Autonomy	Difficulty in Countering
Monolithic	Single state sponsor	Strong and direct	Minimal	Clear options; easiest to counter
Strong multipolar	Several state sponsors	Strong and direct	Minimal	Clear options
Weak multipolar	Dissident groups; several state sponsors	Weak and insecure	Strong	Problematic and unpredictable
Cell-based	Dissident groups	Weak	Strong	Problematic and unpredictable; most difficult to counter

_____ The International Dimension of the New Terrorism

Movement Case: The Afghan Arabs at War

The series of wars in Afghanistan, which began with the 1979 Soviet invasion, produced a large **cadre** of hardened Islamic fighters and led to a sustained guerrilla insurgency that eventually forced the occupying Soviet army to withdraw after losing 15,000 of its men. The war was considered a jihad in the eyes of the insurgents, who declared themselves mujahideen in a holy war against nonbelievers.

Muslims from around the world volunteered to fight alongside or otherwise support the Afghani mujahideen. This created a pan-Islamic consciousness that led to the creation of the Al Qaeda network and domestic jihadi movements as far afield as the Philippines, Malaysia, Central Asia, and Algeria. The Muslim volunteers—Afghan Arabs—became a legendary fighting force among Muslim activists. It is not known exactly how many Afghan Arabs fought in Afghanistan during the anti-Soviet jihad. However, reasonable estimates have been calculated:

> [A former] senior CIA official . . . claims the number is close to 17,000, while the highly respected British publication Jane's Intelligence Review suggests a figure of more than 14,000 (including some 5,000 Saudis, 3,000 Yemenis, 2,000 Egyptians, 2,800 Algerians, 400 Tunisians, 370 Iraqis, 200 Libyans, and scores of Jordanians).[4]

After the Soviet phase of the war, many of the Afghan Arabs carried on their jihad in other countries, becoming **international mujahideen**.

Case: Mujahideen in Bosnia

The first wars that were fought after the breakup of Yugoslavia were between the Serbs, Croatians, and Bosnian Muslims. When Bosnia declared independence, it precipitated a brutal civil war that at times had all three national groups fighting against each other simultaneously. The war between the Bosnians and Serbs was particularly violent and involved genocidal "ethnic cleansing" that was initiated by the Serbs and imitated later by other national groups. From the beginning, the Bosnian Muslims were severely pressed by their adversaries and were nearly defeated as town after town fell to the Serbs. Overt international arms shipments were prohibited, although some Islamic countries did provide covert support. Into this mix came Muslim volunteers who fought as mujahideen on behalf of the Bosnians. Most were Middle Eastern Afghan Arabs. An estimated 500 to 1,000 mujahideen fought alongside the Bosnians, coming from nearly a dozen countries. Many came from Afghanistan, Algeria, Egypt, Tunisia, and Yemen. Although the international mujahideen were motivated by religious zeal, the Bosnians are traditionally secular Muslims, so that they were motivated by nationalist zeal.[5] This was a cause for some friction, but a handful of Bosnians also took an oath to become mujahideen.

Organization Case: Al Qaeda and International Terrorism

Al Qaeda is a transnational movement with members and supporters from throughout the Muslim world. It is, at its very core, an international revolutionary movement that uses terrorism as a matter of routine. Al Qaeda has two overarching goals: to link together Muslim extremist groups throughout the world into a loose pan-Islamic revolutionary network and to expel non-Muslim (especially Western) influences from Islamic regions and countries.

Photo 7.2 Death in Kenya: A Kenyan man is lifted out of the rubble of the American embassy in Nairobi, Kenya. On August 7, 1998, the U.S. embassy in Nairobi was bombed by an Al Qaeda cell, which coordinated the assault with an attack conducted by another cell against the American embassy in Dar es Salaam, Tanzania.

Case: International Cells

Operatives who were trained or inspired by Al Qaeda established cells in dozens of countries and regions. For instance, cells and larger groups became resident in the following predominantly Muslim countries and regions: Afghanistan, Algeria, Bosnia, Chechnya, Indonesia, Iraq, Kosovo, Lebanon, Malaysia, Pakistan, the southern Philippines, Somalia, Sudan, the West Bank, and Yemen. Other cells were covertly positioned in the following Western and non-Muslim countries: Britain; France; Germany; Israel; Spain; the United States; and the border region of Argentina, Brazil, and Paraguay.

Members communicate with each other using modern technologies such as faxes, the Internet, cell phones, and e-mail. Most Al Qaeda cells are small and self-sustaining and apparently receive funding when activated for specific missions. For example, the bombings of the American embassies in Kenya and Tanzania may have cost $100,000.[6] Not all cells are **sleeper cells,** defined as groups of terrorists who take up long-term residence in countries prior to attacks. For example, most of the September 11, 2001, hijackers entered the United States for the express purpose of committing terrorist acts; they were never prepositioned as sleepers to be activated at a later date. Al Qaeda has been rather prolific in its direct and indirect involvement in international terrorism.

The following cases of international Al Qaeda terrorist activity illustrate the scope, skill, and operational design of the network.

The Hijacking of Air France Flight 8969

In December 1994, **Air France Flight 8969** was hijacked in Algeria by an Algerian Islamic extremist movement, the **Armed Islamic Group (GIA).** The GIA apparently has ties to Al Qaeda, and many members of its movement are Afghan Arab war veterans. After hijackers killed three passengers, the plane was permitted to depart the Algerian airfield and made a refueling stop in Marseilles. Intending to fly to Paris, the hijackers demanded three times the amount of fuel needed to make the journey.[7] The reason for this demand was that they planned to blow up the aircraft over Paris—or possibly crash it into the Eiffel Tower—thus killing themselves and all of the passengers as well as raining flaming debris over the city. French commandos disguised as caterers stormed the plane in Marseilles, thus bringing the incident to an end. The assault was played out live on French television.

The Singapore Plot

During December 2001 and January 2002, security officers in Singapore, Malaysia, and the Philippines arrested more than three dozen terrorist suspects. Thirteen were arrested in Singapore, 22 in Malaysia, and 4 in the Philippines. The sweep apparently foiled a significant plot by Al Qaeda to attack Western interests in Singapore. The arrests were made after Northern Alliance forces in Afghanistan captured a Singaporean of Pakistani descent who was fighting for the Taliban. The prisoner gave details about "plans to bomb U.S. warships, airplanes, military personnel and major U.S. companies in Singapore [as well as] American, Israeli, British and Australian companies and government offices there."[8]

Photo 7.3   The 1972 Munich Olympics. A hooded Black September terrorist peers over a balcony during the tragic crisis that ended with the deaths of all of the Israeli hostages.

Members of the Singaporean cell had been trained by Al Qaeda in Afghanistan, and Al Qaeda operatives had traveled to Singapore to advise some of the suspects about bomb construction and other operational matters. During their preparations for the strike, the cell had sought to purchase 21 tons of ammonium nitrate—by comparison, Timothy McVeigh's **ANFO (ammonium nitrate and fuel oil)** truck bomb had used only 2 tons of ammonium nitrate. The Singaporean cell organized itself as the Islamic Group (Jemaah Islamiah). The Malaysian cell called itself the Kumpulan Militan Malaysia. In an interesting twist, the Malaysian group apparently had indirect ties to the October 2000 bombing of the *USS Cole* in Yemen; the September 11 attacks; and perhaps to **Zacarias Moussaoui**, the French Moroccan implicated as affiliated with the September 11 terrorists.[9]

Chapter Perspective 7.4 explores the bona fide threat from terrorist organizations that have been inspired by the Al Qaeda model.

CHAPTER PERSPECTIVE 7.4

Beyond Al Qaeda

It is well known that the idea of Al Qaeda was born in the crucible of the anti-Soviet jihad in Afghanistan. International fighters were brought together under the banner of jihadist solidarity, and those who passed through the Al Qaeda network became imbued with a global and extreme belief system.

Because Al Qaeda's belief system is grounded in fundamentalist religious faith, the organization evolved into an ideology and an exemplar for other radical Islamists.[a] In a sense, Al Qaeda franchised its methods, organizational model, and internationalist ideology.[b] As the original leadership and Afghan Arabs were killed,

(Continued)

captured, or otherwise neutralized, new personnel stepped to the fore, many of whom had minimal contacts with the original network. Thus, new extremists became affiliated with Al Qaeda by virtue of their replication of its model.

Although Al Qaeda remains actively engaged in its war against Western influence and perceived Muslim apostasies, its role has in part become that of an instigator and mentor. For example, the Islamists who joined the anticoalition resistance in Iraq became an Al Qaeda–affiliated presence. Jordanian-born Abu Musab al-Zarqawi became an important symbol of Islamist resistance in Iraq, and Osama bin Laden communicated with him. Al-Zarqawi and his followers eventually claimed credit for terrorist attacks under the banner of a hitherto unknown group called **Al Qaeda Organization for Holy War in Iraq**. This example is not unique, as evidenced by the cells who carried out the July 2005 attacks in Great Britain and Egypt.

In effect, an emerging pattern of terrorist behavior involves claims of responsibility for terrorist attacks around the world by Al Qaeda–inspired or loosely affiliated groups.

One plausible scenario is that the unanticipated resistance encountered during the occupation of Iraq produced a new generation of extremists with a similar internationalist mission as the original Afghan Arabs.[c] Assuming this has occurred, the global jihad has evolved beyond Al Qaeda into a loose web of similar organizations and networks.

Notes

a. Sebastian Rotella and Richard C. Paddock, "Experts See Major Shift in Al Qaeda's Strategy," *Los Angeles Times,* November 19, 2003.

b. Douglas Farah and Peter Finn, "Terrorism, Inc.," *Washington Post,* November 21, 2003.

c. Dana Priest, "Report Says Iraq is New Terrorist Training Ground," *Washington Post,* January 14, 2005.

Incident Case: The Madrid Train Bombings

Spain suffered its worst terrorist attack on March 11, 2004, when terrorists detonated ten bombs on several commuter trains in Madrid[10] after delivering them in backpacks dropped at key locations. The explosions were synchronized, occurring within a span of 20 minutes at the height of rush hour, when the trains were certain to be crowded with commuters. Casualties were high: 191 people were killed and more than 1,500 were injured. Casualties could have been more severe, but fortunately three additional bombs (also in backpacks) failed to explode and were later detonated safely by Spanish authorities.

The **Abu Hafs al-Masri Brigades** are an example of how Al Qaeda–affiliated cells operate as prepositioned sleepers in foreign countries. It was estimated that 12 to 30 terrorists may have participated in the delivery of the Madrid bombs.[11] Fifteen suspects, 11 of them Moroccans, were arrested within several weeks of the attack. It was believed that some of the suspects were members of the Moroccan Islamic Combatant Group, another Al Qaeda–affiliated group. When Spanish police prepared to storm an apartment in the town of Leganes near Madrid, three terrorists and a police officer were killed when the suspects detonated explosives and blew themselves up.

Incident Case: The London Transportation System Attacks

On July 7, 2005, four bombs exploded in London, three on London Underground trains, and one on a double-decker bus, killing more than 50 people and injuring more than 700. The attack was well synchronized by suicide bombers, so that the three Underground bombs exploded within 50 seconds of each other.[12] Several days later, on July 21, an identical attack was attempted but failed when the explosives misfired. Four bombs—like those in the first attack, three on Underground trains and one on a bus—failed to detonate because the explosives had degraded over time. Investigators quickly identified four suspects from video surveillance cameras. All were residents of London. A fifth bomb was found in a London park on July 23, abandoned by a fifth bomber. Authorities were acutely concerned because British-based cells—sympathizers of Al Qaeda—were responsible for both attacks.[13]

Wartime Case: Terrorist Violence in Iraq

The U.S.-led invasion of Iraq, designated Operation Iraqi Freedom, commenced on March 20, 2003 with an early-morning decapitation air strike intended to kill Iraqi leaders. Ground forces crossed into Iraq from Kuwait the same day. During the drive toward Baghdad from March 21 to April 5, coalition forces encountered stiff resistance from regular and irregular Iraqi forces, but this opposition was overcome or bypassed. American armored units swept through Baghdad on April 5, British troops overcame resistance in Basra on April 7, and Baghdad fell on April 9. On May 1, the coalition declared that major combat operations had ended.

Unfortunately for the coalition, the security environment and quality of life in Iraq became progressively poorer in the immediate aftermath of the fall of Saddam Hussein's government. Crime became widespread and basic services such as electricity and water became sporadic. Most ominous, an insurgency took root. It only increased in intensity.

It spread during the first year of the occupation. The insurgents engaged in classic guerrilla attacks, some of them professionally and intricately planned, against occupation troops and Iraqi security forces. Roadside bombs, ambushes, harassing mortar fire, and firefights occurred in many areas. Insurgents also used terrorism and assassinations against foreign contract workers and those they defined as collaborators—police officers, Iraqi soldiers, moderate leaders, election workers, and many others.

This early and politically motivated insurgency was the first phase of violence in postinvasion Iraq. A decidedly communal quality of violence also spread, primarily between Shi'a and Sunni Iraqis, but also within these communities. Bombs were indiscriminately detonated in Shi'a neighborhoods in Baghdad, mosques were attacked, Sunni and Shi'a leaders were assassinated, and hundreds of bodies were found in rivers, mass graves, and other locations. These incidents were the result of intracommunal power struggles and tit-for-tat revenge killings. Thousands of Iraqis died.

The insurgency in Iraq had international implications from the outset because the intensity of the noncommunal resistance was widely admired throughout the Arab and Muslim world, from which international volunteers came to participate in the insurgency. In addition, disaffected individuals in Muslim communities in Europe sympathized with the resistance in Iraq, and a few Western Muslims declared their solidarity. These individuals became active in the United Kingdom Spain, and elsewhere.

The insurgency also became a focal point for debate about the prosecution of the war against terrorism. For example, some leaders and supporters of the occupation reasoned that terrorists were being flushed out and that it would be better to fight them in Iraq than elsewhere. Others maintained that Iraq had never posed a direct threat of Al Qaeda–style terrorism, that resources were needlessly expended in an unnecessary war, and that a significant number of new extremists became inspired by the Iraqi insurgency.

Table 7.4 summarizes the activity profiles of secular, sectarian, and ethno-national stateless revolutions.

TABLE 7.4   THE STATELESS REVOLUTIONS

Some movements operate almost exclusively in the international arena. The reasons for this include ideologies of transnational revolution, global spiritual visions, or simple practicality brought on by their political environment.

Political Orientation	Activity Profile				
	Group	**Constituency**	**Adversary**	**Benefactor**	**Goal**
Secular	Japanese Red Army	Oppressed of the world	Capitalism, imperialism, Zionism	Unclear; probably PFLP, maybe Libya	International revolution
Sectarian	Al Qaeda	Muslims of the world	Foes of Islam	Self-supporting; maybe radical states	International Islamic revolution
Ethno-nationalist	Palestinian nationalists	Palestinians	Israel	Government sponsors	Palestinian nation

Photo 7.4   The remains of Pan Am Flight 103 after it was bombed in the skies over Lockerbie, Scotland. The attack led to economic sanctions against Libya and the trial of two members of the Libyan security service.

Chapter Summary

This chapter provided readers with a critical assessment of international terrorism, which was defined from the motives of the perpetrators and the symbolism of their selected target; terrorists who choose to operate on an international scale select targets that symbolize an international interest. Readers were reintroduced to the terrorist environments discussed in previous chapters. These environments were evaluated from the perspective of the international venue.

Within the framework—ideological, practical, tactical, and historical—of the reasons identified for international terrorism, readers were asked to apply adaptations of revolutionary theory to international operations. Recalling previous discussions about Mao's doctrine for waging guerrilla warfare as fish swimming in the sea of the people and Marighella's strategy for destabilizing established governments, it is no surprise that modern terrorists would carry variations of these themes onto the international stage.

International terrorist networks have been a facet of international terrorism since the 1960s. The models presented in this chapter are useful in understanding the networking features of terrorist environments. The cases of Al Qaeda and the international mujahideen illustrate the nature of modern terrorist networks.

DISCUSSION BOX

This chapter's Discussion Box is intended to stimulate critical debate about whether spillover incidents are a legitimate expression of grievances.

Understanding Terrorist Spillovers: Middle Eastern and North African Spillovers in Europe

Terrorist spillovers have occurred regularly since the 1960s. The most common sources have been from the Middle East and North Africa, and the most frequent venue has been Western Europe. For example, hundreds of Middle Eastern and North African terrorist incidents occurred during the 1980s in more than a dozen countries, causing hundreds of deaths and hundreds of people wounded.

There are several types of spillovers. Some arise from the foreign policies of governments; others are PLO-affiliated groups; and others are attacks by Islamist revolutionaries.

The motives behind these incidents have included

◆ Silencing or intimidating exiles
◆ Attempts to influence European policies
◆ Retaliation against a person or state
◆ Solidarity attacks by indigenous European groups

Europe has been an attractive target for many reasons, including

◆ The presence of large Middle Eastern and North African communities
◆ Many soft targets
◆ Immediate publicity
◆ Open borders and good transportation
◆ Proximity to the Middle East and North Africa

(Continued)

These attacks have been widespread geographically, and many have been indiscriminate.

Discussion Questions

1. Is the international arena a legitimate option for the expression of grievances?
2. Is Europe an appropriate venue for conflicts originating in the Middle East or North Africa?
3. Are some grievances more legitimate, or more acceptable, for spillovers?
4. What is the likely future of spillover attacks in Europe?
5. Because Europe has historically been perceived to be an easy battlefield with soft targets, should the European Union "harden" itself?

Key Terms and Concepts

The following topics were discussed in this chapter and can be found in the glossary:

Air France Flight 8969
Armenian Revolutionary
  Army
Armenian Secret Army
  for the Liberation of
  Armenia (ASALA)
asymmetrical warfare
cell-based terrorist
  environment

imperialism
international
  mujahideen
Islamic jihad
Justice Commandos of the
  Armenian Genocide
monolithic terrorist
  environment
multinational corporations

neocolonialism
Operation Death Trains
sleeper cells
spillover effect
strong multipolar
  terrorist environment
weak multipolar
  terrorist environment
Zionism

Terrorism on the Web

Log on to the Web-based student study site at **www.sagepub.com/martinessstudy** for additional Web sources and study resources.

Web Exercise

Using this chapter's recommended Web sites, conduct an online investigation of international terrorism.

1. From your review of these Web sites, how significant do you think the threat is from international terrorist groups?
2. Which groups do you think pose a serious threat of international terrorism? Which groups do you think present less serious threats?
3. Critique the Web sites of the international monitoring organizations. Are some more helpful than others?

For an online search of international terrorism, readers should activate the search engine on their Web browser and enter the following keywords:

"International Terrorism (or International Terrorists)"
"Global Terrorism"

Recommended Readings

The following publications provide discussions about the nature of international terrorism and cases in point about international terrorists.

Follain, John. *Jackal: The Complete Story of the Legendary Terrorist, Carlos the Jackal.* New York: Arcade, 1998.

Gunaratna, Rohan. *Inside Al Qaeda: Global Network of Terror.* New York: Columbia University Press, 2002.

Herman, Edward S. *The Real Terror Network: Terrorism in Fact and Propaganda.* Boston, MA: South End, 1982.

Kegley, Charles W., Jr., ed. *International Terrorism: Characteristics, Causes, Controls.* New York: St. Martin's, 1990.

Pillar, Paul R. *Terrorism and U.S. Foreign Policy.* Washington, DC: Brookings Institution, 2001.

Reeve, Simon. *The New Jackals: Ramzi Yousef, Osama bin Laden and the Future of Terrorism.* Boston, MA: Northeastern University Press, 1999.

8

Domestic Terrorism in the United States

Photo 8.1 Communal terrorism in America. The lynching of Tommy Shipp and Abe Smith in Marion, Indiana, on August 7, 1930. The crowd is in a festive mood, including the young couple holding hands in the foreground.

The quality of postwar ideological extremism in the United States reflects the characteristics of the classical ideological continuum. Readers may recall that the classical ideological continuum, discussed earlier, incorporates political tendencies that range from the fringe left to the fringe right, but that many examples of nationalist and religious terrorism do not fit squarely within the continuum categories. However, the United States is an idiosyncratic subject, and most terrorism in the postwar era did originate from the left- and right-wing spectrums of the continuum.

The discussion in this chapter will review the following:

♦ Extremism in America
♦ Leftist terrorism in the United States
♦ Right-wing terrorism in the United States

Figure 8.1 shows the number of terrorist incidents in the United States by group class, from 1980 to 2001. Table 8.1 summarizes and contrasts the basic characteristics of contemporary left-wing, right-wing, and international political violence in the United States.[1] This is not an exhaustive profile, but it is instructive for purposes of comparison.

Figure 8.1   Terrorism in the United States by Group Class, 1980–2001

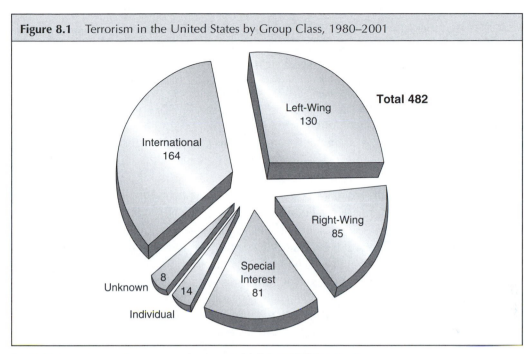

SOURCE: United States Department of Justice, Federal Bureau of Investigation.

TABLE 8.1   ATTRIBUTES OF TERRORISM IN THE UNITED STATES

In the United States, terrorism has typically been conducted by groups and individuals espousing leftist or rightist ideologies or those who engage in international spillover conflicts. These interests are motivated by diverse ideologies, operate from different milieus, possess distinctive organizational profiles, and target a variety of interests.

Environment	Activity Profile			
	Ideological Profile	Bases of Operation	Organizational Profile	Typical Targets
Leftist	Marxist; left-wing nationalist	Urban areas; suburbs	Clandestine groups; movement-based	Symbolic structures; avoidance of human targets
Rightist	Racial supremacist; antigovernment; religious	Rural areas; small towns	Self-isolated groups; cells; lone wolves	Symbolic structures; human targets
"Old" international terrorism	Ethno-nationalist	Urban areas	Clandestine groups	Symbols of enemy interest
"New" international terrorism	Religious	Urban areas	Cells	Symbolic structures; human targets

CHAPTER PERSPECTIVE 8.1

American Cults and Terrorist Violence

Cults are self-isolated mystical communities that claim to have definitive answers for societal and personal problems. They are usually tightly knit groups of people who follow a spiritual belief, often based on extreme interpretations of existing religions, that is a product of some higher mystical insight. Many cults are led by a single charismatic leader. The Tokyo subway terrorist attack by the Japanese cult Aum Shinrikyō was discussed in Chapter 6.

Most American cults have not directed violence against nonmembers. However, some—such as the apocalyptic **People's Temple of the Disciples of Christ**—have committed acts of extreme violence against their own members, as well as nonmembers who challenged their authority. Founded and led by **Jim Jones**, the People's Temple was originally based in California. In the late 1970s, hundreds of its members migrated to Guyana to establish a compound called Jonestown after Jones and others became obsessed with conspiracy theories about government oppression and the imminence of the Apocalypse (a time of global trial and warfare). In November 1978, during a visit to Jonestown by a delegation led by U.S. Congressman Leo Ryan, a number of cult members decided to leave the People's Temple. They accompanied Ryan's party to an airfield, where they were assaulted by heavily armed members of the People's Temple; Ryan and several others were killed. Jones and hundreds of his followers committed suicide—mostly by drinking flavored sugar water that had been poisoned, although some shot themselves. A number of members were apparently murdered by gunshots or injections of the poison. Jim Jones audiotaped the episode, in which 914 people died, including more than 200 children.

Another cult—the Black Hebrew Israelites—is a counterpart to the neo-Nazi Christian Identity faith, with the distinction that it promotes African racial supremacy. Branches of the cult have called themselves the Tribe of Judah, Nation of Yahweh, and Temple of Love. Black Hebrew Israelites believe that the ancient Hebrews were Africans, and that these chosen people migrated to West Africa during the Diaspora (scattering) of the ancient Jewish people. The descendents of these migrants were brought to the western hemisphere as slaves by Europeans during the 400-year African slave trade. Thus, Africans in general—and African Americans in particular—are the true Chosen People of God. This "truth" has been rediscovered by followers of the Black Hebrew Israelite movement. One branch was established in Miami in 1979 by **Hulon Mitchell, Jr.**, who adopted the name of Yahweh Ben Yahweh (God, Son of God). This branch became highly insular, and members followed Mitchell's authority without question. Disobedience was dealt with harshly, so that when some members tried to leave the group they were assaulted, and some were beheaded. Mitchell also taught that whites are descendants of the devil and worthy of death. Mitchell and some of his followers were eventually prosecuted and imprisoned after the group dispatched death squads known as the Death Angels to murder whites in the Miami area. The group was linked to at least 14 murders.

In the Pacific Northwest, followers of the **Bhagwan Shree Rajneesh** committed at least one act of biological terrorism. The group occupied a rural area in Oregon where their leader—the Bhagwan—was revered as the source of ultimate spiritual truth and enlightenment. Their mystical belief system was loosely based on Hinduism. Cult members, known as **Rajneeshees**, were expected to renounce their previous lifestyles and totally devote themselves and their livelihoods to their group and the Bhagwan. For example, they expressed great pride in the Bhagwan's fleet of Rolls Royce sedans. In September 1984, the group poisoned the salad bars of 10 Oregon restaurants with *Salmonella* bacteria. More than 700 people fell ill. The incident had been an experiment to test options for how the group could influence a local election in which the Rajneeshees hoped to elect members of the cult.

Extremism in America

Unlike many terrorist environments elsewhere in the world, where the designations of left and right are not always applicable, most political violence in the United States falls within them. Even nationalist and religious sources of such incidents have tended to reflect the attributes of leftist or rightist movements.

Left-Wing Extremism in the United States

The modern American left is characterized by several movements that grew out of the political fervor of the 1960s. They were fairly interconnected, so that understanding their origins provides instructive insight into the basic issues of the left. One should bear in mind that none was fundamentally violent, and that they were not terrorist movements. However, extremist trends within them led to factions that sometimes espoused violent confrontation, and a few engaged in terrorist violence.

The Rise of Black Power[2]

The modern civil rights movement initially centered on the struggle to win equality for African Americans in the South. During the early 1950s, the movement—at first led by the National Association for the Advancement of Colored People—forced an end to segregation on trains and interstate buses by successfully appealing several federal lawsuits to the U.S. Supreme Court. Despite these victories, southern state laws still allowed segregation on intrastate transportation.

In 1955 and 1956 in Birmingham, Alabama, Reverend Martin Luther King, Jr., led a bus boycott in Birmingham that lasted for 13 months. A Supreme Court decision forced the bus company to capitulate. This was the beginning of the application of civil disobedience using a strategy known as collective nonviolence. The theory was that massive resistance, coupled with moral suasion and peaceful behavior, would lead to fundamental change. Unfortunately, official and unofficial violence was directed against the movement. There were numerous anti–civil rights bombings, shootings, and beatings in the South during this period. For this reason, not every member of the civil rights movement accepted collective nonviolence as a fundamental principle, and the strategy was not particularly effective outside of the southern context.

As a direct result of the violence directed against the nonviolent movement, an emerging ideology of African American empowerment took root among many activists. It began in June 1966, when civil rights activist **James Meredith** planned to walk through Mississippi to demonstrate that African Americans could safely go to polling places to register to vote. When he was ambushed, shot, and wounded early in his walk, Martin Luther King and other national civil rights activists traveled to Mississippi to finish Meredith's symbolic march. One of the leaders was **Stokely Carmichael**, chairman of the Student Nonviolent Coordinating Committee (SNCC).

Carmichael renounced collective nonviolence. At a rally in Mississippi, he roused the crowd to repeatedly shout "Black power!" and adopted the clenched fist as a symbol of defiance. The slogan caught on, as did the clenched-fist symbol, and the **Black Power** movement began. It is important to note that the movement occurred at a time when the violence in the South was paralleled by urban activism, unrest, and rioting in the impoverished African American ghettos of the North, Midwest, and West. The ideology of Black Power advocated political independence, economic self-sufficiency, and a cultural reawakening. It was expressed in Afrocentric political agendas; experiments in economic development of African American communities; and cultural chauvinism that was expressed in music, art, and dress (the **Black Pride** movement).

Growth of the New Left

The so-called old left was characterized by orthodox Marxist ideologies and political parties dating from the Russian Revolution. However, new issues galvanized a new movement among educated young activists, primarily on the nation's university campuses. The New Left arose in the mid-1960s when a new generation of activists rallied around the antiwar movement, the civil rights movement, women's rights, and other political and social causes. New student organizations such as **Students for a Democratic Society (SDS)** advocated a philosophy of **direct action** to confront mainstream establishment values. In the fall of 1964, participants in the Free Speech Movement at the University of California in Berkeley seized an administration building on the campus. This was a wakeup call for adopting direct action as a central tactic of the fledgling New Left.

The New Left adapted its ideological motivations to the political and social context of the 1960s. It championed contemporary revolutionaries and movements, such as the Cuban, Palestinian, and Vietnamese revolutionaries. At its core, "the [American] New Left was a mass movement that led, and fed upon, growing public opposition to U.S. involvement in Vietnam."[3] Many young Americans experimented with alternative lifestyles, drugs, and avant-garde music. They also challenged the values of mainstream American society, questioning its fundamental ideological and cultural assumptions. This component of the New Left was commonly called the **counterculture**. The period was marked by numerous experiments in youth-centered culture.

Right-Wing Extremism in the United States

Extremist tendencies within the modern American right are characterized by self-defined value systems. This tendency is rooted in emerging trends such as antigovernment and evangelical religious activism, as well as in historical cultural trends such as racial supremacy. The following discussion surveys the modern (postwar) characteristics of these trends, which provide a useful background to contemporary terrorism on the right.

Religious Politics and the Christian Right

The movement commonly termed the **Christian Right** is a mostly Protestant fundamentalist movement that links strict Christian values to political agendas. Its modern origins lie in the conservative political environment of the 1980s. It is not an inherently violent movement, and some activists have practiced variations of **collective nonviolence** and direct action by blockading and protesting at the offices of abortion providers. The ultimate goal of the Christian Right is to make Christian religious values (primarily evangelical Christian values) an integral part of the nation's social and political framework.

Far- and fringe-right members of the Christian Right have adopted a highly aggressive and confrontational style of activism. One significant aspect of the more **reactionary** tendency within the movement is the promotion of a specifically evangelical Christian agenda, thus rejecting agendas that are secular, non-Christian, or nonfundamentalist Christian.

Rise of the Antigovernment Patriots

The Patriot movement came to prominence during the early 1990s. The movement considers that it represents the true heirs of the ideals of the framers of the U.S. Constitution. For many Patriots, government in general is not to be trusted, the federal government in particular is to be distrusted, and the United Nations is a dangerous and evil institution. The Patriot movement is not ideologically monolithic, and numerous tendencies have developed, such as the Common Law Courts and Constitutionalists. Conspiracy theories abound within the Patriot movement. Some have long and murky origins, having been developed over decades. Others appear and disappear during periods of political or social crisis. Patriots cite evidence that non-American interests are threatening to take over—or have already taken over—key governmental centers of authority. This is part of an international plot to create a one-world government called the **New World Order**.

Chapter Perspective 8.2 summarizes several conspiracy theories from the right.

CHAPTER PERSPECTIVE 8.2

Conspiracy Theories on the American Right

The modern far and fringe right have produced a number of conspiracy theories and rumors. Although they may seem fantastic to nonmembers of the Patriot (and other) movements, many adherents of these theories live their lives as if the theories were reality.

Communist Invaders During the Cold War

◆ Rumors "confirmed" that Soviet cavalry units were preparing to invade Alaska across the Bering Strait from Siberia.

◆ Thousands of Chinese soldiers (perhaps an entire division) had massed in tunnels across the southwestern border of the United States in Mexico.

◆ Thousands of Viet Cong and Mongolian troops had also massed in Mexico across the borders of Texas and California.

The New World Order Replaces the Communist Menace

◆ Hostile un-American interests (which may already be in power) include the United Nations, international Jewish bankers, the Illuminati, the Council on Foreign Relations, and the Trilateral Commission.

◆ Assuming it is Jewish interests who are in power, the U.S. government has secretly become the Zionist Occupation Government (ZOG).

(Continued)

- ◆ The government has constructed concentration camps that will be used to intern Patriots and other loyal Americans after their weapons have all been seized (possibly by African American street gangs).

- ◆ Invasion coordinates for the New World Order have been secretly stuck to the backs of road signs.
- ◆ Sinister symbolism and codes have been found in the Universal Product Code (the bar lines on consumer goods), cleaning products, cereal boxes, and dollar bills (such as the pyramid with the eyeball).
- ◆ Sinister technologies exist that will be used when ZOG or the New World Order make their move. These include devices that can alter the weather and scanners that read the plastic strips in American paper currency.

With conspiracy theories as an ideological foundation, many within the Patriot movement organized themselves into citizens' **militias**. Scores were organized during the 1990s, and at their peak an estimated 50,000 Americans were members of more than 800 militias, drawn from 5 to 6 million adherents of the Patriot movement.[4] Some joined to train as weekend soldiers, whereas others organized themselves as paramilitary survivalists.

The potential for political violence from some members of the armed, conspiracy-bound Patriot movement has been cited by experts and law enforcement officials as a genuine threat.

Racial Supremacy: An Old Problem With New Beginnings

The history of racial supremacy in the United States began during the period of African chattel slavery and continued with the policy to remove Native Americans from ancestral lands. The racial dimensions of these practices became norms of the early American nation. After the Civil War and prior to World War II, the United States became a highly segregated country. After World War II, though, the tide turned against overt and unquestioned racial supremacy. Supremacist beliefs, however, continued to win adherents.

Modern organized racial supremacist groups include the modern KKK, neo-Nazi movements, racist skinhead youth gangs, and some adherents of the Neo-Confederate movement. New non-Klan groups came into their own during the 1980s, when supremacist groups created their own mythologies and conspiracy theories. For example, many neo-Nazis considered the novel *The Turner Diaries*[5] a blueprint for the Aryan revolution in America. Also on the racist right, the **Fourteen Words** have become a rallying slogan. Originally coined by David Lane, a convicted member of the terrorist group the Order, the Fourteen Words are: "We must secure the existence of our people and a future for White children." The Fourteen Words have been incorporated into the Aryan Nations' "declaration of independence" for the white race, and the slogan is often represented by simply writing or tattooing.[14]

Racial Mysticism

Neofascist movements and political parties in Europe are decidedly secular. They reference religion and the organized Christian church only to support their political agendas; they do not adopt Christian or cult-like mystical doctrines to justify their legitimacy. In the United States, members of far- and fringe-right movements frequently justify their claims of racial supremacy and cultural purity by referencing underlying spiritual values—essentially claiming that they have a racial mandate from God. Racial supremacists in particular have developed mystical foundations for their belief systems, many of which are cult-like. Two of these cultish doctrines follow.

The Creativity Movement. **Creativity** is premised on a rejection of the white race's reliance on Christianity, which adherents believe Jews created in a conspiracy to enslave whites. According to movement adherents, the white race itself should be worshipped.

Ásatrú. A neopagan movement, **Ásatrú** venerates the pantheon of ancient Norse (Scandinavian) religions. In its most basic form, which is not racial in conviction, Ásatrú adherents worship the Norse pantheon of Odin, Thor, Freyr, Loki, and others. A minority of Ásatrú believers have adopted an activist and racist belief system, linking variants of Nazi ideology and racial supremacy to the Nordic pantheon.

Race and the Bible: The Christian Identity Creation Myth

Christian Identity is the Americanized strain of an 18th-century quasi-religious doctrine called **Anglo-Israelism** developed by Richard Brothers. Believers hold that whites are descended from Adam and are the true chosen people of God, that Jews are biologically descended from Satan, and that nonwhites are soulless beasts, also called the Mud People. Christian Identity adherents have developed two cultish creation stories loosely based on the Old Testament. The theories are called **One-Seedline Christian Identity** and **Two-Seedline Christian Identity**.

One-Seedline Identity accepts that all humans regardless of race are descended from Adam. However, only Aryans (defined as northern Europeans) are the true elect of God. Two-Seedline Identity rejects the notion that all humans are descended from Adam. Instead, their focus is on the progeny of Eve. Two-seedline adherents believe that Eve bore Abel as Adam's son but Cain as the son of the Serpent (that is, the devil). Outside of the Garden of Eden lived nonwhite, soulless beasts who are a separate species from humans. They are the modern nonwhite races of the world and are often referred to by Identity believers as Mud People. When Cain slew Abel, he was cast out of the Garden to live among the soulless beasts. Those who became the descendants of Cain are the modern Jews.

Leftist Terrorism in the United States _____

As New Left and Black Power movements and organizations became radicalized, many individuals and groups began to advocate active resistance against **the Establishment**—defined as mainstream American political and social institutions. This resistance included explicit calls for civil disobedience and confrontation with the authorities. Many within these movements referred to themselves as revolutionaries, and some advocated the overthrow of the military-industrial complex. Prototypical revolutionary organizations began to form in the late 1960s, and a few groups produced factions that became terrorist organizations. All of this occurred in a generalized environment of activism and direct action. A large number of politically motivated bombings, shootings, and assaults occurred during this period. The Senate Committee on Government Operations reported the following statistics:[6]

1969, 298 explosive and 243 incendiary bombings
January to July 1970, 301 explosive and 210 incendiary bombing incidents

January 1968 to June 1970, 216 ambushes against law enforcement personnel and headquarters

January 1968 to June 1970, 359 total assaults against police, causing 23 deaths and 326 injuries

Chapter Perspective 8.3 presents two examples of radicalized organizations—one from the New Left (Students for a Democratic Society) and the other from the Black Power movement (the Black Panthers). The story of both groups illustrates the evolutionary process of left-wing revolutionary cadres and factions that eventually advocated political violence.

CHAPTER PERSPECTIVE 8.3

Seeds of Terrorism: Radicals on the American Left

Two militant case studies are discussed here—the Black Panthers and the radicalized Students for a Democratic Society. Within each were groups or cadres who advocated violent revolution.

Case: The Black Panthers

The **Black Panther Party for Self-Defense** was organized in 1966 in Oakland, California. The name was selected from an African American organization founded in Alabama called the Lowndes County Freedom Organization. The Lowndes County group had used the symbol of a black panther on voter ballots, ostensibly so that illiterate voters would know who their candidates were.

The Oakland Black Panthers initially imitated a tactic used by the Los Angeles-based Community Alert Patrol, which had been formed after the Watts riot in August 1965.[a] The Community Alert Patrol would dispatch observers to scenes of suspected harassment by the Los Angeles Police Department and observe the police stop. In Oakland, the Black Panthers took this tactic one step further and arrived on the scene openly carrying law books and shotguns or rifles (legal at the time in California).[b] The symbolism of young African Americans projecting a paramilitary image in poor urban ghettos attracted members to the Black Panthers around the country. More that 40 chapters were formed, with a total of more than 2,000 members. By 1968, the group made worldwide headlines and came to symbolize the Black Power movement. Public demonstrations by the Black Panthers maximized the use of paramilitary symbolism, with members marching and chanting slogans in precision and wearing black berets and black leather jackets.

Ideologically, the Black Panthers were inspired by Malcolm X,[c] Frantz Fanon, and Mao Zedong. They were advocates of Black Nationalism and encouraged economic self-sufficiency and armed self-defense in the black community. Black Panther self-help initiatives included free breakfasts for poor schoolchildren in urban areas. The police, at that time all male and mostly white in most cities, were especially singled out and labeled as a kind of occupation force in African American communities.

The group's militancy attracted the attention of federal and local law enforcement agencies, who considered the organization to be a threat to national security. The revolutionary and antipolice rhetoric of Black Panther leaders and the militant articles in its newspaper *The Black Panther* increased their concern. FBI Director J. Edgar Hoover stated that the Black Panthers were the most significant threat to domestic security in the United States. A series of arrests and shootouts at Black Panther offices occurred. The leadership of the organization was decimated by arrests, police raids, and a successful disinformation campaign that sowed distrust among central figures. Internal feuds between leaders Huey Newton and Eldridge Cleaver also disrupted the group. Although the Black Panther movement continued to be active into the late 1970s—after significantly moderating its militancy by the mid-1970s—its heyday as a paramilitary symbol of Black Nationalism was during the late 1960s and early 1970s. As it declined under relentless internal and external pressures, some of its more radical members joined the revolutionary underground.

Case: Students for a Democratic Society

In June 1962, a group of liberal and mildly leftist students, many from the University of Michigan, met to draft a document that became known as the **Port Huron Statement**. In it, SDS harshly criticized the values of mainstream American society and called for a New Left movement. The statement was a critique and a call for action directed to middle-class students. At the time, SDS was liberal and leftist, but hardly revolutionary. SDS espoused direct action, which originally meant peaceful and nonviolent confrontation.

By 1965, SDS had moved to the radical left, and when the bombing of North Vietnam began, its national membership soared. By 1966, its focal point was the war in Vietnam and support for the Black Power movement (SDS's membership was mostly white students). In 1967, in a classic Marcuse-like interpretation, it cast activist American youth as a new working class oppressed by the military-industrial complex. By 1968, SDS leadership was revolutionary. In an SDS-led takeover of Columbia University during the 1968 spring term, students seized five buildings for 5 days. When the police were called in, a riot ensued, more than 700 people were arrested, and nearly 150 were injured. A student strike—again led by SDS—closed Columbia. The group led dozens of other campus disturbances in 1968.

In June 1968, ideological tensions within the group led SDS to fragment. Some members formed a prototypical Revolutionary Youth Movement, others aligned themselves with developing world revolutionary heroes, and still others (sometimes called Crazies) espoused violent revolution. At its next meeting, in June 1969 in Chicago, SDS split along doctrinal and tactical lines into the Maoist **Progressive Labor Party** (also known as Worker-Student Alliance), the **Revolutionary Youth Movement II**, and the violent revolutionary **Weatherman** group.

Notes

a. The toll for the Watts disturbance was high; 34 people were killed, more than 1,000 injured, and nearly 4,000 arrested. Approximately 200 businesses were destroyed and about 700 damaged. For a study of the Watts riot, see Robert Conot, *Rivers of Blood, Years of Darkness* (New York: Bantam Books, 1967).

b. The armed patrols ended when California passed a law prohibiting the open display of firearms.

c. For more information about Malcolm X, see Malcolm X, *The Autobiography of Malcolm X* (New York: Grove Press, 1964).

The following discussion evaluates four trends on the violent left:

◆ Generational rebellion: New Left terrorism   ◆ The revolution continues: Leftist hard cores
◆ Civil strife: Ethno-nationalist   ◆ Single-Issue violence on the Left
  terrorism on the Left

Generational Rebellion: New Left Terrorism

The New Left was deeply affected by the war in Vietnam, the civil rights movement, and the turmoil in inner-city African American communities. A number of terrorist groups and cells grew out of this environment. Although the most prominent example was the Weatherman group, other groups such as the Symbionese Liberation Army also engaged in terrorist violence. The United Freedom Front proved to be the most enduring of all terrorist groups of the era.

The Weatherman/Weather Underground Organization

The Weatherman group—known as the **Weathermen**—jelled at the June 1969 Students for a Democratic Society national convention in Chicago, when SDS splintered into several factions. The Weathermen were mostly young, white, educated members of the middle class. They represented—starkly—the dynamic ideological tendencies of the era, as well as the cultural separation from the older generation. Although they and others were sometimes referred to collectively as the Crazies, they operated within a supportive cultural and political environment.

From the beginning, the Weathermen were violent and confrontational. In October 1969, they distributed leaflets in Chicago announcing what became known as their **Days of Rage** action. The Days of Rage lasted for four days and consisted of acts of vandalism and running street fights with the Chicago police. In December 1969, the Weathermen held what they called a war council in Michigan. Its leadership, calling itself the **Weather Bureau**, advocated bombings, armed resistance, and assassinations. In March 1970, an explosion occurred in a Greenwich Village townhouse in New York City that was being used as a bomb factory. Three Weathermen were killed, several others escaped through the subway system, and hundreds of members went underground to wage war.

By the mid-1970s, the Weathermen—renamed the **Weather Underground Organization**—had committed at least 40 bombings. Aside from these actions, the Weather Underground also freed counterculture guru Timothy Leary from prison,[7] published a manifesto called *Prairie Fire,* and distributed an underground periodical called *Osawatomie.* By the mid-1970s, members of the Weather Underground began to give up their armed struggle and returned to aboveground activism—a process that they called inversion. Those who remained underground (mostly the East Coast wing) committed acts of political violence into the 1980s. Others joined other terrorist organizations.

The Symbionese Liberation Army

The Symbionese Liberation Army (SLA) was a violent terrorist cell that gained notoriety for several high-profile incidents in the mid-1970s. The core members were led by **Donald DeFreeze**, who took the nom de guerre Cinque (named for the leader of a 19th-century rebellion aboard the slave ship *Amistad*). Members trained in the Berkeley hills of California near San Francisco, rented safe houses, and obtained weapons. In November 1973, the Oakland school superintendent was assassinated when he was shot eight times; five of the bullets were cyanide-tipped. In a communiqué, the SLA took credit for the attack, using a rhetorical phrase that became its slogan: "Death to the fascist insect that preys upon the people!"

In February 1974, newspaper heiress Patricia Hearst was kidnapped by the cell. She was kept bound and blindfolded in a closet for more than 50 days under constant physical and psychological pressure, including physical abuse and intensive political indoctrination. She broke down under the pressure, and a tape recording was released in which she stated that she had joined the SLA. In April 1974, Hearst participated in a bank robbery in San Francisco. This was a classic case of Stockholm syndrome.

In May 1974, five of the SLA's core members, including DeFreeze, were killed in a shootout in a house in the Watts neighborhood of Los Angeles.

Patricia Hearst was a fugitive for approximately a year. She was hidden—probably by the Weather Underground—and traveled across the country with compatriots. By 1975, the SLA had a rebirth with new recruits and was responsible for several bank robberies and bombings in California. They referred to themselves as the **New World Liberation Front**. Hearst was captured in September 1975 in San Francisco, along with another underground fugitive.

Photo 8.2 The Symbionese Liberation Army in action. A bank camera captures Patricia Hearst exiting a bank after an SLA robbery. Hearst, who joined the group after being kidnapped by them, likely suffered from Stockholm syndrome.

Civil Strife: Ethno-Nationalist Terrorism on the Left

Ethno-national violence—which is distinguishable from racial supremacist violence—has been rare in the United States. This is primarily because activist environments have not historically supported nationalist terrorism. Exceptions grew out of the political environment of the 1960s, when nationalist political violence originated in African American and Puerto Rican activist movements.

The Black Liberation Movement

Racial tensions in the United States were extremely high during the 1960s. African Americans in the South directly confronted southern racism through collective nonviolence and the burgeoning Black Power ideology. In the urban areas of the North and West, cities became centers of confrontation between African Americans, the police, and state National Guards. When President Lyndon Johnson and the U.S. Senate organized inquiries into the causes of these disorders, their findings were disturbing. Table 8.2 describes the quality of these findings, which indicate the severity of tensions in urban areas during the mid-1960s.

Within this environment grew cadres of African American revolutionaries dedicated to using political violence to overthrow what they perceived to be a racist and oppressive system.

The Black Liberation Army

The **Black Liberation Army (BLA)** was an underground movement whose membership included Vietnam veterans and former members of the Black Panthers. BLA members were

TABLE 8.2   RACIAL CONFLICT IN AMERICA: THE "LONG HOT SUMMERS" OF THE 1960S

The urban disturbances in the United States during the 1960s caused an unprecedented period of communal discord. Incidents were widespread, violent, and a culmination of many factors. One of these was the deeply rooted racial polarization in American society. The presidentially appointed National Advisory Commission on Civil Disorders (known as the **Kerner Commission**) reported in 1968 that

> segregation and poverty have created in the racial ghetto a destructive environment totally unknown to most white Americans. What white Americans have never fully understood—but what the Negro can never forget—is that white society is deeply implicated in the ghetto. White institutions created it, white institutions maintain it, and white society condones it.[a]

These data are from a Senate Permanent Subcommittee on Investigations inquiry into urban rioting after the serious disturbances in the summer of 1967.[b]

	Activity Profile		
Incident Report	**1965**	**1966**	**1967**
Number of urban disturbances	5	21	75
Casualties			
Killed	36	11	83
Injured	1,206	520	1,897
Legal sanctions			
Arrests	10,245	2,298	16,389
Convictions	2,074	1,203	2,157
Costs of damage (in millions of dollars)	$40.1	$10.2	$664.5

a. *Report of the National Advisory Commission on Civil Disorders* (New York: Bantam Books, 1968), 2.

b. In 1967, the Senate passed a resolution ordering the Senate Permanent Subcommittee on Investigations to investigate what had caused the 1967 rioting and to recommend solutions. Senate Permanent Subcommittee on Investigations, data reported in *Ebony Pictorial History of Black America,* vol. 3 (Chicago: Johnson, 1971), 69.

nationalists who were inspired in part by the 1966 film *Battle of Algiers,* a semidocumentary of an urban terrorist uprising in the city of Algiers against the French during their colonial war in Algeria. In the film, Algerian rebels organized themselves into many autonomous cells to wage urban guerrilla warfare against the French. There were at least two cells (or groups of cells) of the BLA—the East Coast and West Coast groups. Although the BLA was likely active in late 1970 and early 1971, both cells became known later, and in similar fashion, to law enforcement agencies and the media.

The BLA is suspected to have committed a number of attacks in New York and California prior to and after these incidents. They are thought to have been responsible for numerous bombings, ambushes of police officers,[8] and bank robberies to "liberate" money to support their cause. Their areas of operation were California and New York City, though some members were apparently trained in the South.

Most members of the BLA were eventually captured or killed. Those who escaped the FBI net re-formed to join other radical organizations. Interestingly, the only known white member of the BLA, Marilyn Buck, was a former member of the radicalized SDS who had disappeared into the revolutionary underground.

Puerto Rican Independencistas

Puerto Rico is a commonwealth of the United States, meaning that it is self-governed by a legislature and an executive (a governor) and has a nonvoting delegate to Congress. Those who desire independence are nationalists called independencistas, most of whom use democratic institutions to promote the cause of independence; they are activists but not prone to violence. The Puerto Rico Independence Party, for example, is a fairly mainstreamed leftist political movement in Puerto Rico.

Some independencistas are revolutionaries, and a few have resorted to violence. Modern violent nationalists pattern themselves after Cuban nationalism and view the United States as an imperial and colonial power. Cuba has, in fact, provided support for violent independencista groups, especially during the 1980s.

The FALN

The FALN[9] was a particularly active terrorist organization that concentrated its activities on the U.S. mainland, primarily in Chicago and New York City. One important fact about the FALN stands out: It was the most prolific terrorist organization in U.S. history. The group became active in 1974, and from 1975 to 1983 approximately 130 bombings were linked to it. Most attacks were symbolic, directed against buildings, but some were deadly. For example, in January 1975 the FALN detonated a bomb at the trendy restaurant Fraunces Tavern in New York, killing four people and wounding more than 50. In another incident in 1983, three New York City police officers were maimed while trying to defuse explosives at the New York police headquarters. FALN was also responsible for armored car and bank robberies.

In 1980, more than a dozen FALN members were convicted of terrorist-related crimes. Sentences were imposed for seditious conspiracy, possession of unregistered firearms, interstate transportation of a stolen vehicle, interference with interstate commerce by violence, and interstate transportation of firearms with intent to commit a crime. None of the charges were linked to homicides. FALN members' sentences ranged from 15 to 90 years, and they considered themselves prisoners of war.

The Revolution Continues: Leftist Hard Cores[10]

The left-wing revolutionary underground re-formed after the decline of groups like the Weather Underground and the BLA. These new groups were made up of die-hard former members of the Weather Underground and BLA, as well as former activists from other organizations such as the radicalized SDS and the Black Panthers. Two cases illustrate the character of the reconstituted revolutionary left in the 1980s.

May 19 Communist Organization

The **May 19 Communist Organization (M19CO)** derives its name from the birthdays of Malcolm X and Vietnamese leader Ho Chi Minh. The symbolism of this designation is obvious—it combines domestic and international examples of resistance against self-defined U.S. racism and imperialism. The group was composed of remnants of the **New Africa Freedom Fighters**, the BLA, the Weather Underground, and the Black Panthers. These cadres included the founders of the **Republic of New Africa** and the most violent members of the Weather Underground. Many of its members were individuals who had disappeared into the revolutionary underground for years.

M19CO was fairly active, engaging in bank and armored car robberies, bombings, and other politically motivated actions. The group adopted several different names when claiming responsibility for its attacks. These aliases included Red Guerrilla Resistance, Revolutionary Fighting Group, and Armed Resistance Unit. M19CO remained active and engaged in several bombings, but the group was finally broken when its remaining members were arrested in May 1985.

The United Freedom Front[11]

Formed in 1975, the **United Freedom Front (UFF)** was underground and active for approximately 10 years. In 1975, it detonated a bomb at the Boston State House under the name of the **Sam Melville-Jonathan Jackson Unit**, named for two politicized inmates. The group was never large but was active, peaking in during the early 1980s. The UFF is suspected to have committed at least 25 bombings and robberies in New York and New England. The attacks were primarily intended to exhibit anticorporate or antimilitary symbolism. A group calling itself the Armed Resistance Unit detonated a bomb to protest the U.S. invasion of Grenada on the Senate side of the U.S. Capitol building on November 6, 1983. It is possible that the Armed Resistance Unit was in fact the UFF operating under another name. The UFF was broken when its members were arrested in late 1984 and early 1985. Few leftist groups had survived by remaining both underground and active for as long as the UFF.

Single-Issue Violence on the Left

The left has produced violent single-issue groups and individuals who focus on one issue to the exclusion of others. To them, their championed issue is the central point—arguably the political crux—for solving many of the world's problems. For example, Ted Kaczynski, also known as the Unabomber, protested the danger of technology by sending and placing bombs that killed 3 people and injured 22 others during a 17-year campaign.

Typical of leftist single-issue extremism is the fringe environmental movement. Groups such as the **Animal Liberation Front (ALF)** and the **Earth Liberation Front (ELF)** have committed numerous acts of violence, such as arson and vandalism.

The ALF and ELF have coordinated their activities. Several joint claims have been made about property damage and other acts of vandalism, and it is likely that the two groups share members. For the most part, both have been nonviolent toward humans, but they are responsible for many incidents of property destruction. ALF and ELF targets include laboratories, facilities where animals are kept, and sport utility vehicles (SUVs). Some of these incidents are vandalism sprees.

Right-Wing Terrorism in the United States _____

Right-wing terrorism in the United States is usually motivated by racial supremacist and antigovernment sentiment. Unlike the violent left, terrorist campaigns by underground rightist organizations and networks have been rare, as have massive bombings such as the Oklahoma City attack. It is more common for the right to be characterized by small-scale, cell-based conspiracies within the Patriot and neo-Nazi movements. In comparison with the left, the violent right has been less organized and less consistent.

Chapter Perspective 8.4 summarizes several examples of racial supremacist activity on the right in the modern era. These examples illustrate how potentially violent members of the right wing can find organizations to provide direction and structure for their underlying animosity toward target groups.

CHAPTER PERSPECTIVE 8.4

Seeds of Terrorism: Reactionaries on the American Right

Three reactionary case studies are discussed here—**White Aryan Resistance** (WAR), Aryan Nations, and the National Alliance. Each group has directly or indirectly influenced activists on the racial supremacist right.

White Aryan Resistance

White Aryan Resistance is an overtly racist organization founded and led by Tom Metzger. Based in California, WAR publishes neo-Nazi propaganda, manages an active Web site, and has tried to recruit and organize racist skinheads. Implicit in its message is the notion that skinheads should be mobilized as Aryan shock troops in the coming Racial Holy War. WAR has used popular culture and music to appeal to potential skinhead recruits, and its Web site is largely marketed to racist youth. In October 1990, WAR lost a $12.8 million verdict after the Southern Poverty Law Center litigated a case on behalf of the family of an Ethiopian immigrant who was beaten to death by WAR-inspired racist skinheads.

Aryan Nations

The "Reverend" **Richard Butler** established the Aryan Nations organization as a political counterpart to his Christian Identity sect, called the Church of Jesus Christ Christian. Aryan Nations established its spiritual and political headquarters in a compound at Hayden Lakes, Idaho. Residents of the compound were overtly neo-Nazi. They adopted a rank hierarchy, established an armed security force, trained as survivalists, worshipped as Identity believers, and took to wearing uniforms. A number of people who passed through the Aryan Nations group eventually engaged in political and racial violence, a pattern that included violence by the Order and **Buford Oneal Furrow**. This pattern led to its financial ruin. In a celebrated lawsuit brought by the Southern Poverty Law Center, Aryan Nations lost its title to the Hayden Lakes property in September 2000 when a $6.3 million verdict was decided. During the trial, the Southern Poverty Law Center successfully linked Aryan Nations security guards to the terrorizing of a family who had driven to the compound's entrance.

National Alliance

The **National Alliance** is historically linked to the now-defunct American Nazi Party, which had been founded and led by George Lincoln Rockwell. **William Pierce**, the founder and leader of the National Alliance, was long considered by experts and members of the neo-Nazi movement to be the most prominent propagandist of the

(Continued)

movement. Before his death in July 2002, Pierce wrote *The Turner Diaries* (under the nom de plume Andrew MacDonald), published a magazine called the *National Vanguard*, made regular radio broadcasts, and managed an active Web site. The National Alliance's original headquarters is a compound in rural Hillsboro, West Virginia, where Pierce's followers try to carry on his tradition. Although some violent neo-Nazis or other reactionaries may have been inspired by the National Alliance's message (recall that *The Turner Diaries* was found in the possession of Timothy McVeigh), no acts of terrorism or **hate crimes** were directly linked to the original group.

Postscript: Aryan Nations and National Alliance in Disarray

Two of the most active and influential neo-Nazi organizations were thrown into disarray when their founders and longtime leaders died in the early years of the new millennium. National Alliance's William Pierce died in July 2002, and Aryan Nations' Richard Butler died in September 2004. With the deaths of these leaders, both organizations engaged in bitter infighting over who would assume leadership and whose ideology most reflected the ideologies of the founding leaders. The infighting led to splits within the organizations, factions claiming to be the heirs of the original groups formed, and membership declined.

The following discussion explores the terrorist right by investigating the following subjects:

◆ Homegrown racism: the legacy of the Ku Klux Klan
◆ Racial mysticism: neo-Nazi terrorism
◆ Patriot threats
◆ Case: moralist terrorism

Homegrown Racism: The Legacy of the Ku Klux Klan

The Ku Klux Klan is a racist movement with no counterpart among international right-wing movements—it is a purely American phenomenon. Its name comes from the Greek word *kuclos*, or circle. The KKK is best described as an enduring movement that developed the following ideology:

◆ Racial supremacy
◆ Protestant Christian supremacy
◆ American cultural nationalism (also known as **nativism**)
◆ Violent assertion of Klan racial doctrine
◆ Ritualistic symbolism, greetings, and fraternal behavior

Klan terminology in many ways is an exercise in racist secret fraternal bonding. Table 8.3 samples the exotic language of the KKK.

There have been several manifestations of the KKK, which most experts divide into five eras.

First-Era Klan

The KKK was founded in 1866 in the immediate aftermath of the Civil War. Some sources date its origin to Christmas Eve 1865, whereas others cite 1866. According to most sources, the KKK was first convened in Pulaski, Tennessee, by a group of Southerners who

---

TABLE 8.3 THE FRATERNAL KLAN

From its inception in 1866, the Ku Klux Klan has used fraternity-like greetings, symbolism, and rituals. These behaviors promote secrecy and racial bonding within the organization. Examples of Klan language include the following greeting: *Ayak*? (Are you a Klansman?), and *Akia*! (A Klansman I am!). The language used for regional offices is also unique, as indicated in the following examples:

♦ Invisible Empire—national
♦ Realm—state
♦ Klavern—local

Klan Official	Duties	Scope of Authority	Symbolic Identification
Imperial Wizard	National leader	Invisible Empire	Blue stripes or robe
Grand Dragon	State leader	Realm	Green stripes or robe
Exalted Cyclops	County leader	Klaverns within county	Orange stripes or robe
Nighthawk	Local security and administration	Klavern	Black robe
Klonsel	General counsel	Invisible Empire	White robe
Citizen	Member	Klan faction	White robe

---

initially formed the group as a fraternal association. Their first Imperial Wizard, or national leader, was former Confederate general and slave trader Nathan Bedford Forrest. Military-style rankings were established, and by 1868 the KKK was a secretive and politically violent underground. Its targets included African Americans, Northerners, and southern collaborators. The KKK was suppressed by the Union Army and the anti-Klan Ku Klux laws passed by Congress. Forrest ordered the KKK to be officially disbanded, and their robes and regalia were ceremoniously burned. It has been estimated that the Klan had about 400,000 members during its first incarnation.

Second-Era Klan

After the Reconstruction era, that is, after the departure of the Union Army from the South and the end of martial law, the KKK re-formed into new secret societies and fraternal groups. It wielded a great deal of political influence and successfully helped restore racial supremacy and segregation in the South. African Americans lost most political and social rights during this period, beginning a condition of racial subjugation that did not end until the civil rights movement in the mid-20th century. The targets of Klan violence during this period were African Americans, immigrants, Catholics, and Jews.

Third-Era Klan

During the early part of the 20th century, and continuing into the 1920s, the KKK became a broad-based national movement. In 1915, members gathered at Stone Mountain, Georgia, and formed a movement known as the Invisible Empire. The Klan was glorified in the novel *The Clansman* and in the 1915 film *Birth of a Nation,* which was shown in the White House during the administration of President Woodrow Wilson. During this period, the Invisible Empire had between 3 and 4 million members. In 1925 in Washington, DC, 45,000 Klansmen and Klanswomen paraded down Pennsylvania Avenue. Also during this period, Klan and Klan-inspired violence was widespread. Thousands of people—mostly African Americans—were victimized by the KKK. Many acts of terrorism were ritualistic communal lynchings.

Fourth-Era Klan

After a decline because of revelations about Third-Era violence and corruption, the Klan was reinvigorated in 1946—once again at Stone Mountain, Georgia. At this gathering, the Invisible Empire disbanded, and new independent Klans were organized at local and regional levels. There was no longer a single national Klan, but rather autonomous Klan factions. During the civil rights movement, some Klan factions became extremely violent. The White Knights of Mississippi and the United Klans of America (mostly in Alabama) committed numerous acts of terrorism to try to halt progress toward racial equality in the American South. This era ended after several successful federal prosecutions on criminal civil rights charges, though the Klan itself endured.

Fifth-Era Klan

Violence during the Fifth Era has been committed by lone wolves rather than as organized Klan action. The modern era of the Ku Klux Klan is characterized by two trends:

The Moderate Klan. Some Klansmen and -women have tried to moderate their image by adopting more mainstream symbolism and rhetoric. Rather than advocating violence or paramilitary activity, they have projected an image of law-abiding activists working on behalf of white civil rights and good moral values. Those who promote this trend have eschewed the prominent display of Klan regalia and symbols.

The Purist Klan. A traditional and "pure" Klan has emerged that hearkens back to the original traditions and ideology of the KKK. This group has held a number of aggressive and vitriolic rallies—many in public at county government buildings. Its rhetoric is unapologetically racist and confrontational. Some factions of the purist trend prohibit the display of Nazi swastikas or other non-Klan racist symbols at KKK gatherings.

KKK membership has ebbed and flowed in the Fifth Era, in part because of competition from other racial supremacist movements, such as the racist skinhead and neo-Nazi groups. There was also fresh competition beginning in the late 1990s from the neo-Confederate movement.

Racial Mysticism: Neo-Nazi Terrorism

In the modern era, most non-Klan terrorism on the right wing has come from the neo-Nazi movement. Neo-Nazi terrorism is predicated on varying mixes of religious fanaticism, political violence, and racial supremacy. Their worldview is predicated on the superiority of the Aryan race, the inferiority of non-Aryans, and the need to confront an evil global Jewish conspiracy. Another common theme is the belief that a **racial holy war**—called **Rahowa**—is inevitable.

Although most violence emanating from these beliefs has been expressed as lone wolf terrorism and hate crimes, several groups have embarked on violent sprees. For example, a group calling itself the **Aryan Republican Army (ARA)** operated in the Midwest from 1994 to 1996.[12]

The Order

The Order was a covert, underground, and violent group that was inspired by a fictional secret organization depicted in the novel *The Turner Diaries*. In the book, the Order was a heroic inner circle and vanguard for the Aryan revolution in the United States. **Robert Jay Mathews**, a racial supremacist activist, was the founder of the actual Order in 1983.

The Order's methods for fighting the war against the Zionist Occupation Government were counterfeiting, bank robberies, armored car robberies, and occasional murders.[13] Its area of operation was primarily in the Pacific Northwest. In April 1984, the group bombed a synagogue in

Photo 8.3 Oklahoma City. The rubble of the Alfred P. Murrah Federal Building in the aftermath of a bomb attack by right-wing members of the Patriot movement. This was the worst terrorist incident on American soil before the September 11 attacks. It remains the worst incident of domestic terrorism carried out by Americans.

Boise, Idaho. In March 1984, members of the Order seized $500,000 from a parked armored car in Seattle; the group detonated a bomb at a theater as a diversion. In June 1984, Alan Berg, a Jewish talk-radio host, was murdered in Denver; he had regularly lambasted the neo-Nazi movement. Also in June, a Brinks armored car was robbed near Ukiah, California, with disciplined precision, and the Order made off with $3.6 million. The end of the Order came when more than 20 of its members were prosecuted and imprisoned in December 1985.

Patriot Threats

Although the Patriot movement attracted a significant number of adherents during the 1990s, and militias at one point recruited tens of thousands of members, no underground similar to that of the radical left was formed. Few terrorist movements or groups splintered from the Patriot movement. Thus, despite many implicit and explicit threats of armed violence from Patriots, terrorist conspiracies were rarely carried to completion.

In 1992, former KKK member Louis Beam began to publicly advocate **leaderless resistance** against the U.S. government. Leaderless resistance is a cell-based strategy based on forming **phantom cells** to wage war against the government and enemy interests. Dedicated Patriots and neo-Nazis believe that leaderless resistance and the creation of phantom cells will prevent infiltration from federal agencies. The chief threat of violence came from the armed militias, which peaked in membership immediately before and after the Oklahoma City bombing. In the wake of that incident, federal authorities broke up at least 25 Patriot conspiracies.

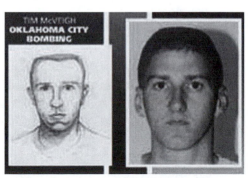

Photo 8.4 A comparison of the sketched witness description of Timothy McVeigh to his mug shot. An excellent example of the value of cooperation between witnesses and law enforcement officials.

The number of militias declined between the April 1995 Oklahoma City bombing and the attacks of September 11, 2001.[14] By 2000, the number of Patriot organizations was only one fourth of the 1996 peak,[15] and this decline continued after September 11, 2001.[16] Experts have noted, however, that the most militant and committed Patriot adherents remain within the movement and that these dedicated members constitute a core of potentially violent true believers.

Case in Point: Moralist Terrorism

Moralist terrorism refers to acts of political violence motivated by a moralistic worldview. Most moralist terrorism in the United States is motivated by an underlying religious doctrine, which is usually a fringe interpretation of Christianity. Abortion clinics and gay bars have been the most frequent targets of moralist violence.

Examples of violent moralist movements include the **Army of God** and the **Phineas Priesthood**. They are both shadowy movements that apparently have little or no organizational structure, operate as lone wolves or cells, and answer to the so-called higher power of their interpretations of God's will. They seem to be belief systems in which like-minded activists engage in similar behavior. The Phineas Priesthood is apparently a calling (divine revelation) for Christian Identity fundamentalists, and the Army of God membership is perhaps derived from fringe evangelical Christian fundamentalists. These profiles are speculative. Both groups may simply be manifestations of terrorist contagion (copycatting). There has also been speculation that both movements are linked.

Chapter Summary

This chapter investigated domestic political violence in the United States, the sources of which were identified as extremist tendencies that grew out of movements and cultural histories.

On the left, modern terrorism originated in the social and political fervor of the 1960s and 1970s. Some members of activist movements became radicalized by their experiences within this environment. A few became dedicated revolutionaries and chose to engage in terrorist violence. Members of New Left and nationalist terrorist groups waged terrorist campaigns until the mid-1980s. Single-issue and nascent anarchist tendencies have replaced the now-defunct Marxist and nationalist movements on the left.

On the right, the long history of racial violence continued into the 21st century. The Ku Klux Klan is a uniquely American racist movement that has progressed through five eras, with terrorist violence occurring in each. Modern Klansmen and -women, neo-Nazis, and moralists have also engaged in terrorist violence. Threats of potential political violence come from antigovernment movements and emerging "heritage" movements. The activity profile of the modern era is primarily either lone wolf or cell-based. It has become rare for racial supremacist and moralist terrorists to act as members of established organizations.

DISCUSSION BOX

This chapter's Discussion Box is intended to stimulate critical debate about the idiosyncratic nature of domestic terrorism in the United States.

Domestic Terrorism in the American Context

The subject of domestic terrorism in the United States is arguably a study in idiosyncratic political violence. Indigenous terrorist groups reflected the American political and social environments during historical periods when extremists chose to engage in political violence.

In the modern era, left-wing and right-wing political violence grew from very different circumstances. Leftist violence evolved from a uniquely American social environment that produced the civil rights, Black Power, and New Left movements. Rightist violence grew out of a combination of historical racial and nativist animosity, combined with modern applications of religious and antigovernment ideologies.

In the early years of the new millennium, threats from right-wing antigovernment and racial supremacist extremists continued. Potential violence from leftist extremists remained low in comparison. When the September 11 attacks created a new security environment, the question of terrorism originating from domestic sources remained uncertain; this was especially true after the anthrax attacks on the East Coast.

Discussion Questions

1. Assume that a nascent anarchist movement continues in its opposition to globalism. How should the modern leftist movement be described? What is the potential for violence originating from modern extremists on the left?
2. Keeping in mind the many conspiracy and mystical beliefs of the American right, what is the potential for violence from adherents of these theories to the modern American environment?
3. As a matter of policy, how closely should hate and antigovernment groups be monitored? What restrictions should be imposed on their activities? Why?
4. Is the American activity profile truly an idiosyncratic profile, or can it be compared with other nations' environments? If so, how? If not, why not?
5. What is the likelihood that the new millennium will witness a resurgence of a rightist movement on the scale of the 1990s Patriot movement? What trends indicate that it will occur? What trends indicate that it will not occur?

Key Terms and Concepts

The following topics were discussed in this chapter and can be found in the glossary:

Animal Liberation Front (ALF)	Armadas de Liberación Nacional, or FALN)	Aryan Republican Army (ARA)
Armed Forces for National Liberation (Fuerzas	Army of God	Ásatrú
	Aryan Nations	Bhagwan Shree Rajneesh

Black Liberation
    Army (BLA)
Black Panther Party
    for Self-Defense
Black Power
Butler, Richard
Christian Identity
Christian Right
counterculture
Creativity
Days of Rage
DeFreeze, Donald
direct action
Earth Liberation
    Front (ELF)
Establishment, the
Fourteen Words
Jewish Defense
    League (JDL)
Ku Klux Klan (KKK)
kuclos
leaderless resistance

long hot summer
lynch mob
Mathews, Robert Jay
May 19 Communist
    Organization (M19CO)
    (Revolutionary Fighting
    Group, Armed Resistance
    Unit)
militia
Mitchell, Hulon, Jr.
Mud People
National Alliance
Nativism
New Africa Freedom
    Fighters
New Left
New World Liberation Front
New World Order
One-Seedline Christian
    Identity
Order, The
Osawatomie

People's Temple of the
    Disciples of Christ
phantom cell
Phineas Priesthood
Pierce, William
Prairie Fire
racial holy war (Rahowa)
Rajneeshees
Sam Melville-Jonathan
    Jackson Unit
Students for a Democratic
    Society (SDS)
Turner Diaries, The
Two-Seedline Christian
    Identity
United Freedom
    Front (UFF)
Weather Bureau
Weather Underground
    Organization
Weathermen
White Aryan Resistance

Terrorism on the Web

Log on to the Web-based student study site at **www.sagepub.com/martinessstudy** for additional Web sources and study resources.

Web Exercise

Using this chapter's recommended Web sites, conduct an online investigation of terrorism in the United States.

1. How would you describe the typologies of groups that predominate in the United States?
2. Conduct a Web search of American monitoring organizations, read their mission statements, and assess their services. Which organizations do you think provide the most useful data? Why?
3. If you were an American dissident extremist (leftist or rightist), how would you design your own Web site?

For an online search of terrorism in the United States, readers should activate the search engine on their Web browser and enter the following keywords:

"Homeland Security"
"Domestic Terrorism"

Recommended Readings

The following publications discuss the nature of terrorism in the United States and the root causes of political violence in American society.

Emerson, Steven. *American Jihad: The Terrorists Living Among Us*. New York: Free Press, 2002.

George, John, and Laird Wilcox. *American Extremists: Militias, Supremacists, Klansmen, Communists, and Others*. Amherst, NY: Prometheus, 1996.

MacDonald, Andrew. *The Turner Diaries*. New York: Barricade, 1978.

McCarthy, Timothy Patrick, and John McMillian. *The Radical Reader: A Documentary History of the American Radical Tradition*. New York: New Press, 2003.

Michel, Lou and Dan Herbeck. *American Terrorist: Timothy McVeigh & the Oklahoma City Bombing*. New York: ReganBooks, 2001

Ridgeway, James. *Blood in the Face: The Ku Klux Klan, Aryan Nations, Nazi Skinheads, and the Rise of a New White Culture*. New York: Thunder's Mouth, 1990.

Sargent, Lyman Tower, ed. *Extremism in America: A Reader*. New York: New York University Press, 1995.

Smith, Brent L. *Terrorism in America: Pipe Bombs and Pipe Dreams*. Albany: State University of New York Press, 1994.

Stern, Kenneth S. *A Force Upon the Plain: The American Militia Movement and the Politics of Hate*. New York: Simon & Schuster, 1996.

Zakin, Susan. *Coyotes and Town Dogs: Earth First! and the Environmental Movement*. Tucson: University of Arizona Press, 2002.

PART III

The Terrorist Battleground

Part III Photo   The war on terrorism. U.S. Army soldiers patrol in Mosul, Iraq, during the U.S.-led occupation.

9

Terrorist Violence and the Role of the Media

This chapter explores the role of the media in a terrorist environment—the frequent interplay between media reporting and the use of violence by extremist movements. If terrorism is a strategy characterized by symbolic attacks on symbolic targets, it is also a strategy characterized by intentional manipulation of the news media: "Terrorist attacks are often carefully choreographed to attract the attention of the electronic media and the international press."[1] The truism that **information is power** is clearly understood by all parties, the media and governments as well as terrorists, their audiences, and their adversaries.

Terrorists understand that instantaneous media exposure for their grievances requires simply a dramatic incident to attract the world's press. Terrorists seeking publicity are likely to garner a large audience if they dramatically carry out targeted hijackings, bombings, hostage takings, assassinations, or other acts of violence. The press also has its own incentives to report major terrorist incidents. From the media's point of view—and aside from its fundamental responsibility to objectively report the news—drama guarantees increased attention from viewers.

The discussion in this chapter will review the following:

- Understanding the role of the media
- Mass communications and the war for information

Understanding the Role of the Media

In societies that champion freedom of the press, tension exists between the media's professional duty to objectively report the news and the terrorists' desire to promote their cause, between the necessity to keep the public informed and deliberate attempts to disseminate **propaganda** through the media. Organizations, movements, and governments use propaganda to spread their interpretations of the truth or to invent a new truth. Propaganda can incorporate truth, half-truths, and lies. Underlying all extremist propaganda is a particular political agenda.

The media sometimes tread a fine line between providing news and disseminating the terrorists' message when they report the details of terrorist incidents, broadcast interviews with terrorists and their extremist supporters, or investigate the merits of the terrorists' grievances. In theory, the media will be mindful of the fine line and carefully weigh what news to report and how to report it. In practice, some media outlets are blatantly sympathetic to one side of a conflict and completely unsympathetic to the other. In authoritarian states, this occurs as a matter of routine, because the government heavily regulates the media. In democracies, the **free press** enjoy the liberty to apply whatever **media spin** is deemed desirable in their reporting practices. Some media purposely use provocative language and photographs to attract an audience.

The following discussion will review several factors one should consider in understanding the role of the media:

◆ Mass communications and the terrorists' message
◆ Mass communications and the new media
◆ The reporting of terrorism

Mass Communications and the Terrorists' Message

The technological capability to convey considerable amounts of information to a large number of people, **mass communications** include printed material, audio broadcasts, video broadcasts, and expanding technologies such as the Internet. Today's revolutionaries consider mass communications an invaluable tool for achieving the goals of their cause. In fact, the theories of armed propaganda are partly technology driven.

For terrorists, efficiency and timeliness are critical components to mass communication. Efficiency is necessary so that delivery of the message will be orderly (as opposed to garbled or chaotic) and the message itself will be intelligible and easily understood. Timeliness is also necessary, because the message must be received while it is still fresh and relevant. It makes little sense to send a message before an issue has had an opportunity to ripen; it likewise makes little sense to send a message after an issue has become moot. Thus, if one's message is efficiently delivered and timely, it will have a stronger impact on the target audience.

Since the advent of printing presses using industrial-age technologies in the 19th century, terrorists and extremist movements have used virtually every available mass communications technology. The following technologies are commonly used by political extremists today:

◆ Print media   ◆ Television
◆ Radio   ◆ The Internet

Print Media

Dissident movements relied on the printed word throughout the 20th century. Sympathetic publishers and clandestine printing presses were instrumental in promulgating propaganda on behalf of dissident causes. Governments readily understood the power of the press to sway public sentiment, and of crackdowns on aboveground newspapers. There are also many examples of the deployment of security forces to locate and shut down clandestine presses. In an interesting example of how political blackmail can be used to promulgate an extremist message, the *New York Times* and *Washington Post* published the political manifesto of Ted Kaczynski, the so-called Unabomber, on September 19, 1995. On the recommendation of the U.S. Department of Justice, who hoped Kaczynski's writing style would be recognized, the *Times* and *Post* published his document, entitled *Industrial Society and Its Future.*

Radio

Radio broadcasts were used by many dissident movements before the advent of television. Many 20th-century movements continued to broadcast over the radio in societies where large numbers of people were unable to receive uncensored television broadcasts and where shortwave radio was widely used. In 1969, for example, the Brazilian groups National Liberation Action and MR-8 kidnapped the American ambassador to Brazil, successfully demanding—in the terms for his release—that their manifesto be broadcast over the radio. Historically, clandestine radio broadcasts have been instrumental in publicizing the cause to selected audiences, including potential supporters; shortwave radio was particularly effective in reaching a broad audience. As with dissident printing presses, governments have used security forces to root out clandestine radio stations.

Photo 9.1   The nature of the job. A British war correspondent travels with southern Sudanese rebels. The rebels, who practice Christianity or traditional religions, fought a long and brutal war against the northern Islamic government.

Television

Today, the medium of choice for terrorists is television, especially in the era of cable and digital feeds. It provides immediate visibility and increases the size of the audience. It also allows for dramatic images, many of which are relatively uncensored in sympathetic markets. Especially significant, of course, is that televised news breaks very quickly—often within minutes of an incident—and is broadcast worldwide. Satellite feeds can be linked from almost anywhere in the world. Television has thus become quite useful for promulgating the terrorists' message both visually and with dramatic, on-the-scene audio. All they need to do is manipulate the media into broadcasting a sympathetic spin for their grievances.

If successful, terrorists can bring images of their war into the homes of hundreds of millions of people worldwide nearly instantaneously—possibly with content that might sway large audiences to their cause.

The Internet

Computer technology is now an invaluable tool, used extensively by many terrorist groups and extremist movements. It is not uncommon for terrorist Web sites to be visually attractive, user-friendly, and interactive; to include music, photographs, videos, and extensive written materials; to portray a sense of the peaceful and rich culture of the downtrodden group. One apt example of the anonymity and scope of the Internet is the activities of a purported member of the Iraqi resistance who called himself Abu Maysara al-Iraqi. Al-Iraqi regularly posted alleged updates and communiqués about the Iraqi resistance on sympathetic Islamic Web sites. It proved very difficult to verify his authenticity, or even whether he (or they) was based in Iraq, given how accomplished he was in one online account for another.[2] As a counterpoint to such online materials, organizations independently monitor extremist Web sites for their origin and content.

Mass Communications and the New Media

The traditional and new resources just discussed are not the only media outlets. In the United States, there is a growing market in, and consumer demand for, the so-called **New Media**. New Media use existing technologies and alternative broadcasting formats to analyze and disseminate information. These alternative formats include talk-show models, tabloid (sensational) styles, celebrity status for hosts, blatant entertainment spins, and strong and opinionated political or social commentary.[3] Some extremist groups have appeared in the New Media, but terrorists have not been particularly active in attempting to manipulate it as a resource.

Reporting Terrorism

The propensity of those in print and broadcast media has been to give priority to terrorist incidents in their news reports. This is understandable, given the influence terrorism can have on policy making and either domestic or international political environments. However, the media have not been consistent about which incidents they report or how they report them. They frequently cover some acts of political violence extensively but provide little if any information about others.[5] For example, during invasions of Afghanistan and Iraq after September 11, 2001, American cable news outlets focused on the military campaign, lacing their broadcasts with on-the-scene reporting from embedded journalists advancing with the troops. In contrast, Qatar's **Al Jazeera** cable news outlet regularly broadcasted images of injured civilians or destruction from the fighting, lacing its broadcasts with on-the-scene reporting from journalists on the street and inside hospitals. Similarly, media reports are inconsistent in labeling perpetrators of political violence, and media interest in the sheer violent nature of terrorism can be disproportionate to exploration of the underlying causes of this violence.

Because of such disparities, organizations such as the Middle East Media Research Institute (MEMRI) were established to bridge the gap between Western and Middle Eastern media outlets.

Market Competition

The news media are owned and controlled by large corporations largely motivated by market share and profit. This affects their style, content, and reporting practices. Objective reporting is often outweighed by other factors, such as trying to acquire a larger share of the viewing market. The scoop and the news exclusive are prized objectives. Coverage can thus be quite selective, often allowing public opinion and government pronouncements to set the agenda for how the news will be spun. In such a political and market environment, the media will often forgo criticism of counterterrorist policies. Coverage can also be quite subjective, with the biases of executives, editors, and commentators reported to the public as if they were the most salient features of an issue.

Deciding Which Incidents to Report

The process for deciding which events to report (news triage) is often driven by evaluating what kind of news is likely to attract an audience. If it is decided that dramatic incidents will bring in sizable shares of viewers, the media will not hesitate to prioritize such incidents for the day's editions or broadcasts. The media can be highly selective about which terrorist incidents to report. The ultimate decision tends to weigh in favor of what most affects readers or the viewing public.

The personal stories of participants in a terrorist environment are particularly appealing to the media. Strong emotions such as outrage, grief, and hatred play well to many audiences. Certain kinds of terrorist incidents are particularly susceptible to the human interest spin. The key task for the media is to find personal stories that resonate well with their readers or viewers. Stories that do not resonate are likely to be left out of the mix.

Deciding How to Report Incidents

Deciding how to report terrorist incidents is, from the readers' or viewers' perspective, seemingly a subjective exercise. Media reports have never been consistent in their descriptions of the perpetrators of terrorist incidents, nor have they been consistent in characterizing examples of extremist violence as terrorism per se. **Labeling** vacillates from the pejorative to the somewhat positive, from *terrorist* to *commando*, for example. This point is further demonstrated by the following sequence of reporting that occurred in 1973:

> One *New York Times* leading article . . . [described] it as "bloody" and "mindless" and [used] the words "terrorists" and "terrorism" interchangeably with "guerrillas" and "extremists." . . . The *Christian Science Monitor* reports of the Rome Pan Am attack . . . avoided "terrorist" and "terrorism" in favor of "guerrillas" and "extremists"; an Associated Press story in the next day's *Los Angeles Times* also stuck with "guerrillas," while the two *Washington Post* articles on the same incident opted for the terms "commandos" and "guerrillas."[6]

These labels reflect a tendency to use indirect or vague language to describe what might otherwise be appalling behavior.[7] Euphemisms that apply words outside of their normal meaning to mask or soften the language of violence are also common among governments, policy makers, and others. This practice is deliberately media oriented so that the press and general public will more easily accept an incident or policy.

Terrorist-Initiated Labeling

Terrorist groups also use labeling and **euphemistic language**, most often in two circumstances: first, when they label enemy interests as potential targets, and second, when they label themselves. This language is promulgated in communiqués to supporters and journalists.

Labeling Enemies and Targets. Using symbolism to dehumanize potential targets is a universal trait among violent extremists, regardless of ideology. For example, leftists might recast Western business travelers as imperialists. Armenian nationalists might symbolically hold Turkish diplomats to account for the Armenian genocide of the early 20th century. Anti-Semitic and religious terrorists might label a Jewish Community Center as a Zionist interest. Al Qaeda and its sympathizers denounce Western culture and values as contrary to Islam and the values of the faithful; they also denounce secular Arab governments as apostasies. This labeling process creates important qualifiers for acts of extreme violence, allowing terrorists to justify their behavior even though their victims are often noncombatants.

Self-Labeling. Choosing an organization or movement name is a significant decision. Those who engage in political violence consider themselves an elite—a vanguard—waging war against an implacable foe. They consciously use labels and euphemisms to project their self-image. Members of the cause become self-described martyrs, soldiers, or freedom fighters. An image of freedom, sacrifice, or heroism is a given. This pattern is universal among groups on the extremist fringe and is likely to continue. Table 9.1 surveys a few examples of self-labeling and euphemistic language.

_____ Mass Communications and the War for Information

Mass communications technologies can become weapons of war in both conventional and asymmetrical conflicts. Adversaries in a terrorist environment frequently try to shape the character of the environment by manipulating the media. For terrorists, the media serve several useful purposes: first, dissemination of information about the cause; second, the delivery of messages to supporters and adversaries; and, third, as a "front" in the war to shape official governmental policies or influence the hearts and minds of the audience. For governments, the media can be a powerful tool in suppressing terrorist propaganda and manipulating the opinions of large segments of society. This is why every regime will be aggressive in delivering selective information to the media or, as in authoritarian regimes, officially suppress some stories.

Revisiting the Participants in a Terrorist Environment

Depending on their role when an incident occurs, participants in a terrorist environment often provide different assessments of the motives, methods, and targets of violent extremists.[8] It is instructive to summarize each participant group's role within the context of the informational battlefield.

The Terrorist

Terrorists seek attention and legitimacy for their cause by engaging in media-oriented violence. Propaganda by the deed, if properly carried out, carries powerful symbolic messages to the target audience and to large segments of an onlooker audience. One message, for example, might be to "show their power preeminently through deeds that embarrass their more powerful opponents."[9]

TABLE 9.1	PUBLIC RELATIONS: ORGANIZATIONAL TITLES OF VIOLENT EXTREMISTS		

Terrorists seek to project an image that casts them in the role of liberators and soldiers. They are often conscious of their public image and can become quite media savvy.

Organizational Title	Self-Designation	Purpose	Championed Group
Popular Front for the Liberation of Palestine	A united front; a forward position in a war	Liberation of a people	Palestine and Palestinians
Al Qaeda Organization for Holy War in Iraq	A clandestine paramilitary group	Transnational religious warfare on the Iraq "front"	The Faithful
Irish Republican Army	An army; members are soldiers	Republican unification	Northern Irish Catholics
New People's Army (Philippines)	An army; members are soldiers	The continuation of a people's liberation movement	The people
Party of God (Hezbollah)	Movement representing God's will	Carry out God's will on earth	The faithful
Liberation Tigers of Tamil Eelam (Sri Lanka)	Warriors possessing the fierceness of tigers	Liberation	Tamils
Québec Liberation Front	A united front; a forward position in a war	Liberation of a people	Québec and people of Québec

Terrorists also, as mentioned, attempt to cast themselves as freedom fighters, soldiers, and martyrs. If successful, their image will be of a vanguard movement representing the just aspirations of an oppressed people. When this occurs, political and moral pressure can be brought against their adversaries, possibly forcing them to grant concessions to the movement.

The Supporter

Supporters and patrons of terrorists often help with spinning the terrorists' cause and manipulating how incidents are reported. Supporters with sophisticated information departments—such as Northern Ireland's **Sinn Féin** or Lebanon's Hezbollah—can successfully use the media to deliver their message to a wide audience. In societies with a free press, or with supportive authoritarian regimes, sympathetic reporters and editors might lend a hand in portraying the terrorists as freedom fighters. Supporters will always defend the underlying grievances of the extremists and will often allude to these as the reason for the group's decision to use terrorist methods. The key for activist supporters is to convey the impression that the terrorists' methods are understandable under the circumstances. If they can do this successfully, public opinion "can provide the movement with a feeling of legitimacy."[10]

The Victim

Media-oriented terrorist violence can be used to spin incidents to symbolize punishment or chastisement for injustices. From the terrorists' point of view, high-profile attacks that victimize an audience are useful as wake-up calls for the victims to understand the underlying grievances of the movement. Terrorists believe that though victims rarely sympathize with those who cause their suffering, propaganda by the deed can help educate them. Because they are the innocent collateral damage of a conflict, victims—with help from media commentators—will often question why they have become caught up in a terrorist environment. This process can theoretically cause public opinion shifts.

The Target

Media-oriented attacks select targets that symbolize the interests of the terrorists' adversaries. Of course, attacks on some targets—such as symbolic buildings—frequently risk inflicting casualties on large numbers of people. With the appropriate spin, terrorists can garner sympathy, or at least a measure of understanding, if the media convey their reasons for selecting the target. Assessing targeted interests is not unlike assessing the impact on victims and media commentators assist with both. The difference is that the investigatory process is conducted with the understanding that targeted interests have been specifically labeled as an enemy interest. In many circumstances, targeted audiences can have a significant impact on public opinion and government policy.

The Onlooker

Onlookers to terrorist incidents observe the dynamics of the attack, public reactions to the event, and political and media analyses of the incident. They can be directly or indirectly affected, and the media play a significant role in how they receive information. Depending on who is more successful in the battle for information, the onlooker will sympathize with the terrorists' grievances, oppose them, or remain indifferent. If the government is repressive and terrorists or their supporters spin this to their advantage, "one positive effect of repression is that it can supply the movement with new volunteers."[11]

The Analyst

The media play a strong role as interpreters of the terrorist incident. They also play a role in how nonmedia analysts will have their views broadcast to a larger audience. Political leaders, experts, and scholars all rely on the media to promulgate their expert opinions. Aside from contact with these analysts, journalists are prominently—and consistently—in communication with other participants in the terrorist environment. Journalists and other media analysts investigate perspectives, interpret incidents, and have significant input on the labeling process. Media analysts often define who is—or is not—a terrorist.

Practical Considerations: Using the Media

Terrorists and their supporters use time-honored techniques to attract media attention. In the tradition of mainstream media-savvy organizations (and aside from acts of dramatic violence), terrorists have invited the media to press conferences, issued press releases, granted interviews, released audio and video productions, and produced attractive photographic essays. There is a tendency for the media to sensationalize information, so that broadcasts of terrorist audio and video recordings, news conferences, or written statements often take on an entertainment quality.

Terrorists' Manipulation of the News Scoop

News outlets compete in trying to preempt the newsworthiness of their competitors stories—known colloquially as **media scooping,** that is, being the first to report breaking news. "This is a situation that, however unwittingly, is tailor-made for terrorist manipulation and contrivance."[12] Terrorists and other radicals have in fact successfully manipulated this propensity for scooping and sensationalizing news on a number of occasions. Several examples follow:

◆ Ilich Ramirez Sanchez, also known as Carlos the Jackal, did not make his escape during the December 1975 OPEC hostage crisis until the television cameras arrived. Only then did he dramatically and publicly make his getaway from Vienna to Algeria with 35 hostages in tow.[13]

◆ During the November 1979 to January 1981 seizure of the American embassy in Iran, Iranian crowds played to the cameras several times. Crowds would come alive when the cameras were on them, while parts of the crowd not under scrutiny were rather quiescent.[14]

◆ In May 1986, ABC News broadcast a short interview with Abu Abbas, the leader of the Palestine Liberation Front. The PLF was notorious at the time because a PLF terrorist unit had carried out the October 1985 hijacking of the *Achille Lauro* cruise ship, in which American passengers were terrorized and one was murdered. During the interview, Abbas threatened to carry out acts of terrorism in the United States and referred to President Ronald Reagan as Enemy Number One.

Points of Criticism

Because of these and other examples of overt (often successful) manipulation, critics have identified a number of problems in news reporting:

1. First, critics argue that journalists sometime cross the line between reporting the news and disseminating terrorist propaganda.
2. Second, critics argue that the media's behavior sometimes shifts from objectivity to sensational opinion during particularly intense incidents.
3. Third, critics argue that the ability of the mass media to reach large audiences, when combined with the first two factors, can lead to realignments within the political environment.

Thus, in the aftermath of the September 11 attacks, political and media critics hotly debated whether the media should continue to broadcast feeds from Al Jazeera news service. Al Jazeera is one of the rare independent news services in the Middle East, and it broadcast extensive footage of injured Afghani and Iraqi civilians, whom the U.S. classifies as the unfortunate collateral damage of war. Al Jazeera also aired film clips of Al Qaeda leaders Osama bin Laden and Ayman al-Zawahiri that Al Qaeda had delivered to the station. The fear in the United States was that uninterpreted broadcasts of these images could spread enemy propaganda or send messages to sleeper cells. For this reason, American news services were asked to limit Al Jazeera coverage.

Counterpointing the Criticism

As a counterpoint to these criticisms, and in defense of journalistic reporting of terrorist incidents, some proponents of the free press argue that full exposure of terrorism and the terrorists' grievances should be encouraged. This way, the public can become fully informed about the nature of terrorism in general and about the motives of specific terrorists. Thus, "defenders of media coverage feel that it enhances public understanding of terrorism and reinforces public hostility toward terrorists."[15]

Information Is Power: The Media as a Weapon

For terrorists and other extremists, information can be wielded as a weapon of war, making **media as a weapon** an important concept. Because symbolism is at the center of most terrorist incidents, the media are explicitly identified by terrorists as potential supplements to their arsenal. When terrorists successfully—and violently—manipulate important symbols, relatively weak movements can influence governments and societies. Even when a terrorist unit fails to complete its mission, intensive media exposure can lead to a propaganda victory. For example, during the 1972 Munich Olympics attack by Black September terrorists, "an estimated 900 million persons in at least a hundred different countries saw the crisis unfold on their television screens."[16] As one PLO leader later observed, "World opinion was forced to take note of the Palestinian drama, and the Palestinian people imposed their presence on an international gathering that had sought to exclude them."[17]

Case: Hezbollah and the Hijacking of TWA Flight 847

Lebanon's Hezbollah has demonstrated its skill at conducting extraordinary strikes, some of which ultimately affected the foreign policies of France, Israel, and the United States. It regularly markets itself to the media by disseminating grievances as press releases, filming and photographing moving images of its struggle, compiling human interest backgrounds of Hezbollah fighters and Shi'a victims, and packaging its attacks as valiant assaults against Western and Israeli invaders and their proxies. This has been done overtly and publicly, and incidents are manipulated to generate maximum publicity and media exposure. For example, the January 1987 kidnapping in Beirut of Terry Waite, the Archbishop of Canterbury's envoy, was broadcast globally. He was released in November 1991.

On June 14, 1985, three Lebanese Shi'a terrorists hijacked **TWA Flight 847** as it flew from Athens to Rome. It was diverted to Beirut, Lebanon, and then to Algiers, Algeria. It was then flown back to Beirut, made a second flight to Algiers, and then flew back to Beirut a second time. During the odyssey, the terrorists released women, children, and those who were not American, until 39 American men remained on board the aircraft. At the final stop in Beirut, the hostages were offloaded and dispersed throughout the city.

Photo 9.2 TWA Flight 847. A terrorist waves a gun to cut short a press interview with TWA pilot John Testrake during the hijacked airliner's odyssey around the Mediterranean. The terrorists skillfully manipulated the international press.

As the hijacking unfolded, the media devoted an extraordinary amount of airtime to the incident. The television networks ABC, CBS, and NBC broadcast almost 500 news reports, or 28.8 per day, and devoted two thirds of their evening news programs to the crisis.[18] "During the 16 days of the hijacking, CBS devoted 68% of its nightly news broadcasts to the event while the corresponding figures at ABC and NBC were 62% and 63% respectively."[19]

The hijackers were masterful as puppeteers of the world's media. They granted carefully orchestrated interviews, held press conferences, and selected the information they permitted the news outlets to broadcast. It was reported later that they had offered to arrange tours of the airliner for the networks for a $1,000 fee and an interview with the hostages for $12,500.[20] After the hostages were dispersed in Beirut, **Nabih Berri**, the leader of Lebanon's Syrian-backed Shi'a **Amal** movement (an ally and occasional rival of the Shi'a Hezbollah movement), was interviewed by news networks as part of the negotiations to trade the hostages for concessions. In the end, the terrorists' media-oriented tactics were effective. They broadcast their grievances and demands to the world community and achieved their objectives. "[The] media exposure of the hostages generated enough pressure for the American president to make concessions."[21]

The hostages were released on June 30, 1985.

The Contagion Effect

The **contagion effect** refers to the theoretical influence of media exposure on the future behavior of other like-minded extremists.[22] This concept can also be applied "to a rather wide range of violent behavior [other than terrorism], including racial disturbances."[23] In theory, when terrorists successfully garner wide exposure or a measure of sympathy from the media and their audience, other terrorists may be motivated to replicate the tactic or strategy. This may be especially true if concessions have been forced from the targeted interest. Assuming that contagion theory has merit (the debate on this point continues), the question becomes the extent to which the contagion effect influences behavior.

Assessments of the contagion effect have produced some consensus that the media do have an effect on terrorist cycles. For example, empirical studies have indicated a correlation between media coverage and time lags between terrorist incidents.[24] These studies have not definitively proven that contagion is a behavioral fact, but do suggest that the theory may have some validity.

The era of the New Terrorism arguably presents an unprecedented dynamic for contagion theory, because transnational cell-based movements are a new model for—and may suggest new assessments of—the theory. Transnational organizations such as Al Qaeda engage in a learning process from the lessons of attacks by their operatives around the world. The advent of communications technologies such as faxes, cellular telephones, e-mail, text messaging, and the Internet—especially in combination with focused manipulation of the media—means that the terrorists' international learning curve can be quick and efficient. Hence, in theory, the contagion effect may be enhanced within New Terrorist movements on a global scale.

Problems on the New Battleground: The Risk of Backlash

Unfortunately for terrorists, this widespread exposure does not always work to their advantage. Governments are also experts at spinning the nature of violence to the media. When the violence is truly horrific, and when the victims, targets, or onlooker audiences recoil in popular disgust, the terrorists significantly diminish their influence over their adversary. They can, in effect, actually strengthen the adversary's resolve.

Public opinion among victims, targets, and onlooker audiences is critical to the success of **media-oriented terrorism**. However, one should bear in mind that terrorists often play to their supporter audiences, so that success is always a relative term in the battle for the media.

Freedom of the Press and Regulating the Media

Freedom of the press is an ideal standard—and arguably an ideology—in many democracies. The phrase embodies a conceptual construct that suggests that the press should enjoy the liberty to independently report information to the public, even when the information might be unpopular. News editors and journalists, when criticized for their reports, frequently cite the people's right to know as a justification. The counterpoint to freedom of the press is regulation of the press. This issue arises when the media publish unpleasant facts (often in lurid prose) about subjects that the public or the government would rather not consider. Regulation is also a genuine option when matters of national security are at stake.

The Free Press

The international media operate under many rules inherent to and defined by their cultural environments. Some media have few if any codes of professional self-regulation, whereas others have adopted rather strict self-standards.[25] The Netherlands Broadcasting Corporation has traditionally had no formal code of operations, for example, whereas its British counterpart

Photo 9.3 Disseminating information. A **Rewards for Justice** program poster disseminated by the U.S. Department of State's **Diplomatic Security Service**. It depicts the aftermath of the 1993 bombing of the World Trade Center in New York City.

operates under a detailed set of rules.[26] Consensus indicates that ethical standards should be observed when reporting terrorist incidents.

Gatekeeping. In societies that champion freedom of the press, one model professional environment is **journalistic self-regulation,** sometimes referred to as **media gatekeeping.** If conducted under established standards of professional conduct, self-regulation obviates the need for official regulation and censorship. In theory, moral arguments brought to bear on the press from political leaders and the public will pressure them to adhere to model standards of fairness, accuracy, and objectivity.

This is, of course, an ideal free press environment; in reality, critics argue that journalistic self-regulation is a fluid and inconsistent process. Chapter Perspective 9.1 illustrates this criticism by contrasting different standards of gatekeeping by the American media when reporting news about several U.S. presidents.

CHAPTER PERSPECTIVE 9.1

Delivering the Message

Extremist movements will often use coded language to convey their message. Such language is often peculiar to the group and is not easily interpreted by those who are not members. From the outsider's point of view, the message is sometimes incoherent. The following example demonstrates how extremist propaganda can become lost in rhetoric.[a]

Against Social-Democracy and Liquidationism—For Steadfast Revolutionary Work!

Reformism does not mean improving the conditions of the masses; on the contrary, the vital role that reformism has played in the capitalist offensive shows that reformism means collaborating with the **bourgeoisie** in suppressing the mass struggle and implementing the capitalist program. . . .

The work of the Marxist-Leninist Party has been a beacon against the opportunism of the liquidationist and social-democratic trends. The Marxist-Leninist Party has persevered in steadfast revolutionary struggle, while the opportunists, as fair-weather "revolutionaries," are reveling in despondency and renegacy, are denouncing the revolutionary traditions from the mass upsurge that reached its height in the 1960s and early 1970s, and are cowering behind the liberals, labor bureaucrats and any bourgeois who is willing to throw them a crumb.[b]

Communiqué on the Second Congress of the Marxist-Leninist Party, USA

Fall 1983

Notes

a. See Lyman Tower Sargent, ed., *Extremism in America: A Reader* (New York: New York University Press, 1995), 85.
b. *Communiqué on the Second Congress of the MLP, USA*, 88–89.

Regulation of the Free Press. Many governments occasionally regulate or otherwise influence their press community while advocating freedom of reporting. Governments selectively release information, or release none, during terrorist incidents. The rationale is that proper investigation, by definition, requires limitations. This occurs as a matter of routine during wartime or other national crises.

A number of democracies have state-run and semiprivate radio and television stations—for example, Great Britain, France, Germany, and other European democracies. These networks are expected to promote accepted standards of professional conduct and to practice self-regulation for the sake of good taste and national security. In some democracies, the law permits the government to suppress the reporting of news.

The State-Regulated Press

State-regulated media exist in environments in which the state routinely intervenes, even in societies that otherwise have a measure of democratic freedoms. For example, the **state-regulated presses** of most countries in the Middle East led many people in those countries to believe that the September 11 attacks either were not the work of Al Qaeda or were the work of Zionists. Some disseminated far-fetched rumors. As noted earlier, some Middle Eastern mainstream commentators reported and supported a popular conspiracy theory that anonymous telephone calls warned thousands of Jewish workers in the World Trade Center to leave the buildings before the attack, and that therefore no Jews were among the casualties.

Scales of intervention range from permitting independent but regulated newspapers to creating government-controlled propaganda enterprises. In **authoritarian** and **totalitarian regimes,** terrorists have no chance to rely on the media to sensationalize their deeds. In these societies, the media serve to promote the government's interests and often to disseminate government propaganda. There are no gripping stories that might sway an audience. The general public is never privy to sympathetic human interest stories or to an extremist manifesto's call to arms. When terrorist incidents occur, they are either underreported by the government or manipulated to the absolute advantage of the regime.

In very restrictive societies, the media are used as outlets for propaganda on behalf of the existing regime. For example, the late Iraqi dictator Saddam Hussein created an extensive **cult of personality** not unlike that of dictators Joseph Stalin in the Soviet Union and Kim Il Sung (and son Kim Jong Il) in North Korea. Cults of personality are used by dictatorial regimes to promote the leader or ruling party as the source of absolute wisdom, truth, and benevolence. Likenesses of the leader are widely distributed, usually in a variety of symbolic poses. Saddam Hussein, for example, was regularly depicted as a visionary, a warrior, the good father, a common citizen, a devout Muslim, and the medieval leader Saladin. Hussein, Stalin, and Kim promoted themselves, their regimes, and their policies by completely controlling the dissemination of information.

Chapter Summary

This chapter investigated the role of the media in terrorist environments. Particular attention was given to terrorists' efforts to publicize the cause, their manipulation of mass communications, and the potential impact of the New Media. Issues related to reporting terrorism include questions about which incidents to report and how to report them. The concepts of information as power and media-oriented terrorism were defined and explored as critical considerations in understanding the role of the media.

Evaluation of the new battleground for information requires that readers first explore the issue from the perspectives of participants in terrorist environments. Practical considerations for terrorists' treatment of the media include their manipulation of the media's desire to scoop their competitors. The contagion effect and the example of the hijacking of TWA Flight 847 demonstrate how media exposure can become an effective weapon in a terrorist arsenal.

Regulation of the media is a challenge for every government. For authoritarian and totalitarian regimes, this challenge is easily resolved by simply prohibiting certain kinds of reporting practices. It is a more complex issue in most democratic systems, though most have adapted by adopting laws and practices that restrict media access to operational information.

DISCUSSION BOX

This chapter's Discussion Box is intended to stimulate critical debate about how to balance freedom of the press with the protection of national security.

Freedom of Reporting and Security Issues

During times of crisis, governments restrict media access to information about matters that affect security policy. The logic is understandable: Governments believe that the war effort (or counterterrorism policy) requires limitations to be imposed to prevent information from helping the enemy and to prevent the enemy from spreading its propaganda. For example, the British **Official Secrets Act** was designed to manage the flow of information both from and to adversaries.

The challenge for democracies is to strike a balance between government control over information—for the sake of national security—and unbridled propaganda. The following examples illustrate how the United States and Great Britain managed the flow of information during international crises:

During the Vietnam War, journalists had a great deal of latitude in the field to visit troops and observe operations. Vietnam was the first television war, so violent and disturbing images were broadcast into American homes on a daily basis. These reports were one reason American public opinion turned against the war.

During the 1982 Falklands War, the British government tightly controlled and censored news about operations. Press briefings were also strictly controlled, under the rationale that useful information could otherwise be received by the Argentines and jeopardize the war effort.

During the 1991 Persian Gulf War, news was likewise tightly controlled. The media received their information during official military press briefings and were not permitted to travel into the field except under highly restricted conditions.

During the late 2001 Afghan phase of the war on terrorism, news was again highly restricted. Official press briefings were the norm, and requests were made for cooperation in not broadcasting enemy propaganda.

During the 2003 conventional phase of the invasion of Iraq, reporters were embedded with military units and reported events as they unfolded. Official press briefings were the norm.

Discussion Questions

1. Should the United States adopt information-control regulations similar to Britain's Official Secrets Act?
2. What are the policy implications of permitting journalists to have the same degree of access to information as occurred during the Vietnam War?
3. What are the policy implications of permitting journalists to have the same degree of access to information as occurred during the Gulf War?
4. Under what circumstances should the state increase restrictions on the media? How would you justify these restrictions?
5. Do you think that the media in democracies are more prone to manipulation by terrorists? Is this a myth?

Key Terms and Concepts

The following topics were discussed in this chapter and can be found in the glossary:

Al Jazeera	information is power	New Media
Amal	journalistic	Official Secrets Act
authoritarian regimes	self-regulation	print media
Berri, Nabih	labeling	propaganda
contagion effect	mass communications	state-regulated press
cult of personality	media as a weapon	totalitarian regime
free press	media gatekeeping	TWA Flight 847

Terrorism on the Web

Log on to the Web-based student study site at **www.sagepub.com/martinessstudy** for additional Web sources and study resources.

Web Exercise

Using this chapter's recommended Web sites, conduct an online investigation of the reporting of terrorism by the media.

1. Compare and contrast the reporting of political violence by the referenced media services. What patterns of reporting can you identify?
2. To what extent are the media services biased in their reporting? How so?
3. To what extent are the media services objective in their reporting? How so?

For an online search of media reporting of terrorism, readers should activate the search engine on their Web browser and enter the following keywords:

"Media and Terrorism"
"Propaganda and Terrorism"

Recommended Readings

The following publications provide discussions for evaluating the role of the media in the reporting of terrorism, national conflict, and political dissent.

Davis, Richard, and Diana Owen. *New Media and American Politics*. New York: Oxford University Press, 1998.

Gitlin, Todd. *The Whole World Is Watching: Mass Media in the Making & Unmaking of the New Left*. Los Angeles: University of California Press, 1980.

Herbst, Philip. *Talking Terrorism: A Dictionary of the Loaded Language of Political Violence*. Westport, CT: Greenwood, 2003.

Knightley, Phillip. *The First Casualty: The War Correspondent as Hero and Myth Maker From the Crimea to Kosovo*. Baltimore: Johns Hopkins University Press, 2002.

Paletz, David L., and Alex P. Schmid, eds. *Terrorism and the Media*. Newbury Park, CA: Sage, 1992.

Weimann, Gabriel, and Conrad Winn. *The Theater of Terror: Mass Media and International Terrorism*. New York: Longman, 1994.

10

Tactics and Targets of Terrorists

In this chapter, readers will investigate terrorist objectives, methods, and targets. The discussion focuses on the rationale behind the calculation of terrorists' ends and means—what terrorists are trying to do and how they try to do it. Weaponry is, of course, an integral factor in evaluating ends and means, so attention will also be given to the terrorists' arsenal.

Ironically, to most onlookers, many methods appear senseless and random; however, from the perspective of terrorists, they are neither. Two commonalities must be remembered:

> *Terrorist violence is rarely senseless.* It is usually well thought out, and regardless of the ultimate scale of violence applied or the number of civilian casualties, these are considered to be logical and sensible consequences of waging a just war.
>
> *Terrorist violence is rarely random.* Targets are specifically selected and the outcome of careful deliberation. An element of randomness occurs when targets of opportunity are attacked without a period of careful planning.

Photo 10.1 A dramatized depiction of a 19th-century scientific anarchist constructing a bomb in his apartment.

Among extremists, acceptance is almost universal that terrorist violence is a kind of poor man's warfare the weak use against stronger opponents. As a matter of practicality, extremists adopt terrorist methods for several reasons:

- ◆ Terror tactics are relatively easy to use and therefore commend themselves to an organization without sophisticated weapons or popular support.
- ◆ Terrorism produces disproportionate publicity, which is highly prized by separatist movements or political factions that may feel they have no other way of seizing the world's attention.
- ◆ Spectacular atrocities illustrate a government's inability to rule. If a government is perceived to be weakened, exasperated security forces may be provoked to overreaction.[1]

Based on such justifications and practical considerations, terrorists have selected methods and targets from a menu of options derived from their interpretation of their environment.

Many terrorists in the past were known to discriminate in selecting methods and targets. Terrorists today, on the other hand, are apt to wield any available weapon against broadly defined enemy interests.

The discussion in this chapter will review the following:

◆ Understanding terrorist objectives
◆ The terrorists' arsenal
◆ Terrorist targets

Understanding Terrorist Objectives

Objectives and goals are parts of the process toward a final outcome. An objective is an incremental step in the overall process that leads to an ultimate goal. A goal is the final result of the process, the terminal point of a series of objectives. Thus, an objective in a revolutionary campaign could be the overthrow of an enemy government or social order; the goal could be to establish a new society. During a revolutionary campaign, many objectives have to be achieved to reach the final goal.

Typical Objectives

Politically violent groups and movements show certain similarities in their objectives. The following discussion identifies a few such commonalities. These are by no means either common to all violent extremists at all phases of their campaigns, or exhaustive analyses.[2] However, it is instructive to review a few central objectives:[3]

◆ Changing the existing order   ◆ Social disruption
◆ Psychological disruption   ◆ Creating a revolutionary environment

Changing the Existing Order

At some level, all terrorists seek to change an existing order, even if it is simply a short-term objective to disrupt the normal routines of society by inflicting maximum casualties. When evaluating what it means to change an existing order, one must take into consideration the different profiles of terrorist movements, their motives, and the idiosyncrasies of individual terrorists. Several examples follow:

◆ Ethno-nationalist terrorists seek to win recognition of their human rights, or a degree of national autonomy, from the present order.
◆ Nihilists wish to destroy systems and institutions without regard for what will replace the existing order.
◆ Religious terrorists act on behalf of a supernatural mandate to bring about a divinely inspired new order.
◆ Lone wolves have a vague and sometimes delusional assumption that their actions will further a greater cause against a corrupt or evil social order.

Psychological Disruption

An obvious objective is to inflict maximum psychological damage by applying dramatic violence against symbolic targets. "From the terrorists' perspective, the major force of terrorism comes not from its physical impact but from its psychological impact."[4] When terrorist violence is applied discerningly, the weak can influence the powerful, and the powerful can intimidate the weak. Cultural symbols, political institutions, and public leaders are examples of iconic (nearly sacred) targets that can affect large populations when attacked.

Social Disruption

Social disruption is an objective of propaganda by the deed. The ability of terrorists and extremists to disrupt the normal routines of society demonstrates both the weakness of the government and the strength of the movement; it provides terrorists with potentially very effective propaganda. When governments fail to protect the normal routines of society, discontent may spread throughout society, thus making the population susceptible to manipulation by a self-styled vanguard movement. For example, bombing attacks on public transportation systems certainly cause social disruption. A group might be attacked specifically to deter those individuals from traveling through a region or territory. Tourists, for example, have been targeted repeatedly in Egypt, such as the bombing incident in July 2005 in the resort city of Sharm el Sheikh on the Sinai Peninsula, which killed approximately 90 people.

Creating a Revolutionary Environment

Dissident extremists understand that they cannot hope to win in their struggle against the state without raising the revolutionary consciousness of the people. For many terrorists, propaganda by the deed is considered the most direct way to create a broad-based revolutionary environment, so that "the destruction of one troop transport truck is more effective propaganda for the local population than a thousand speeches."[5] Revolutionary theorists predicted that terrorism would force the state to overreact, the people would understand the true repressive nature of the state, and a mass rebellion would occur—led by the revolutionary vanguard.

Playing to the Audience: Objectives, Victims, and Constituencies

Terrorists adapt their methods and selection of targets to the characteristics of their championed group and the idiosyncrasies of their environment. Targets are selected for specific symbolic reasons, with the objectives of victimizing specific groups or interests and sending symbolic messages to the terrorists' constituency. In a sense, the targeted groups or interests serve as conduits to communicate the extremist movement's message.

If skillfully applied, propaganda by the deed can be manipulated to affect specific audiences. These audiences can include the following segments of society:[6]

Politically apathetic people. The objective of terrorist violence directed toward this group is to force an end to their indifference and, ideally, to motivate them to petition the government for fundamental changes.

The government and its allied elites. Terrorists seek to seriously intimidate or distract a nation's ruling bodies to force them to deal favorably with the underlying grievances of the dissident movement.

Potential supporters. An important objective of propaganda by the deed is to create a revolutionary consciousness in a large segment of society. This is more easily achieved within the pool of those who are sympathetic to the extremists' objectives but do not yet approve of their methods.

Confirmed supporters. Terrorists seek to assure their members and confirmed supporters that the movement continues to be strong and active. They communicate this through acts of symbolic violence.

Depending on whom they claim to champion, extremist movements adapt their tactics to their environment as a way to communicate with (and attract) their defined constituency. With a few exceptions, terrorists and extremists usually direct their appeals to specific constituencies. These appeals are peculiar to the environment and idiosyncrasies of the movement, although leftists and ethno-nationalists have sometimes championed the same groups out of a sense of revolutionary solidarity. Table 10.1 illustrates the relationship between several extremist groups and their constituencies, objectives, methods, and targeted interests.

The New Terrorism and New Objectives

The New Terrorism is different from previous models because it is characterized by vaguely articulated political objectives, indiscriminate attacks, attempts to achieve maximum psychological and social disruption, and the potential use of weapons of mass destruction. It also includes an emphasis on building horizontally organized, semiautonomous cell-based networks.

TABLE 10.1   CONSTITUENCIES AND ENEMIES: SELECTING TACTICS AND TARGETS

Terrorists select their methods within the context of their social and political environments. They appeal to specific constituencies and justify their choice of methods by championing the political cause of their constituencies. Their targeted interests (that is, enemy interests) can be defined narrowly or broadly, so that civilian populations can be included as legitimized targets.

Group or Movement	Activity Profile			
	Constituency	**Objectives**	**Methods**	**Targeted Interest**
Al Aqsa Martyr Brigades	Palestinians	Palestinian state	Suicide bombings; small-arms attacks	Israeli civilians; Israeli military
Secular Iraqi Insurgents	Iraqi People	Expulsion of occupation troops; establishment of nationalist government	Terrorist attacks; guerrilla warfare	Occupation troops; foreign contractors; Iraqi collaborators
Al Qaeda	Devout Muslims	Worldwide Islamic revolution	Well-planned bombings; indigenous insurrections	The West; secular Islamic governments
Provos	Irish Catholics	Union with the Irish Republic	Small-arms attacks; bombings	British; Ulster Protestants
Bosnian Serb Militias	Bosnian Serbs	Serb state	Ethnic cleansing; communal terrorism	Bosnian Muslims; Bosnian Croats
Tamil Tigers	Sri Lankan Tamils	Tamil state	Terrorist attacks; guerrilla warfare	Sri Lankan government; Sinhalese

Weapons of Mass Destruction and the Objectives of the New Terrorism

The redefined morality of the New Terrorism opens the door for methods to include high-yield weapons and for targets to include large populations. Symbolic targets and enemy populations can now be hit much harder than in the past; all that is required is the will to do so.

Why would terrorists deliberately use high-yield weapons? What objectives would they seek? Depending on the group, many reasons have been suggested, including the following general objectives:[7]

- ◆ Attracting attention
- ◆ Pleasing God
- ◆ Damaging economies
- ◆ Influencing enemies

Terrorists may strike with the central objective of killing as many people as possible. They are not necessarily interested in overthrowing governments or changing policies as their primary objectives. Rather, their intent is simply to deliver a high body count and thereby terrorize and disrupt large audiences. For example, the 1993 and 2001 attacks on the World Trade Center in New York City by radical Islamists, Aum Shinrikyō's 1995 sarin nerve gas attack in Tokyo, and the 1995 bombing in Oklahoma City by American terrorists were all intended to kill as many civilians as possible and to demonstrate the vulnerability of society. Little if any consideration was given to changing government policies.

The Terrorists' Arsenal

The terrorist environment today is shaped by advances in technology, information, and transnational interconnectivity. This globalized environment has given rise to new possibilities in terrorist methodology.[8] Two factors in particular are believed to contribute significantly to the distinctiveness of New Terrorism methodologies. The first is "the diffusion of information technology and advanced communications."[9] A second is the "increased movement, and ease of movement, across international boundaries."[10]

The following discussion reviews common methods and weapons terrorists use to achieve their objectives.

Concept: Asymmetrical Warfare

The concept of **asymmetrical warfare** refers to the use of unconventional, unexpected, and nearly unpredictable methods of political violence. Terrorists intentionally strike at unanticipated targets and apply unique and idiosyncratic tactics. This way, they can seize the initiative and redefine the international security environment and overcome the traditional protections and deterrent policies that societies and the international community use.

The Appeal of Asymmetrical Conflict

Dissident terrorists are quantitatively and qualitatively weaker than conventional security forces. In today's intensive security environment, they simply cannot prevail or last indefinitely in an urban-based guerrilla campaign. Because of this, state-level rivals must resort to unconventional and subversive methods to confront U.S. and Western interests. At the same time, they must also deliver maximum propaganda and symbolic blows against the seemingly overwhelming power of enemy states or societies. New Terrorism is thus characterized by a new doctrine that allows for the use of weapons of mass destruction, indiscriminate attacks, maximum casualties, technology-based terrorism, and other exotic and extreme methods.

In adopting asymmetrical methods, "the weaker forces are seeking total war, encompassing all segments of society."[11] They are trying to break the enemy's will to resist by whatever means are at their disposal.

Netwar: A New Organizational Theory

The New Terrorism incorporates maximum flexibility into its organizational and communications design. Semiautonomous cells are either prepositioned around the globe as sleepers (such as the 2004 Madrid terrorists) or they travel to locations where an attack is to occur (such as the September 11 hijackers in the United States). They communicate using new cyber and digital technologies. An important concept in the new terrorist environment is the notion of **netwar**, which refers to

> an emerging mode of conflict and crime . . . in which the protagonists use network forms of organization and related doctrines, strategies, and technologies attuned to the information age. These protagonists are likely to consist of dispersed small groups who communicate, coordinate, and conduct their campaigns in an internetted manner, without a precise central command.[12]

The new "internetted" movements have made a strategic decision to establish virtual linkages via the Internet and other technologies.

Case: The Martyr Nation as an Asymmetrical Strategy

The application of asymmetrical warfare is evident in the conflict between the Palestinians and Israelis during 2001 and 2002. The doctrine of engagement used by Palestinian nationalists called for incessant confrontation using guerrilla, terrorist, and suicidal martyrdom tactics to strike unexpectedly at soft civilian targets. As one Palestinian leader said, "Our ability to die is greater than the Israelis' ability to go on killing us." Nationalists contended that "Israel is confronting a martyr 'nation,' not just individual fanatics or militant groups."[13] Thus, the Maoist concept of people's war arguably has been applied asymmetrically in the martyrdom tactics of Palestinian extremists, because it suggests that an entire people is willing to sacrifice a great deal to achieve its goals.

Weapons Old and New

Twenty-first century weaponry can be classified along a sliding scale of technological sophistication and threat potential. This scale includes a high, medium, and low range, summarized as follows:[14]

High Range. The New Terrorism is defined in part by the threatened acquisition of **chemical agents**, **biological agents**, or **nuclear weapons**. This threat includes the development of **radiological agents** that spread highly toxic radioactive materials by detonating conventional explosives. The first case of widespread use of a biological agent by terrorists occurred when **anthrax** was deliberately sent through the mails in the United States in the aftermath of the September 11 attacks.

Medium Range. Terrorists currently have extensive access to military-style weaponry. These include automatic weapons, rocket launchers, and military-grade explosives of many varieties. Sympathetic state sponsorship and the international arms black market facilitate procurement of a virtually unlimited array of conventional small arms and munitions. These arms have been the weapons of choice for terrorists in innumerable incidents.

Low Range. Often forgotten in the discussions about the threat from medium- and high-range weaponry are the powerful homemade weapons that can be manufactured from commercial-grade components. For example, ammonium nitrate and fuel oil (ANFO) bombs can be easily manufactured from readily available materials. Iraqi insurgents became quite adept at deploying **improvised explosive devices (IEDs),** commonly referred to as **roadside bombs,** against U.S.-led occupation troops.

Firearms

Small arms and other handheld weapons have been, and continue to be, the most common types of weapon that terrorists use. These are light and heavy infantry weapons and include pistols, rifles, submachine guns, assault rifles, machine guns, **rocket-propelled grenades**, mortars, and precision-guided munitions. Typical firearms include the following:

Submachine Guns. Originally developed for military use, **submachine guns** are now used primarily by police and paramilitary services. Although new models have been designed, such as the famous Israeli Uzi and the American Ingram, World War II–era models are still on the market and have been used by terrorists.

Assault Rifles. Usually capable of both automatic (repeating) and semiautomatic (single-shot) fire, **assault rifles** are military-grade weapons that are used extensively by terrorists and other irregular forces. The **AK-47**, invented by **Mikhail Kalashnikov** for the Soviet Army, is the most successful assault rifle in terms of production numbers and widespread adoption by standing armies, guerrillas, and terrorists. The American-made **M-16** has likewise been produced in large numbers and adopted by a range of conventional and irregular forces.

Rocket-Propelled Grenades (RPGs). Light self-propelled munitions are common features of modern infantry units. The **RPG-7** has been used extensively by dissident forces throughout the world, particularly in Latin America, the Middle East, and Asia. Manufactured in large quantities by the Soviets, Chinese, and other communist nations, it is an uncomplicated and powerful weapon that is useful against armor and fixed emplacements such as bunkers or buildings.

Precision-Guided Munitions (PGMs). Less commonly found among terrorists, but extremely effective when used, **PGMs** are weapons that can be guided to their targets by using infrared or other tracking technologies. The American-made **Stinger** is a shoulder-fired surface-to-air missile that uses an infrared targeting system. It was delivered to the Afghan mujahideen during their anti-Soviet jihad and used very effectively against Soviet helicopters and other aircraft. The Soviet-made **SA-7**, also known as the **Grail**, is also an infrared-targeted surface-to-air missile. Both the Stinger and the Grail pose a significant threat to commercial airliners and other aircraft.

Common Explosives

Terrorists regularly use explosives to attack symbolic targets. Along with firearms, explosives are staples of the terrorist arsenal. The vast majority of terrorist bombs are self-constructed improvised weapons rather than premanufactured military-grade bombs. The one significant exception to this rule is the heavy use of military-grade **mines**. These are buried in the soil or rigged to be detonated as booby traps. Some improvised bombs are constructed from commercially available explosives such as **dynamite** and **TNT**, whereas others are manufactured from military-grade compounds. Examples of compounds found in terrorist bombs include the following:

Plastic Explosives. **Plastic explosives** are putty-like explosive compounds that can be easily molded. The central component of most plastic explosives is a compound known as **RDX**. Nations that manufacture plastic explosives often use chemical markers to tag each batch manufactured. The explosives can thus be traced back to their source.

Semtex. **Semtex** is a potent plastic explosive of Czech origin. During the Cold War, Semtex appeared on the international market, and Libya obtained a large supply. It is popular among terrorists.

Composite-4 (C-4). Invented in the United States, **Composite-4 (C-4)** is a high-grade and powerful plastic explosive. It is more expensive and more difficult to obtain than Semtex.

Ammonium Nitrate and Fuel Oil (ANFO) Explosives are manufactured from common ammonium nitrate fertilizer that has been soaked in fuel oil. Using ammonium nitrate as a base for the bomb, additional compounds and explosives can be added to intensify the explosion. These devices require hundreds of pounds of ammonium nitrate, so are generally constructed as car or truck bombs.

Triggers

Regardless of the type of explosive that is used, some bomb makers construct sophisticated triggering devices and are able to shape explosive charges to control the direction of the blast. Examples of triggering devices include the following:

Timed Switches. Time bombs are constructed from acid-activated or electronically activated triggers. They are rigged to detonate after a period of time.

Fuses. A very old and low-technology method to detonate bombs is to light a fuse that detonates the explosive. It can be timed by varying the length of the fuse. Shoe bomber Richard Reid was overpowered after a flight attendant smelled burning matches as he tried to light a fuse in his shoe.

Pressure Triggers. Using **pressure triggers**, weapons such as mines are detonated when physical pressure is applied to a trigger. Car bombers in Iraq apparently attached broom handles or other poles to the front of their vehicles as plungers and then rammed their target with the plunger. A variation on physical pressure triggers is trip-wire booby traps. More sophisticated pressure triggers react to atmospheric (barometric) pressure, such as changes in pressure when an airliner ascends or descends.

Electronic Triggers. Remotely controlled bombs are commonly employed by terrorists. **Electronic triggers** are activated by a remote electronic or radio signal.

High-Technology Triggers. Some sophisticated devices may use triggers that are activated by motion, heat, or sunlight.

Types of Bombs

Gasoline Bombs. The most easily manufactured (and common) explosive weapon dissidents use is nothing more than a gasoline-filled bottle with a flaming rag for its trigger. A **gasoline bomb** is thrown at targets after the rag is stuffed into the mouth of the bottle and ignited. Tar, Styrofoam, or other ingredients can be added to create a gelling effect for the bomb, which causes the combustible ingredient to stick to surfaces. These weapons are commonly called **Molotov cocktails,** named for **Vyacheslav Molotov**, the Soviet Union's foreign minister during World War II.

Pipe Bombs. These devices are easily constructed from common pipes and filled with explosives (usually gunpowder) and then capped on both ends. Nuts, bolts, screws, nails, and other shrapnel are usually taped or otherwise attached, to be released as projectiles on detonation. Terrorists have used many hundreds of **pipe bombs**.

Vehicular Bombs. Ground vehicles that have been wired with explosives are a frequent weapon in the terrorist arsenal. **Vehicular bombs** can include car bombs and truck bombs; they are mobile, covert in the sense that they are not readily identifiable, capable of transporting large amounts of explosives, and rather easily constructed. They have been used on scores of occasions throughout the world.

Barometric Bombs. These sophisticated devices use triggers that are activated by changes in atmospheric pressure. An altitude meter can be rigged to become a triggering device when a specific change in pressure is detected. Thus, an airliner can be blown up in midair as the cabin pressure changes.

Case: Weapons of Mass Destruction

Within the context of threats from high-range weapons, it is important to understand the basic differences between four types of weapons: biological agents, chemical agents, radiological agents, and nuclear weapons.

Biological Agents. These weapons are "living organisms . . . or infective material derived from them, which are intended to cause disease or death in man, animals, and plants, and which depend on their ability to multiply in the person, animal, or plant attacked."[15] Once biological components are obtained, the problem of converting them into weapons can be difficult.[16] However, experts generally agree that the most likely biological agents terrorists might use would be the following:

♦ *Anthrax.* A disease that afflicts livestock and humans, anthrax can exist as spores or suspended in aerosols. Humans contract anthrax either through cuts in the skin (cutaneous anthrax), through the respiratory system (inhalation anthrax), or by eating contaminated meat. Obtaining lethal quantities of anthrax is difficult, but not impossible.

♦ *Smallpox.* Eradicated in nature, **smallpox** is a virus that is very difficult to obtain because samples exist solely in laboratories, apparently only in the United States and Russia. Its symptoms appear after about 12 days of incubation and include flu-like symptoms and a skin condition that eventually leads to pus-filled lesions. It is highly contagious and can be deadly if it progresses to a hemorrhagic (bleeding) stage known as the black pox.

♦ *Botulinum toxin (botulism).* Also known as **botulism, botulinum toxin** is a rather common form of food poisoning. It is a bacterium rather than a virus or fungus and can be deadly if inhaled or ingested even in small quantities.

♦ *Bubonic plague.* A bacterium that leads to the disease known as the Black Death in medieval Europe, **bubonic plague** is highly infectious, often fatal, and spread by bacteria-infected fleas that infect hosts when bitten.

Chemical Agents. As weapons, these are "chemical substances, whether gaseous, liquid, or solid, . . . used for hostile purposes to cause disease or death in humans, animals, or plants, and . . . depend on direct toxicity for their primary effect."[17] Some chemical agents, such as pesticides, are commercially available. Others can be manufactured using available instruction guides. Because of many plausible threat scenarios, experts believe that chemical weapons in the possession of terrorists pose a more likely possibility than do biological, radiological, or nuclear weapons.[18]

Examples of possible weaponized chemical agents in the arsenals of terrorists could include the following:

- ◆ **Phosgene gas** causes the lungs to fill with water, choking the victim.
- ◆ **Chlorine gas** destroys the cells that line the respiratory tract.
- ◆ **Mustard gas** is actually a mist rather than a gas. It is a blistering agent that blisters skin, eyes, and the nose and can severely damage the lungs if inhaled.
- ◆ **Nerve gases**, such as **sarin**, tabun, and VX, block (or short-circuit) nerve messages in the body. A single drop of a nerve agent, whether inhaled or absorbed through the skin, can shut down the body's neurotransmitters.

Radiological Agents. To become threatening to life or health, these radioactive substances must be "ingested, inhaled, or absorbed through the skin" in sufficient quantities.[19] Non-weapons-grade radiological agents could theoretically be used to build a toxic **dirty bomb** that would use conventional explosives to release a cloud of radioactive contaminants. Absent large quantities of radioactive materials, this type of weapon would likely cause minimal casualties outside of the blast radius of the bomb, but its psychological effect could be quite disruptive.

Nuclear Weapons. These high-explosive military weapons use high-grade plutonium and uranium. Explosions from nuclear bombs devastate the area within their blast zone, irradiate an area outside the blast zone, and are capable of sending dangerous radioactive debris into the atmosphere that falls to earth as toxic **fallout**. Nuclear devices are sophisticated and difficult to manufacture, even for highly motivated governments. Although it is conceivable that terrorists could do the same, the challenge is technically and logistically formidable.[20] Most threat scenarios thus envision that terrorists have acquired tactical nuclear weapons such as artillery shells.

Case: The Suicide Bombers

"Human bombs" have become an accepted method of political violence in a number of conflicts. Although some examples of suicidal behavior by ideological extremists can be cited, most incidents have been committed by ethno-national and religious terrorists. When considering the tactical and symbolic value of suicide attacks, it is instructive to recall the words of the Chinese military philosopher Wu Ch'i: "One man willing to throw away his life is enough to terrorize a thousand."[21]

In some conflicts, suicide bombings have occurred only rarely. For example, the IRA, ETA, and European leftists and rightists did not use suicidal violence. In others, suicide attacks became common. The Tamil Tigers in Sri Lanka, Hezbollah (Islamic Jihad) in Lebanon, several Palestinian groups in Israel, and Al Qaeda are all examples. In other conflicts, suicide operations became a **signature method**, as with Chechen rebels and Iraqi insurgents. The following cases illustrate this behavior:

Religion-Motivated Suicide and the Lebanon Model. Lebanon descended into anarchy for approximately 15 years during the 1970s and 1980s. The fighting was mostly religious, among contending paramilitaries drawn from the Shi'a, Sunni, Druze, and Christian communities.

The group that pioneered suicide bombing as an effective method of terrorist violence in the Middle East was Lebanon's Hezbollah.[22] The group conducted a series of suicide bombings in 1983 through 1985 against Israeli, American, and French interests. The October 1983 attacks against the French and American peacekeeping troops in Beirut were particularly effective—the attackers killed 58 French paratroopers and 241 American Marines, forcing the peacekeepers to withdraw. This tactic continued through the 1990s during Hezbollah's campaign against the Israeli occupation of Lebanon's southern border region. An important aspect of the Lebanese example is that each suicide bomber was later glorified as a martyr. This concept of martyrdom is an important motivation behind the recruitment of young suicide bombers.

Intifada-motivated Suicide in Israel. Israel has experienced more suicide attacks than perhaps any other nation. Readers may recall that the Islamic Resistance Movement, better known as Hamas,[23] is a Palestinian Islamic fundamentalist movement founded in December 1987 when the first Palestinian intifada broke out. Hamas's military wing is the Izzedine al-Qassam Brigade, which first appeared in January 1992.

Hamas made a concerted effort from 1994 to 1996 to establish itself as the preeminent Palestinian liberation organization. It was during this period that it set the precedent—and honed the methodology—for suicide bombings. In 1995 and 1996, its campaign became more deadly as its bombs became increasingly sophisticated. This was the handiwork of an electrical engineer named **Yehiya Ayyash**, the master bomb maker better known as The Engineer.

Hamas had launched the campaign in retaliation for the February 1994 Hebron massacre when Baruch Goldstein killed and wounded scores of Muslim worshippers at the Ibrahim Mosque on the holy site of the Cave of the Patriarchs (for a discussion of Goldstein's attack, see Chapter 6). Hamas recruited human-bomb candidates into its Izzedine al-Qassam Brigade cells, with the specific mission to attack Israeli civilian targets—primarily at commuter transportation sites. The suicide bombers used shrapnel-laden vehicular bombs, satchel charges (bagged bombs), and garment-strapped bombs, attacks that inflicted significant damage on Israel in terms of the number of Israeli casualties. Four Hamas bombers, for example, killed 59 Israelis in 1996.[24]

Beginning in 2001, suicide bombers from sectarian Hamas and the secular PLO-affiliated Al Aqsa Martyr Brigades carried out dozens of attacks against civilian targets, killing scores of people. The targets, often buses, were selected to disrupt everyday life in Israel. This was not the first suicide bombing campaign in Israel, but it was by far the most sustained and lethal campaign. During 2001 to 2004, approximately 125 suicide bombings occurred, many carried out by young women.

Al Qaeda and Martyrdom in the New Era of Terrorism. Operatives of Osama bin Laden's Al Qaeda network have demonstrated a propensity for suicidal violence. Highly destructive attacks are most frequently targeted against American interests, but not unknown against rival factions.

Interestingly, Al Qaeda apparently designed an internal consensus about how to conduct their operations. Members of the network committed to writing what are best described as operational protocols, discovered during searches of Al Qaeda hideouts in the aftermath of the September 11 attacks. Manuals—including a six-volume, 1,000-page CD-ROM version—were found in locations as diverse as Chechnya, the United States, Afghanistan, and Manchester, England. Chapter Perspective 10.1 presents sample guidelines designed for Al Qaeda operatives.

CHAPTER PERSPECTIVE 10.1

The Al Qaeda Terrorist Manual

In May 2000, a document written in Arabic was found during the search of a home of an alleged Al Qaeda member in Manchester, England. The document was a manual, approximately 180 pages long, entitled *Military Studies in the Jihad Against the Tyrants*. It is essentially an operations blueprint for engaging in cell-based terrorist activities in foreign countries. The following passages are excerpts:

Goals and Objectives

The confrontation that we are calling for with the apostate regimes does not know Socratic debates . . . Platonic ideals . . . nor Aristotelian diplomacy. . . . But it knows the dialogue of bullets, the ideals of assassination, bombing and destruction, and the diplomacy of the cannon and machine gun.

Missions Required: The main mission for which the Military Organization is responsible is: the overthrow of the godless regimes and the replacement with an Islamic regime. Other missions consist of the following:

1. Kidnapping enemy personnel, documents, secrets and arms.
2. Assassinating enemy personnel as well as foreign tourists.
3. Freeing the brothers who are captured by the enemy.
4. Blasting and destroying the places of amusement, immorality, and sin; not a vital target.
5. Blasting and destroying the embassies and attacking vital economic centers.
6. Blasting and destroying bridges leading into and out of the cities.

The following security precautions should be taken:

1. All documents of the undercover brother, such as identity cards and passport, should be falsified.
2. The brother who has special work status . . . should have more than one identity card and passport.
3. The photograph of the brother in these documents should be without a beard. . . .

Operations

Cell or cluster methods should be adopted by the Organization. It should be composed of many cells whose members do not know one another . . .

Facsimile and wireless: . . . Duration of transmission should not exceed five minutes in order to prevent the enemy from pinpointing the device location . . .

Measures that should be taken by the undercover member:

1. Not reveal his true name to the Organization's members who are working with him. . . .

> 2. Have a general appearance that does not indicate Islamic orientation (beard, toothpick, book, long shirt, small Koran).
>
> 3. Be careful not to mention the brothers' common expressions or show their behaviors. . . .
>
> 4. Avoid visiting famous Islamic places. . . .
>
> 5. Not park in no-parking zones and not take photographs where it is forbidden.
>
> Important note: Married brothers should observe the following: Not talking with their wives about Jihad work.

Table 10.2 summarizes the scale of violence experienced during the suicide bombing campaign waged by the Palestinians during a short period of the intifada.[25] It is important that the targets were almost exclusively civilians and that the death toll was acceptable from the terrorists' perspective—200 people killed at a cost of 13 human bombs.

Figure 10.1 is an FBI representation of terrorist events carried out in the United States during a 21-year period through the 2001 airplane attacks. It is a good representation of the methods terrorists used during that time.

Terrorist Targets

Terrorists select their targets with the expectation that any moral ambiguities of the deed will be outweighed by the target's propaganda value. Terrorists must calculate that they can manipulate the incident into a positive propaganda context. In many campaigns, the objective has been to disrupt society to the point where the routines of life cannot be managed and the government cannot maintain order. To accomplish this, some terrorist movements have incrementally adapted their methods to new targets.

The following sampling of typical targets indicates that terrorists and extremists must rely on a process of redefining who constitutes an enemy group, thereby turning them into a legitimate target.

The Symbolism of Targets

In light of our previous discussions about terrorist groups, environments, and incidents, one conclusion should now be readily apparent: Terrorists select their targets because of their symbolic and propaganda value. High-profile, sentimental, or otherwise significant targets are chosen with the expectation that the terrorists' constituency will be moved and that the victims' audience will in some way suffer. On occasion, terrorists attempt to demonstrate the weakness of an enemy and terrorize those who trust in that enemy. The following targets are often selected for their anticipated high return in propaganda value.

Embassies and Diplomatic Personnel

The symbolism of embassy attacks and operations against diplomats can be profound. Embassies represent the sovereignty and national interests of nations. Diplomatic personnel

TABLE 10.2 THE INTIFADA SUICIDE BOMBERS

The Palestinian intifada increased in scale and ferocity during 2001 and 2002. Fighting in Gaza and the West Bank became pitched battles between Palestinian guerrillas and the Israeli military. Street fighting broke out in Bethlehem, Nablus, Ramallah, and other ancient cities. At the same time, the Palestinians began extensively using a deadly and unpredictable new weapon—the human bomb.

Having first been used by radical Islamic movements such as Hamas and Palestine Islamic Jihad, suicide bombing became a regular weapon of secular Palestine Liberation Organization fighters. The Al Aqsa Martyr Brigades was a secular martyrdom society linked to the mainstream Fatah organization of the PLO.

Martyr Profile	Activity Profile		
	Date	Target	Fatalities
22-year-old man, Jordanian	June 1, 2001	Tel Aviv discothèque	20 killed
23-year-old man, Hamas activist	August 9, 2001	Jerusalem pizzeria	15 killed, including 7 children
48-year-old man, first known Arab Israeli bomber	September 9, 2001	Train depot	3 killed
21-year-old man, engaged to be married	December 2, 2001	Haifa passenger bus	15 killed
28-year-old woman, first female suicide bomber	January 27, 2002	Jerusalem shopping district	1 killed, an elderly man
21-year-old woman, English student	February 27, 2002	Israeli roadblock	3 hurt
19-year-old man	March 2, 2002	Bar Mitzvah celebration	9 killed
20-year-old man	March 9, 2002	Café near Prime Minister Ariel Sharon's residence	11 killed
20-year-old man	March 20, 2002	Commuter bus	7 killed
23-year-old man, wanted fugitive	March 27, 2002	Passover celebration Seder	21 killed
18-year-old woman, engaged to be married	March 29, 2002	Jerusalem supermarket	2 killed
23-year-old man	March 30, 2002	Tel Aviv restaurant	32 hurt
22-year-old man	March 31, 2002	Haifa restaurant	15 killed

SOURCE: Primarily from Amanda Ripley, "Why Suicide Bombing Is Now All the Rage," *Time,* April 15, 2002. Used with permission.

are universally recognized as official representatives of their home countries, and attacks on embassy buildings or embassy personnel are conceptually the same as direct attacks on the nations they represent. Assaults on embassies also guarantee a large audience.

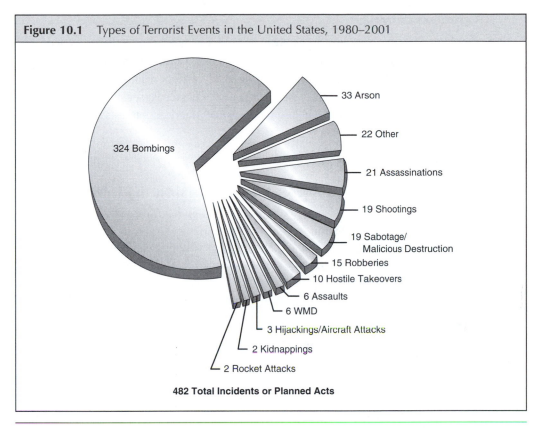

Figure 10.1   Types of Terrorist Events in the United States, 1980–2001

324 Bombings

33 Arson

22 Other

21 Assassinations

19 Shootings

19 Sabotage/
Malicious Destruction

15 Robberies

10 Hostile Takeovers

6 Assaults

6 WMD

3 Hijackings/Aircraft Attacks

2 Kidnappings

2 Rocket Attacks

482 Total Incidents or Planned Acts

SOURCE: United States Department of Justice, Federal Bureau of Investigation.

International Symbols

Many nations deploy military representatives to other countries. They also encourage international investment by private corporations, which consequently set up offices and other facilities. These interests are understandable targets for terrorists because they can be manipulated to depict exploitation, imperialism, or other representations of repression. Thus, terrorists and extremists redefine military facilities, corporate offices, military personnel, and company employees as enemy interests and legitimate targets.

Symbolic Buildings and Sites

Buildings and sentimental sites often represent the prestige and power of a nation or the identity of a people and can evoke strong psychological and emotional reactions from people who revere them. Terrorists and extremists select these cultural symbols because they know that the target audience will be affected. Interestingly, the target audience can be affected even without violence—their perception of these sites as having been desecrated can involve nothing more than a show of strength at a site.

Symbolic People

Terrorists frequently assault individuals because of the symbolic value of their status—security personnel, political leaders, journalists, and business executives are typical targets. Kidnappings and physical violence are common. In hostage situations, videotapes and photographs are sometimes released for propaganda value.

Passenger Carriers

From the terrorists' perspective, passenger carriers are logical targets. If the carrier is big, such as an airliner, it provides a large number of potential victims or hostages who are confined inside a mobile prison. International passenger carriers readily lend themselves to immediate international media and political attention. Chapter Perspective 10.2 applies the foregoing discussion to several symbolic attacks against American interests.

CHAPTER PERSPECTIVE 10.2

The Symbolism of Targets: Terrorist Attacks Against the United States

Many targets are selected because they symbolize the interests of a perceived enemy. This selection process requires that extremists redefine these interests as representations of the forces against whom or which they are waging war. This redefinition process, if properly communicated to the target audience and constituency, can be effectively used as propaganda.

The following attacks were launched against American interests.

Embassies

♦ June 1987. A car bombing and mortar attack were launched against the U.S. embassy in Rome, most likely by the Japanese Red Army.

♦ February 1996. A rocket attack was launched on the American embassy compound in Greece.

International Symbols

♦ April 1988. A USO club in Naples, Italy, was bombed, most likely by the Japanese Red Army. Five people were killed.

♦ November 1995. Seven people were killed when anti-Saudi dissidents bombed an American military training facility in Riyadh, Saudi Arabia.

Symbolic Buildings and Sites

♦ January 1993. Two were killed and three injured when a Pakistani terrorist fired at employees outside the CIA headquarters in Langley, Virginia.

♦ February 1993. The World Trade Center in New York City was bombed, killing six and injuring more than 1,000.

Symbolic People

♦ May 2001. The Filipino Islamic revolutionary movement Abu Sayyaf took three American citizens hostage. One of them was beheaded in June 2001.

♦ January 2002. An American journalist working for the *Wall Street Journal* was kidnapped in Pakistan by Islamic extremists. His murder was later videotaped.

Passenger Carrier Attacks

♦ August 1982. A bomb exploded aboard Pan Am Flight 830 over Hawaii. The Palestinian group 15 May committed the attack. The plane was able to land.

♦ April 1986. A bomb exploded aboard TWA Flight 840. Four were killed and nine injured, including a mother and her infant daughter who fell to their deaths when they were sucked out of the plane. Flight 840 landed safely.

The Effectiveness of Terrorist Violence

Terrorism is arguably effective—however defined—in some manner to someone.[26] The key for terrorists is to establish a link between methods used and desirable outcomes. Of course, success and effectiveness are subjective considerations. In this regard, terrorists tend to use unconventional factors as measures. For example, terrorists have been known to declare victory using the following criteria:

♦ Acquiring global media and political attention
♦ Having an impact on a target audience or championed constituency
♦ Forcing concessions from an enemy interest
♦ Disrupting the normal routines of a society
♦ Provoking the state to overreact

The following review of these criteria is not an exhaustive evaluation of measures of effectiveness, but it demonstrates commonalities among today's terrorists.

Media and Political Attention

At times, focusing world attention on the terrorists' cause is itself a measure of success. One central fact in this age of instantaneous media attention is that

> for the terrorist, success . . . is most often measured in terms of the amount of publicity and attention received. Newsprint and airtime are thus the coin of the realm in the terrorists' mindset: the only tangible or empirical means they have by which to gauge their success and assess their progress. In this respect, little distinction or discrimination is made between good or bad publicity.[27]

Having an Impact on an Audience

Terrorists use propaganda by the deed to affect audiences, hoping to rouse them to action or incite a society-level response. Victim audiences, neutral audiences, and championed groups can all be affected by a terrorist event. When an incident occurs, extremists and their supporters assess reactions from these audiences. From the terrorists' perspective, an attack is effective only if the audience reacts as the terrorists planned.

Chapter Perspective 10.3 explores a hostage-taker tactic of videotaping victims and distributing the images in the media and on the Internet.

CHAPTER PERSPECTIVE 10.3

Tactical Horror: Digital, Video, and Audio Terrorism

With the advent of the Internet and cable news networks, terrorists now have unprecedented access to global audiences. Communications technologies quickly and inexpensively bring symbolic images and extremist messages to the attention of policy makers and civilians around the world. Terrorists have adapted their tactics to these new technologies, and many use them to broadcast their messages and operations.

(Continued)

Early in the millennium, hostage-takers discovered that the plight of their victims would garner intensive global attention as long as their images were distributed to noteworthy cable news networks.

The typical pattern was for an international figure—usually a foreign worker—to be seized and a communiqué claiming credit for the abduction publicized. A video or series of videos would be delivered to a news outlet, with images of the victim pleading for his or her life while seated before a flag and surrounded by hooded and armed terrorists. The outcome was sometimes satisfactory, with the hostage being granted freedom, and other times horrific.

The first noted incident was the kidnapping and videotaped murder of American journalist Daniel Pearl in Pakistan in January 2002. Since then, insurgents in Iraq, terrorists in Saudi Arabia, and violent jihadists elsewhere have either issued Internet and cable news communiqués, or videotaped their hostages, or executed them, or committed all of these actions. In particular, a gruesome cycle of beheadings occurred, as illustrated by the following data from Iraq:

♦ Victims represent the international community, and have included citizens from Bulgaria, Pakistan, South Korea, Nepal, Norway, the United States, Great Britain, Turkey, and Iraq.
♦ Al Qaeda in Iraq and other Islamist or other sectarian movements appeared to be responsible for most of the kidnappings and murders.
♦ A number of hostages were beheaded, sometimes on video recordings later posted on the Internet.
♦ Other videotaped incidents have also been promulgated, such as attacks by insurgents and terrorists in Afghanistan and Iraq.

Forcing Concessions From an Enemy Interest

Enemy interests—broadly defined—will sometimes concede to the demands of a politically violent movement. Concessions vary in magnitude. They can be made as short-term and immediate concessions or as long-term and fundamental, whereby an entire society essentially concedes to a movement. At the immediate level, accommodations might include ransoms paid for the release of hostages. At the societal level, laws might be changed or autonomy granted to a national group. One repeated method used by terrorists to force concessions is kidnapping/hostage-taking.

Disruption of Normal Routines

An obvious measure of effectiveness is whether the normal routines of society can be affected or halted by a terrorist incident or campaign. Some targets—such as the commercial transportation industry—can be selectively attacked to the point where their operations will be disrupted. When this happens, the daily habits of individuals and routines of society will change. In this way, large numbers of people in the broader society in essence respond to the tactics of a relatively weak movement.

Photo 10.2 Daniel Pearl. An American journalist in Pakistan, Pearl was kidnapped and later murdered by Pakistani Islamic terrorists, but not before they disseminated photographs and videotapes to the international media.

Provoking the State to Overreact

One outcome that terrorists allude to as a measure of effectiveness is the state's imposition of violent security countermeasures in response to a terrorist environment. This notion of "enraging the beast" is common across the spectrum of terrorist environments and has been discussed in previous chapters. Terrorists, of course, anticipate that the state will become violently repressive, the people will suffer, and the masses will rise up in rebellion after experiencing the true nature of the enemy. This theory has had only mixed success.

Table 10.3 summarizes measures of effectiveness, the cited cases, and outcomes.

TABLE 10.3 MEASURES OF EFFECTIVENESS		
When extremist movements adopt terrorism as a methodology, they measure its effectiveness of by linking the incident to identifiable outcomes. These measures of effectiveness are unconventional in the sense that they are frequently media oriented and audience oriented.		
Measure	**Activity Profile**	
	Incident	**Outcome**
Media and political attention	Hijacking of TWA Flight 847	Global media and political attention
Impact on an audience	PAGAD's moralist terrorist campaign	Failed campaign to bring about a societal response
Concessions from an enemy interest	Kidnapping of U.S. ambassador to Brazil	Broadcast of terrorists' political manifesto
Disruption of societal routines	Suicidal hijackings of four airliners on September 11, 2001	Fewer Americans traveled by airline; industry suffered revenue losses
Provoke the state	Viable: Provocations by IRA and PLO	Government methods failed to eradicate opposition
	Unviable: Provocations by **Montoneros** and Tupamaros	Violent military governments crushed the opposition

Chapter Summary

This chapter discussed terrorist objectives, methods, targets, and the effectiveness of terrorism.

Typical objectives included the terrorists' desire to change the existing order, to promote the psychological and social disruption of a society, to publicize their cause through propaganda by the deed, and to create a generalized revolutionary environment. To accomplish their objectives, terrorists have traditionally directed their attention to the manipulation of specific audiences. In New Terrorism, objectives have become characterized by vagueness, and methods have included indiscriminate attacks and the possibility of the use of weapons of mass destruction.

Contemporary terrorist methods reflect the changing global political environment and are characterized by asymmetrical warfare and new, cell-based organizational models. However,

Photo 10.3   The rubble of the King David Hotel in Jerusalem. The hotel, which housed British administrative offices for their Palestine mandate, was bombed by the Jewish terrorist group the Irgun.

most terrorists rely on age-old methods that can be accomplished using conventional weapons such as firearms and explosives. Modern technologies such as rocket-propelled grenades, precision-guided munitions, and barometric bombs are updated variations on the same theme. Nevertheless, threats from biological, chemical, radiological, and nuclear weapons are unprecedented in the possible arsenals of terrorists.

Terrorist targets are selected because of their symbolic value and the impact they will have on affected audiences. Typical targets include embassies, international symbols, symbolic buildings and sites, symbolic people, and passenger carriers. These targets are selected because they represent the interests of a defined enemy.

The effectiveness of terrorist attacks is measured by unconventional criteria. From the terrorists' perspective, these include gaining media and political attention, affecting targeted audiences, gaining concessions from an enemy interest, disrupting normal routines, and provoking the state to overreact.

Photo 10.4   Peacekeeping in Beirut. Two U.S. Marines survey the rubble of the Marine Corps barracks, which had been destroyed by a Lebanese suicide bomber. More than 200 of their comrades died in the attack.

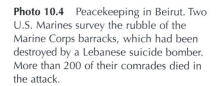

DISCUSSION BOX

This chapter's Discussion Box is intended to stimulate critical debate about how determined terrorist attacks can affect the policies of nations.

Attacks Against the U.S. Marine and French Paratrooper Headquarters in Beirut

In September 1982, 5,000 elite French paratroopers, Italian Bersaglieri, and American Marines were sent into Beirut, Lebanon, as members of the peace-keeping

Multinational Force (MNF). The purpose of the MNF was to restore order to the city in the midst of a civil war and an Israeli invasion that had been launched to drive the Palestine Liberation Organization from Lebanon.

Members of radical Lebanese Islamic militia movements, specifically the Sunni Amal and Shi'ite Hezbollah groups, viewed the MNF as an invasion force. From their perspective, the West supported the Lebanese Christian Phalangists and the Israelis. The Lebanese groups at first waged low-intensity resistance against the Western presence. The situation then gradually escalated, with Amal and Hezbollah fighters becoming more aggressive in their opposition. In response to casualties incurred by Marines and French paratroopers, the United States began shelling Syrian-controlled positions from naval vessels.

On October 23, 1983, two suicide bombers driving vehicular bombs simultaneously struck the headquarters of the U.S. Marines and French paratroopers in Beirut; 241 Marines and 58 paratroopers were killed. The terrorist group Islamic Jihad—probably the Lebanese Shi'ite movement Hezbollah—claimed credit for the attacks. The bombings were hailed by Amal and Hezbollah leaderships, who were careful to deny any responsibility for the attacks, as legitimate resistance by patriots against occupying armies.

After the attacks, the United States began using air power and naval artillery to shell hostile positions. However, public opinion had turned against the increasingly complicated "peace-keeping" mission, and MNF troops were withdrawn in early 1984.

Discussion Questions

1. Were the Lebanese militia fighters terrorists or freedom fighters?
2. Is terrorism poor man's warfare and therefore a legitimate option for waging war?
3. Were the suicide bombings acceptable methods for opposing the deployment of the MNF?
4. Was the presence of Western soldiers indeed an understandable precipitating cause of Amal's and Hezbollah's resistance?
5. Were the targets—the French and American headquarters—logical targets for relatively weak opposition forces?

Key Terms and Concepts

The following topics were discussed in this chapter and can be found in the glossary:

AK-47	chlorine gas	Kalashnikov, Mikhail
anthrax	Composite-4 (C-4)	kidnapping/hostage-taking
assault rifle	dirty bomb	M-16
asymmetrical warfare	dynamite	mine
Ayyash, Yehiya "The Engineer"	electronic trigger	Molotov cocktail
biological agent	gasoline bomb	mustard gas
botulinum toxin (botulism)	Grail	nerve gas
bubonic plague	improvised explosive	netwar
chemical agent	device (IED)	nuclear weapon

phosgene gas	roadside bomb	signature method
plastic explosive	rocket-propelled	smallpox
precision-guided	grenade	Stinger
munitions	RPG-7	submachine gun
pressure trigger	SA-7	suicide bombing
radiological agent	sarin	TNT
RDX	Semtex	vehicular bomb

Terrorism on the Web

Log on to the Web-based student study site at **www.sagepub.com/martinessstudy** for additional web sources and study resources.

Web Exercise

Using this chapter's recommended Web sites, conduct an online investigation of terrorist objectives, methods, and targets.

1. What common patterns of behavior and methods can you identify across regions and movements?
2. Conduct a search for other Web sites that offer advice for organizing terrorist cells and carrying out terrorist attacks. Do you think that the online terrorist manuals and weapons advice are a danger to global society?
3. Compare the Web sites for the monitoring organizations. How would you describe the quality of their information? Are they providing a useful service?

For an online search of terrorist tactics and targets, readers should activate the search engine on their Web browser and enter the following keywords:

"Terrorist Weapons"
"Terrorist Manuals"

Recommended Readings

The following publications provide discussions on terrorist objectives and methods.

Bergen, Peter L. *Holy War, Inc.: Inside the Secret World of Osama bin Laden.* New York: Simon & Schuster, 2001.

Katz, Samuel M. *The Hunt for the Engineer: How Israeli Agents Tracked the Hamas Master Bomber.* New York: Fromm International, 2001.

Powell, William. *The Anarchist Cookbook.* New York: Lyle Stuart, 1971; assigned to Barricade Books, 1989.

Stern, Jessica. *The Ultimate Terrorists.* Cambridge, MA: Harvard University Press, 1999.

Tucker, Jonathan B., ed. *Toxic Terror: Assessing Terrorist Use of Chemical and Biological Weapons.* Cambridge, MA: MIT Press, 2000.

11 Counterterrorism and the War on Terrorism

This chapter reviews policy options for responding to acts of political violence. **Counterterrorism**, as it has come to be known, refers to proactive policies that specifically seek to eliminate terrorist environments and groups. Regardless of which policy is selected, the ultimate goal is clear: to save lives by preventing or decreasing the number of terrorist attacks. There is consensus among policy makers that several basic counterterrorist options and suboptions are available. For the purposes of our discussion, options and suboptions will be classified as follows:

Use of Force

This policy classification allows policymakers to use the force of arms against terrorists and their supporters. The following are examples of military and **paramilitary repressive options**:

- **Suppression campaigns** are military strikes against targets affiliated with terrorists.
- **Covert operations (coercive)** are secretive operations that include assassinations, sabotage, kidnapping, **extraordinary renditions**, and other quasi-legal methods.

Operations Other Than War

Repressive Options

Repressive responses include the following examples of nonmilitary repressive options:

- **Covert operations (nonviolent)** are secretive operations that include a number of possible counterterrorist measures, such as infiltration, **disinformation**, and **cyberwar**.
- **Intelligence** refers to collecting data about terrorist movements and to predict their behavior.
- **Enhanced security** refers to the hardening of targets to deter or prevent terrorist attacks.
- **Economic sanctions** are used to punish or disrupt state sponsors of terrorism.

Conciliatory Options

Conciliatory responses allow policy makers to develop a range of options that do not involve force or other repressive methods, such as the following:

- **Diplomacy** refers to different degrees of capitulation to the terrorists, which refers to engaging with the terrorists to negotiate an acceptable resolution to a conflict. sweeping conditions that may completely resolve the conflict.
- **Social reform** is an attempt to address the grievances of the terrorists and the group they champion.
- **Concessionary options.** Concessions can be incident specific, in which immediate demands are met, or generalized, in which broad demands are accommodated.

Legalistic Options

The overall objective of **legalistic responses** is to promote the rule of law and regular legal proceedings. The following are examples of these responses:

- **Law enforcement** refers to the use of law enforcement agencies and criminal investigative techniques in the prosecution of suspected terrorists.
- Counterterrorist laws attempt to criminalize terrorist behavior.
- International law relies on cooperation among states who are parties to international agreements.

Table 11.1 summarizes activity profiles for these counterterrorist policy categories.

TABLE 11.1 COUNTERTERRORIST OPTIONS: GENERAL POLICY CLASSIFICATIONS

Experts have developed categories for counterterrorist options. These policy classifications have been given a variety of labels, but they refer to similar concepts. Each of the categories has been divided into subcategories.

Counterterrorist Policy Classification	Activity Profile		
	Rationale	**Practical Objectives**	**Typical Resources Used**
Use of force	Symbolic strength	Punish or destroy the terrorists	◆ Military assets ◆ Paramilitary assets ◆ Covert operatives
Operations other than war: repressive options	◆ Deterrence ◆ Prediction ◆ Destabilization	◆ Disruption of the terrorists ◆ Intelligence ◆ Coercion of supporters	◆ Technologies ◆ Intelligence operatives ◆ Covert operatives
Operations other than war: conciliatory options	Resolve underlying problems	◆ End immediate crises ◆ Forestall future crises	◆ Social resources ◆ Economic assets ◆ Negotiators
Legalistic options	Rule of law	◆ International cooperation ◆ Prosecution, conviction, and incarceration of terrorists	◆ International organizations ◆ Law enforcement agencies ◆ Domestic legal establishments

The discussion in this chapter will review the following responses to terrorism:

◆ The use of force
◆ Operations other than war
◆ Legalistic responses

The Use of Force

The use of force is a hard-line policy approach that states and their proxies use to violently suppress terrorist environments. The goals of this approach are case specific, so that the decision to use violent suppression is based on policy calculations peculiar to each terrorist environment. When states decide to use force against terrorists and their supporters, coercion and violence are considered to be justifiable and desirable policy options.

This policy option requires the deployment of military or paramilitary assets to punish, destabilize, or destroy terrorists and their supporters. Military assets are defined as the recognized and official armed forces of a nation. Paramilitary assets are defined as irregular armed units or individuals supported or organized by regimes.

States have waged military and paramilitary counterterrorist operations domestically and internationally. Domestic operations involve coercive use of military, police, and other security forces against domestic threats. International operations involve overt or covert deployment of security assets abroad. These deployments can include ground, air, or naval forces in large or very small operational configurations.

The following discussion explores several use-of-force options that nations commonly adopt.

Maximum Use of Force: Suppression Campaigns

Military and paramilitary forces can undertake counterterrorist campaigns, that is, long-term operations against terrorist cadres, their bases, and support apparatuses. Suppression campaigns are uniquely adapted to the conditions of each terrorist environment. They are launched within the policy contexts of war or quasi-war and are generally waged with the goal of defeating rather than simply suppressing the terrorist movement.

For example, in 2002 and 2003, Israel began a concerted effort to disrupt Hamas and destroy its capability to indefinitely sustain its trademark suicide bombing campaign. Israeli policy included assassinations, military incursions, and a series of raids that resulted in arrests and gunfights with Hamas operatives, as well as many deaths. By 2004, the Israeli government credited these operations and other measures with weakening Hamas's infrastructure and reducing the incidence of Hamas bombings.

Military Suppression Campaigns

Nations sometimes resort to using conventional units and special operations forces to wage war against terrorist movements. The goal is to destroy their ability to use terrorism to attack the nation's interests. Military campaigns require that troops be deployed to bases in friendly countries or (in the case of nations with significant seaborne capability)[1] offshore aboard naval vessels. The following case illustrates the nature of military suppression campaigns:

Case: Operation Enduring Freedom. The United States declared itself at war against global terrorism after the attacks of September 11, 2001. It was joined by a number of allies who agreed to commit their armed forces and domestic security services to the new war. Dubbed **Operation**

Enduring Freedom, the operation began with the October 2001 invasion of Afghanistan and was defined from the beginning as a long-term suppression campaign. It was also made clear from the outset that the war against terrorism would require extended deployments of U.S. and allied troops around the world, as well as an intensive use of special operations forces, commandos, Marines, and other elite units. The immediate objectives were to destroy Al Qaeda's safe havens in Taliban-controlled Afghanistan, collect intelligence, disrupt the terrorist network around the world, and capture or kill as many cadres as possible. The long-term goal of the campaign—which policy makers stated would take years to achieve—was to degrade or destroy the operational capabilities of international terrorists. Interestingly, Operation Enduring Freedom was allegedly first called **Operation Infinite Justice**, but renamed after a negative reaction in the Muslim world (the Muslim stance being that only God can provide infinite justice).[2]

Paramilitary Suppression Campaigns

In some suppression campaigns, military and paramilitary units actively coordinate their operations. A typical government policy has been to arm and support paramilitaries in areas where the military does not have a strong presence. The following example illustrates the nature of such campaigns:

Case: Paramilitary Suppression in Algeria. During the 1990s, Algeria was beset by intensive terrorist violence from jihadis belonging to several radical Islamic groups. The Algerian government was unable to suppress the rebels and was accused of committing human rights violations. Many thousands of civilians were killed, mostly at the hands of the jihadis. During the suppression campaign, the government organized and armed paramilitary home guard units.[3] These units, which were used as local self-defense forces, effectively staved off attacks from the rebels, thus reducing their operational effectiveness in the countryside. In part because of the paramilitary policy, as well as government-initiated programs and brutal suppression, the rebellion eventually ended in a negotiated amnesty settlement in 1999.

Punitive and Preemptive Strikes

Short-duration military and paramilitary operations are usually conducted for a specific purpose under specific rules of engagement. These actions are designed to send a clear symbolic message to terrorists. The following two types have been used:

Punitive strikes are launched as reprisals against terrorists for incidents that have already taken place. Successful punitive strikes require the attacker to symbolically and politically link the attacks to the terrorist incident.

Preemptive strikes are undertaken to hurt terrorists before a terrorist incident. Preemptive operations are launched as a precautionary measure to degrade the terrorists' ability to launch future attacks. Symbolic and political linkage between the attacks and a real threat is often difficult.

Punitive strikes are more common than preemptive strikes. These attacks can be justified to some extent, as long as links can be established between the attacks and a terrorist incident. Linkage must also be made between the specific targets of the reprisal and the alleged perpetrators of the incident. This latter consideration sometimes goes awry. Preemptive strikes are less frequent than punitive strikes, partly because preemption by definition means that the attack is a unilateral action conducted in response to a perceived threat. Such attacks are therefore not as easily justified as punitive operations, which is an important consideration for regimes concerned about world opinion. Unless a threat can be clearly defined, it is unlikely that preemptive actions will receive widespread support. Nevertheless, some nations have adopted preemptive operations as a regular counterterrorist method. Israel has for some time preemptively attacked neighboring countries that harbor anti-Israeli terrorists and dissidents. Chapter Perspective 11.1 presents an example of a large punitive operation that the United States launched against Libya in 1986.

CHAPTER PERSPECTIVE 11.1

Operation El Dorado Canyon

On April 14, 1986, the United States bombed targets in Libya using Air Force bombers based in Great Britain and Navy carrier aircraft based in the Mediterranean. The attacks occurred at the height of tensions between the United States and Libya, which were precipitated by a radically activist Libyan government and its partnership with the Soviet Union.

Background

During the 1980s, Libya established strong ties with the Communist Eastern Bloc (Soviet-dominated Eastern Europe). It also declared itself a champion of people oppressed by the West and Israel. Libyan leader Muammar el-Qaddafi permitted the Libyan government to build camps to train terrorists from around the world, to stockpile weapons far in excess of Libya's needs, to acquire large amounts of plastic explosives (including powerful Czech-made Semtex), and to directly engage in state-sponsored international terrorism.

In March 1986, the U.S. Mediterranean fleet sailed into a disputed exclusionary zone off the coast of Libya. Qaddafi responded to this show of force by declaring that what he called a line of death had been drawn in the disputed waters—which the U.S. Sixth Fleet then purposefully crossed. Two terrorist bombings blamed on Libya subsequently occurred in Europe. The first bomb killed four Americans aboard a TWA airliner in Greece, and the second killed one U.S. serviceman at La Belle Discotheque in West Berlin. The American punitive raids—dubbed Operation El Dorado Canyon—were ostensibly in retaliation for these bombings, but they probably would have occurred in any event.

Aftermath

More than 100 Libyans were killed in Benghazi and Tripoli, including Qaddafi's young adopted daughter. American policy makers considered the attacks successful because Libya thereafter temporarily scaled back its international adventurism and Qaddafi reduced his inflammatory rhetoric. Libya then entered a second, shorter period of international activism and then pulled back—this time for an extended period.

It should be noted that the strike force based in Britain was forced to detour around France and the Iberian Peninsula (Spain and Portugal) and enter the Mediterranean through the Straits of Gibraltar, which separate Europe and Africa. This was necessary because the French and Spanish governments refused to permit the bombers to fly over their airspace. The American air strike was unpopular in Europe. After the attack, demonstrations occurred in several countries, and many Europeans expressed outrage. In Beirut, terrorists murdered one American and two British hostages in retaliation for the attack on Libya.

War in the Shadows, Part 1: Coercive Covert Operations

Coercive covert operations seek to destabilize, degrade, and destroy terrorist groups. Targets include individual terrorists, terrorist networks, and support apparatuses. Although covert assets have been developed by nations to wage so-called shadow wars using special operatives, conventional forces have also been used to surreptitiously resolve terrorist crises.

Case: Assassinations

The underlying rationale for using assassination as a counterterrorist option is uncomplicated: The terrorism will end or diminish if the terrorists and their supporters are eliminated. The argument is that those targeted for assassination will, as a threat, be permanently removed. It is debatable whether this assumption is accurate because little evidence supports the supposition that assassinations have ever had an appreciable deterrent effect on determined terrorists—new recruits continue to enlist in movements worldwide. As discussed in previous chapters, symbolic martyrdom is actively manipulated by terrorist movements to rally their followers and recruit new fighters. Israel has used assassinations repeatedly in its war against terrorism, primarily targeting Palestinian nationalists. For example, a covert Israeli unit known as **Wrath of God** (Mivtzan Elohim) was responsible for tracking and assassinating Black September terrorists after the 1972 Munich Olympics massacre. At least 20 Palestinians were assassinated in Europe and the Middle East. The Wrath of God program went awry in 1973 when Israeli agents mistakenly shot and killed a North African waiter in Norway while his wife looked on; they had mistaken him for a Black September operative.

Surgical Use of Force: Special Operations Forces

Special operations forces are defined here as highly trained military and police units that specialize in unconventional operations. These units are usually not organized the same way as conventional forces because their missions require them to operate quickly and covertly in hostile environments. Operations are frequently conducted by small teams of operatives, though fairly large units can be deployed if circumstances require. Depending on their mission, special operations forces are trained for long-range reconnaissance, surveillance, surgical punitive raids, hostage rescues, abductions, and liaisons with allied counterterrorist forces. Their training and organizational configurations are ideally suited to counterterrorist operations.

Special Operations Forces: Military Units

Special operations forces today have become fully integrated into the operational commands of national armed forces worldwide. Their value in counterterrorist operations has been proven many times in the postwar era. The following examples illustrate the mission and configuration of special operations forces commands.

United Kingdom. The **Special Air Service (SAS)** is a secretive organization in the British army that has been used repeatedly in counterterrorist operations. Organized at a regimental level, but operating in very small teams, the SAS is similar to the French 1st Navy Parachute Infantry Regiment and the American Delta Force (discussed later in the chapter). The SAS has been deployed repeatedly to assignments in Northern Ireland and abroad, as well as to resolve domestic terrorist crises such as rescuing hostages in May 1980 inside the Iranian embassy in London. The **Special Boat Service (SBS)** is a special unit under the command of the Royal Navy. It specializes in operations against seaborne targets and along coastlines and harbors. The SBS is similar to the French Navy's special assault units and the American SEALs (discussed later in the chapter). The **Royal Marine Commandos** are rapid-reaction troops that deploy in larger numbers than the SAS and SBS. They are organized around units called commandos, which are roughly equivalent to a conventional battalion.

France. The 1st Navy Parachute Infantry Regiment (1RPIMa) is similar to the British SAS and American Delta Force. Within the 1RPIMa, small RAPAS (intelligence and special operations) squads are trained to operate in desert, urban, and tropical environments. Along with the special police unit GIGN (discussed later in the chapter), they form the core of French counterterrorist special operations forces. 1RPIMa has been deployed to crises around the world, particularly in Africa. Five French Navy Special Assault Units have been trained for operations against seaborne targets, coastlines, and harbors. Their mission is similar to those of the British SBS and American SEALs. When large elite combat forces must be deployed, the French use their all-volunteer 11th Parachute Division (Paras) and the commando or parachute units of the famous French Foreign Legion.[4]

Israel. The Sayaret are reconnaissance units that were organized early in the history of the Israel Defense Force (IDF). Several Sayaret exist within the IDF, the most noted of which is Sayaret Matkal, a highly secretive formation that is attached to the IDF General Headquarters. The Parachute Sayaret has been deployed in small and large units, often using high mobility to penetrate deep into hostile territory. They participated in the Entebbe operation and were used extensively in Lebanon against Hezbollah. The IDF has also deployed undercover agents against suspected terrorist cells. When large elite combat forces must be deployed, the Golani Brigade and its Sayaret are frequently used. The Golani Brigade was used extensively against Hezbollah in South Lebanon.

United States. U.S. Special Operations Forces are organized under the U.S. Special Operations Command. The Delta Force (1st Special Forces Operational Detachment—Delta) is a secretive unit that operates in small teams. Similar in mission to the British SAS and French 1RPIMa, Delta Force operates covertly outside of the United States. Its missions probably include abductions, reconnaissance, and punitive operations. Green Berets (Special Forces Groups) usually operate in units called A Teams. These teams comprise specialists whose skills include languages, intelligence, medicine, and demolitions. The traditional mission of the A Team is force multiplication; that is, they are inserted into regions to provide military training to local personnel, thus multiplying their operational strength. They also participate in reconnaissance and punitive raids. U.S. Navy Sea Air Land Forces (SEALs) are similar to the British SBS and the French Navy's Special Assault Units. Their primary mission is to conduct seaborne, riverine, and harbor operations, though they have also been used extensively on land. When large elite combat units must be deployed, the United States relies on the army's 75th Ranger Regiment and units from the U.S. Marine Corps. The Marines have formed their own elite units, which include force reconnaissance and long-range reconnaissance units (referred to as Recon). These units are organized into teams, platoons, companies, and battalions.

Photo 11.1 French commandos at Orly Airport outside Paris. The group had just secured the release of hostages held by terrorists in January 1975.

Special Operations Forces: Police Units

Many nations have special units within their police forces that participate in counterterrorist operations. In several examples, these units have been used in a semimilitary role.

France. The **GIGN (Groupe d'Intervention Gendarmerie Nationale)** is recruited from the French gendarmerie, the military police. GIGN is a counterterrorist unit with international operational duties. In an operation that foiled what was arguably a precursor to the September 11 attacks, the GIGN rescued 173 hostages from an Air France Airbus in December 1994. Four Algerian terrorists had landed the aircraft in Marseilles, intending to fly to Paris to crash or blow up the plane over the city.

Germany. **GSG-9 (Grenzschutzgruppe 9)** was organized after the disastrous failed attempt to rescue Israeli hostages taken by Black September at the 1972 Munich Olympics. It is a paramilitary unit that has been used domestically and internationally as a counterterrorist and hostage rescue unit. GSG-9 first won international attention in 1977 when the group freed hostages held by Palestinian terrorists in Mogadishu, Somalia.

Israel. The **Police Border Guards** are an elite force that is frequently deployed as a counterterrorist force. Known as the **Green Police**, it operates in two subgroups. **YAMAS** is a covert group that has been used extensively during the Palestinian intifada to neutralize terrorist cells in conjunction with covert IDF operatives. **YAMAM** was specifically created to engage in counterterrorist and hostage rescue operations.

United States. At the national level, the United States has organized several units that have counterterrorist capabilities, all paramilitary groups that operate under the administrative supervision of federal agencies and perform traditional law enforcement work. Perhaps the best-known is the FBI's **Hostage Rescue Team**. Not as well known, but very important, is the Department of Energy's **Emergency Search Team**. Paramilitary capabilities have also been incorporated into the Treasury Department's Bureau of Alcohol, Tobacco and Firearms and Secret Service. At the local level, American police forces also deploy units that have

Photo 11.2 U.S. Army troops search a building for insurgents in the Iraqi city of Fallujah during heavy fighting.

counterterrorist capabilities. These units are known by many names, but the most commonly known designation is Special Weapons and Tactics (SWAT).

Both military and police special units have been deployed to resolve hostage crises. Chapter Perspective 11.2 summarizes several famous hostage incidents in which special operations forces were used—usually with success, but not always.

Table 11.2 summarizes the activity profile of counterterrorist options that sanction the use of force.

CHAPTER PERSPECTIVE 11.2

Hostage Rescues

When hostage rescue operations succeed, they seem to be almost miraculous victories against terrorism. In a number of hostage rescue operations, well-trained elite units have dramatically resolved terrorist crises. In other cases, however, elite units have failed because of poor planning or overly complicated scenarios. When they have failed, the consequences have been disastrous.

The following cases illustrate the inherently risky nature of hostage rescue operations.

The Munich Olympics

In September 1972, members of the Black September terrorist organization captured nine Israeli Athletes and killed two others at the Olympic Village during the Munich Olympics. They demanded the release of Palestinian and Red Army Faction prisoners, as well as one Japanese Red Army member. West German officials permitted the terrorists to transport their hostages to an airport, using the ruse that they would be flown out of the country. In reality, the plan called for five Bavarian police snipers to shoot the terrorists when they were in the open. At the airport, five terrorists, one police officer, and all of the hostages were killed in a firefight with the Bavarian police. Three terrorists were captured and imprisoned in West Germany, but were later released and flown to Libya. Israel's Wrath of God program later hunted down and assassinated two of them.

(Continued)

Entebbe

In June 1977, an Air France Airbus was hijacked in Athens while en route from Tel Aviv to Paris. Seven terrorists—two West Germans and five PFLP members—forced the plane to fly to Entebbe in Uganda. There were 248 passengers, but 142 were released. The remaining 106 were Israelis and Jews, and were kept as hostages. Israeli troops, doctors, and nurses flew 2,620 miles to Entebbe, attacked the airport, killed at least seven of the terrorists and 20 Ugandan soldiers, and rescued the hostages. Three hostages and one Israeli commando died. One British-Israeli woman who had become ill was moved to a Ugandan hospital, where she was murdered by Ugandan personnel after the rescue.

Mogadishu

In October 1977, a Lufthansa Boeing 737 was hijacked in mid-air while en route from Mallorca, Spain, to Frankfurt. The hijackers took the aircraft on an odyssey to Rome, the Persian Gulf, the Arabian Peninsula, and East Africa. The four Palestinian terrorists demanded a ransom[a] and the release of imprisoned Palestinian and West German terrorists. The plane eventually landed in Mogadishu, Somalia. Throughout its odyssey, the aircraft had been shadowed by a plane bearing elite West German GSG-9 police commandos. In Mogadishu, the unit landed, attacked the hijacked aircraft, killed three of the terrorists, and rescued the hostages. None of the hostages or rescuers was killed, and only one person was injured. As a postscript, several imprisoned West German terrorists committed suicide when they learned of the rescue—suicides that came to be known as **Death Night**.

Force 777

In 1977, Egypt organized **Force 777** as a small elite counterterrorist special operations unit. Soon after its creation, Force 777 was twice used to assault aircraft hijacked to other countries by Palestinian terrorists. In the first incident, the PFLP landed a hijacked Cyprus Airways airliner in Nicosia, Cyprus. The Egyptian government dispatched Force 777 but neglected to inform Cypriot authorities. When the Egyptians landed and rushed the hijacked airliner, Cypriot police and other security personnel opened fire, thinking that the commandos were reinforcements for the terrorists. During an 80-minute firefight, more than 15 Egyptians and Cypriots were killed. In the second incident, Abu Nidal Organization terrorists landed Egyptair Flight 648 in Malta in retaliation for the Egyptian government's failure to protect the *Achille Lauro* terrorists.[b] The Egyptian government again dispatched Force 777, this time in larger numbers and after notifying Maltese officials. Unfortunately, the assault plan called for explosive charges to blow a hole in the aircraft's roof so that Force 777 commandos could jump into the cabin. The charges were much too strong, and the explosion immediately killed approximately 20 passengers. During the ensuing 6-hour firefight, a total of 57 passengers were killed. Reports alleged that Egyptian snipers shot at passengers as they fled the aircraft.

Operation Eagle Claw

During the consolidation of the Iranian Revolution in November 1979, Iranian radicals seized the American embassy in Tehran. Some hostages were soon released, but more than 50 were held in captivity. In April 1980, the United States launched a rescue attempt led by the Delta Force that included units from all branches of the military. The plan—known as **Operation Eagle Claw**—was to establish a base in the Iranian desert, fly commando teams into Tehran, assault the embassy compound, ferry the hostages to a soccer field, have helicopters land to pick them up, shuttle them to an airfield secured by Army Rangers, and then fly everyone to safety. Gunships and other aircraft would provide air cover during the operation. The operation did not progress beyond its first phase. As helicopters approached the desert site—dubbed Desert One—they flew into a massive dust storm. Because they were under orders to not exceed 200 feet, they tried to fly through the storm; two helicopters were forced out of the operation, as was a third that later malfunctioned. On the ground, a helicopter drove into one of the airplanes, and both exploded. Eight soldiers were killed, and the mission was cancelled.

Note

a. The ransom demand was directed to the son of industrialist Hanns-Martin Schleyer, who was in the hands of West Germany's Red Army Faction.
b. During the *Achille Lauro* crisis, Flight 648 was the same aircraft that had transported the PLF terrorists when it was diverted by U.S. warplanes to Sicily.

TABLE 11.2 COUNTERTERRORIST OPTIONS: THE USE OF FORCE

The purpose of violent responses is to attack and degrade the operational capabilities of terrorists. This can be done by directly confronting terrorists or destabilizing their organizations.

Counterterrorist Option	Activity Profile		
	Rationale	Practical Objectives	Typical Resources Used
Suppression campaigns	• Symbolic strength • Punitive measures • Preemption	• Destruction of the terrorists • Disruption of the terrorists	• Military assets • Paramilitary assets
Coercive covert operations	• Symbolic strength • Destabilization • Preemption	• Disruption of the terrorists • Deterrent effect on potential terrorists	• Military assets • Paramilitary assets
Special operations forces	• Coercive covert operations • Destabilization • Preemption	• Disruption of the terrorists • Deterrent effect on potential terrorists	• Military and police assets

---

Operations Other Than War

Nonmilitary and nonparamilitary **repressive responses** can be effective in suppressing terrorist crises and terrorist environments. The purpose is to disrupt and deter terrorist behavior. They are unlikely to bring a long-term end to terrorism, but can reduce the scale of violence by destabilizing terrorist groups and forcing them to be on the run.

War in the Shadows, Part 2: Nonviolent Covert Operations

Nonviolent covert operations encompass a number of options, are often quite creative, and are inherently secretive. Their value lies in manipulating terrorist behavior and taking terrorist groups by surprise. For example, covert operations can create internal distrust, fear, infighting, and other types of discord. The outcome might also be a reduction in operational focus, momentum, and effectiveness. Typical covert operations include the following options:

Infiltration

Ideally, infiltration will increase the quality of intelligence prediction, as well as the possible betrayal of cadres to counterterrorist forces. However, in New Terrorism, infiltration of terrorist organizations is more difficult because of their cell-based organizational profiles.

Disinformation

Disinformation uses information to disrupt terrorist organizations. It is used to selectively create and deliver data that are calculated to create disorder within the terrorist group or its support apparatus. It can also be used to spread damaging propaganda about terrorist organizations and cadres.

Cyberwar

Bank accounts, personal records, and other data are no longer stored on paper, but instead in digital databases. Terrorist movements that maintain or send electronic financial and personal information run the risk of having that information intercepted and compromised. Thus, new technologies have become imaginative counterterrorist weapons.

Knowing the Enemy: Intelligence

Intelligence agencies involve themselves with collecting and analyzing information. The underlying goal is to develop an accurate activity profile of terrorists. Data are collected from overt and covert sources and evaluated by expert intelligence analysts. This process is at the heart of counterterrorist intelligence. The outcome can range from having profiles of terrorist organizations to tracking the movements of terrorists. An optimal outcome is to be able to anticipate the behavior of terrorists and thus to predict terrorist incidents.

SIGINT—Signal Intelligence

Intelligence collection and analysis in the modern era require the use of sophisticated technological resources. These technological resources are used primarily for the interception of electronic signals—known as SIGINT. **Signal intelligence** is used for a variety of purposes, such as intercepting financial data, monitoring communications such as cell phone conversations, and reading e-mail messages.

HUMINT—Human Intelligence

Collecting **human intelligence**, also referred to as HUMINT, is often a cooperative venture with friendly intelligence agencies and law enforcement officials. This sharing of information is a critical component of counterterrorist intelligence gathering. Circumstances may also require covert manipulation of individuals affiliated with terrorist organizations or their support groups, with the objective of convincing them to become intelligence agents. One significant problem with finding resources for human intelligence is that most terrorist cells are made up of individuals who know each other very well. Newcomers are not openly welcomed, and those who may be potential members are usually expected to commit an act of terrorism or other crime to prove their commitment to the cause. In other words, intelligence agencies must be willing to use terrorists to catch terrorists.

Intelligence Agencies

In many democracies, intelligence collection is traditionally divided between agencies that are separately responsible for domestic and international intelligence collection. This separation is often mandated by law. For example, the following agencies roughly parallel each other's missions:

- In Great Britain, **MI-5** is responsible for domestic intelligence, and **MI-6** is responsible for international collection.
- In the United States, the **Federal Bureau of Investigation (FBI)** performs domestic intelligence collection, and the **Central Intelligence Agency (CIA)** operates internationally.
- In Germany, the **Bureau for the Protection of the Constitution** shares a mission similar to MI-5 and the FBI, and the **Military Intelligence Service** roughly parallels MI-6 and the CIA.

In December 2004, the U.S. **intelligence community** was reorganized with the passage of the Intelligence Reform and Terrorism Prevention Act. Members of the community were subsumed under the direction of a new Office of the Director of National Intelligence (ODNI), and include the following agencies:

National Security Agency. The primary mission of the **National Security Agency (NSA)** is to collect communications and other signal intelligence. It also devotes a significant portion of its technological expertise to code-making and code-breaking activities. Much of this work is done covertly from secret surveillance facilities positioned around the globe.

Central Intelligence Agency. The CIA is an independent federal agency and the theoretical coordinator of the intelligence community. It is charged to collect intelligence outside of the borders of the United States, which is done covertly using human and technological assets. The CIA is legally prohibited from collecting intelligence inside the United States.

Defense Intelligence Agency. The **Defense Intelligence Agency (DIA)** is a bureau within the Department of Defense and is the central intelligence bureau for the U.S. military. Each branch of the military coordinates its intelligence collection and analysis with the other branches through the DIA.

Figure 11.1 presents the organizational chart of the ODNI in early 2007.

Problems of Intelligence Coordination

Collecting and analyzing intelligence are covert processes that do not lend themselves easily to absolute cooperation and coordination between countries or members of domestic intelligence communities. National intelligence agencies do not readily share intelligence with allied countries and usually do so only after careful deliberation. The same is true of intelligence communities. For example, before the September 11 attacks on the United States, dozens of federal agencies were involved in collecting intelligence about terrorism. This led to overlapping and competing interests. One example is the apparent failure by the FBI and CIA to collaborative in processing, sharing, and evaluating important intelligence. This and other

Figure 11.1 Organizational Chart for the Office of the Director of National Intelligence

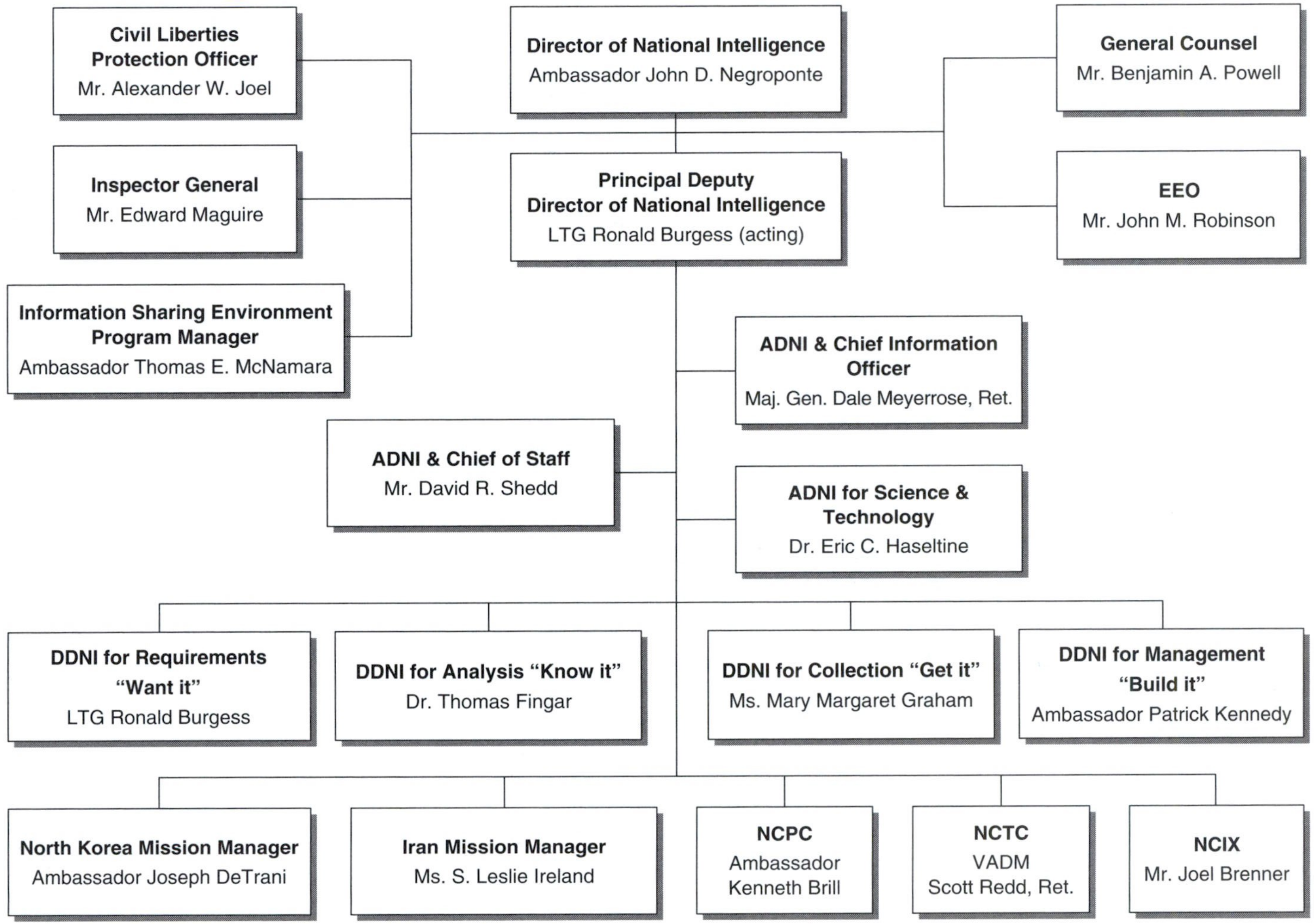

Note: Lieutenant General Burgess was confirmed by the Senate as the Deputy Directory of National Intelligence/Customer Outcomes. The DNI changed the title of the office in September 2006 to more accurately reflect the mission and duties of the position, which have not changed.

shortcomings advanced the call for an intelligence umbrella office, which eventually led to the creation of the ODNI.

Hardening the Target: Enhanced Security

Target hardening is an antiterrorist measure that makes potential targets more difficult to attack. This is a key component of **antiterrorism**, which attempts to deter or prevent terrorist attacks. Enhanced security is also intended to deter would-be terrorists from selecting hardened facilities as targets. These measures are not long-term solutions for ending terrorist environments, but do serve to provide short-term protection for specific sites. Target hardening includes increased airport security, the visible deployment of security personnel, and the erection of crash barriers at entrances to parking garages beneath important buildings. In the United States, technologies permit digital screening of fingerprints and other physical features as one enhancement at ports of entry.[5]

Typical examples of target hardening include the following:

◆ Vehicular traffic was permanently blocked on Pennsylvania Avenue in front of the White House because of the threat from high-yield vehicular bombs in the aftermath of the 1993 World Trade Center and 1995 Oklahoma City bombings. The area became a pedestrian mall.

◆ During the 1990s, Great Britain began to make widespread use of closed-circuit surveillance cameras in its urban areas. The police maintain these cameras on city streets, at intersections, and on highways. The purpose of this controversial policy is to monitor suspicious activities, such as abandoned packages or vehicles.

Long-Term Coercion: Economic Sanctions

Economic sanctions are a precedented counterterrorist method directed against governments. They are defined as trade restrictions and controls that are imposed to pressure sanctioned governments to moderate their behavior. Used as a counterterrorist option, sanctions serve several purposes:

◆ Sanctions symbolically demonstrate strong condemnation of the behavior of the sanctioned regime.
◆ Sanctions are an exercise of the power of the sanctioning body.
◆ Sanctions potentially bring to bear considerable pressure on the sanctioned regime.

Unlike many other counterterrorist options, sanctions inherently require sanctioning nations to commit to a long timeline to ensure success. The reason for this commitment is straightforward: Economic pressure is never felt immediately by nations, unless they are already in dire economic condition. Trade restrictions require time to be felt in a domestic market, particularly if the nation produces a commodity that is desired on the international market.

Problems can include the following:

◆ Sanctioned regimes rarely suffer—it is their people who suffer. Sanctions coalitions do not always remain firm.
◆ Leaks in trade embargoes are difficult to control, and sanctioning policies sometimes become quite porous.

Table 11.3 summarizes the conditions for success, and problems that commonly arise, when attempting to impose economic sanctions. Table 11.4 summarizes the activity profile of repressive counterterrorist options other than war.

TABLE 11.3   ECONOMIC SANCTIONS: CONDITIONS FOR SUCCESS AND PROBLEMS

Economic sanctions can theoretically pressure state sponsors of terrorism to moderate their behavior. Successful sanctions, however, require certain conditions to be in place and to remain firm during the sanctioning process.

Conditions for Success	Common Problems	Cases
Bring pressure to bear on sanctioned regime.	Economic pressure is passed on to a politically powerless population.	Sanctions against Iraq beginning in the early 1990s; regime remained strong
Maintain strong international cooperation.	International cooperation can weaken or dissolve.	U.S. sanctions against Cuba; perceived by world community to be a relic of the Cold War
Control leaks in the sanctioning policy.	When leaks occur, they cannot be controlled.	Uncoordinated sanctions movement against South Africa during apartheid era; failed to affect South African policy

TABLE 11.4   OPERATIONS OTHER THAN WAR: REPRESSIVE OPTIONS

The purpose of repressive responses other than war is to degrade the operational capabilities of terrorists.

Counterterrorist Option	Activity Profile		Typical Resources Used
	Rationale	Practical Objectives	
Nonviolent covert operations	• Deterrence • Destabilization • Prediction	• Deterrent effect on potential terrorists • Disruption of the terrorists	• Covert operatives • Technology
Intelligence	Prediction	Calculating the activity profiles of terrorists	• Technology • Covert operatives
Enhanced security	Deterrence	Hardening of targets	• Security personnel • Security barriers • Security technology
Economic sanctions	• Deterrence • Destabilization	Long-term destabilization and deterrence of sanctioned states	• National economic resources • Coalitional diplomacy

OOTW Conciliatory Options

Conciliatory responses are soft-line approaches for ending terrorist environments. They apply policies designed to resolve underlying problems that cause people to resort to political violence. Diplomatic options such as negotiations and social reform are typical policy options and can be very effective. Concessionary options are also an alternative, but these are more problematic.

Reasoned Dialogue: Diplomatic Options

Diplomatic options refer to the use of channels of communication to secure a counterterrorist objective. These channels range in degree from direct talks with dissidents to formal diplomatic overtures with nations that can influence the behavior of terrorist groups. The characteristics of these overtures are case specific, so the style of interaction and communication between the parties is unique to each example. Peace processes and negotiations are typical diplomatic options used to establish channels of communication.

Peace Processes

In regions with ongoing communal violence, long-term diplomatic intervention has sought to construct mutually acceptable terms for a cease-fire. **Peace processes** often involve long, arduous, and frustrating proceedings. Contending parties are always suspicious of each other and do not always represent all of the factions within their camps. The Northern Ireland peace process is instructive.

Northern Ireland Peace Process. The peace process in Northern Ireland enjoyed a significant turning point in May 1998 when the so-called **Good Friday Agreement** of April 10, 1998 (also known as the Belfast Agreement), was overwhelmingly approved by voters in the Irish Republic and Northern Ireland in May 1998. It signaled the mutual acceptance of a Northern Ireland assembly, and the disarmament, or **decommissioning,** of all paramilitaries. In July 2002, the Irish Republican Army surprised all parties in the peace process by issuing a formal apology for all of the people it killed during the Troubles.[6] Although the IRA and paramilitaries failed to disarm by the May 2000 deadline, the scale of violence has markedly decreased. Nevertheless, in May 2005, the Independent Monitoring Commission—a watchdog group established by the British and Irish governments—reported that the Irish Republican Army and Protestant paramilitaries continued to maintain weapons caches and recruit members. These caches, which contained tons of weapons, became a major hindrance to the peace process. The peace process did not collapse, however, because the IRA maintained as its official position a shift from violence to political struggle.[7] In July 2005, the leadership of the IRA announced an end to armed struggle, and ordered its paramilitary units to cease engaging in political violence. For the first time in three decades, and after more than 3,000 deaths, The Troubles were officially declared ended.[8]

Negotiations

Conventional wisdom in the United States and Israel holds that one should never negotiate with terrorists, never consider their grievances as long as they engage in violence, and never concede to any demand. Nevertheless, history has shown that case-specific negotiations can resolve immediate crises. Not all negotiations end in complete success for either side, but sometimes do provide a measure of closure to terrorist crises. The following cases are examples of negotiations between states, terrorists, and third parties that successfully secured the release of hostages:

Photo 11.3 A lost chance for peace? U.S. President Bill Clinton with Israeli Prime Minister Benjamin Netanyahu and Palestine Liberation Organization (PLO) Chairman Yasir Arafat. Clinton had hoped to establish a framework for an ongoing peace process, but the process collapsed amid the fighting of the Palestinian intifada.

The OPEC hostage crisis of December 1975 in Vienna, Austria, was resolved when the terrorists were permitted to fly to Algiers, Algeria, with a few hostages in tow. The hostages were released when a $5 million ransom was paid for Palestinian causes, and the terrorists were permitted to escape.

The odyssey of TWA Flight 847 in June 1985 ended with the release of the remaining hostages after negotiations were conducted through a series of intermediaries. The Lebanese Shi'a hijackers used the media and the Lebanese Shi'a Amal militia as intermediaries to broadcast their demands. The hostages were freed after the United States negotiated with Israel the release of more than 700 Shi'a prisoners.

The following cases are examples of negotiations (and bribery) between states that successfully secured the capture of terrorist suspects:

Ilich Ramirez Sanchez, better known as Carlos the Jackal, was "purchased" by the French government from the government of Sudan in August 1994. When the French learned that the Sudanese had given Carlos refuge, they secretly negotiated a bounty for permission to capture him, which they did from a Khartoum hospital room. Sanchez was eventually sentenced to life imprisonment for murdering two French gendarmes. **Johannes Weinrich**, a former West German terrorist, was also purchased—this time by the German government from the government of Yemen. Weinrich was sent to Germany in June 1995 to stand trial for the 1983 bombing of a French cultural center in Berlin, in which one person was killed and 23 others were wounded. Weinrich, who had been a very close associate of Carlos, was convicted in January 2000 and sentenced to life imprisonment. The bounty paid for his capture was presumed to be $1 million.

Responding to Grievances: Social Reform

Social reforms attempt to undercut the precipitating causes of national and regional conflicts. Reforms can include improved economic conditions, increased political rights, government recognition of ethno-nationalist sentiment, and public recognition of the validity of grievances.

The following case is an example of social reforms that successfully reduced the severity of a terrorist environment.

Case: Spain and ETA

Ethno-national sentiment in the Basque region had been repressed during the regime of Francisco Franco.9 After his death in 1975, Spanish democracy was restored, and Basque semi-autonomy was granted. Nationalism was permitted to be openly discussed, political groups were recognized, labor organizations became independent, and civil liberties were protected. ETA's political party was legalized. Nevertheless, ETA's most violent period occurred immediately after the restoration of democracy. ETA has since continued to strike from time to time, but Basque sentiment turned against the group during the 1980s. In addition, many ETA members returned to civilian life after accepting the terms of a reintegration program during the mid-1980s. Over time, Spanish commitment to social reforms undercut Basque support for ETA.

Giving Them What They Want: Concessionary Options

Granting concessions to terrorists is widely viewed as a marginally optimal counterterrorist response. The reason for this is obvious: Giving terrorists what they want is likely to encourage them to repeat their successful operation or perhaps to increase the stakes in future incidents. In other words, many extremists and those in the general population should be expected to conclude that concessions simply reward extremist behavior. Concessions include the following policy decisions:

◆ Payment of ransoms

◆ Releases of imprisoned comrades

◆ Broadcasts or other publications of extremist propaganda

◆ Political amnesty for dissidents

An interesting example of concessionary behavior is the Iran-Contra scandal. It is a study in how a nation with a publicly strong counterterrorist position (the United States) secretly tried to appease a country that it had defined as a state sponsor of terrorism (Iran).[10]

Case: Iran-Contra Revisited

The United States was embarrassed in November 1986 when a Lebanese magazine revealed that high-ranking officials in the Reagan administration had secretly agreed to sell arms to Iran. The operation, which was under way in August 1985, involved the sale of arms to Iran in exchange for help from the Iranian government to secure the release of American hostages held by Shi'a terrorists in Lebanon. Profits from the sales (reportedly $30 million) were used to support Nicaraguan contras in their war against the Sandinista government. This support was managed by National Security Adviser John Poindexter and Lt. Col. Oliver North. The administration's embarrassment was aggravated by the fact that the United States had previously adopted a get-tough policy against what it deemed to be terrorist states and Iran had been included in that category. Soon after American captives were released (apparently as a result of the weapons deal), Lebanese terrorists seized more hostages.

Table 11.5 summarizes the activity profile of conciliatory counterterrorist options.

Legalistic Options

The law enforcement approach to combating terrorism has had some success in disrupting conspiratorial networks, and has brought closure to criminal cases arising out of terrorist attacks. For example, in the aftermath of the September 11 attacks, European and Southeast Asian police uncovered a number of Al Qaeda and other jihadi cells in Spain, Germany, Singapore, and elsewhere.

TABLE 11.5   OPERATIONS OTHER THAN WAR: CONCILIATORY OPTIONS

The purpose of conciliatory options is to resolve the underlying grievances of the terrorists. This can be done by addressing immediate or long-term threats.

Counterterrorist Option	Activity Profile		
	Rationale	Practical Objectives	Typical Resources Used
Diplomatic options	Resolve terrorist crises	• Negotiate case-specific agreements • Negotiate long-term agreements	• Direct contacts • Intermediary contacts
Social reform	Degrade terrorist environments	• Win support from potential terrorist supporters • Decrease effectiveness of terrorist propaganda	• Targeted economic programs • Intensive political involvement
Concessionary options	• Resolve terrorist crises • Degrade terrorist environments	Satisfy the demands that motivate the terrorists	• Negotiators • Economic concessions • Political concessions

Law Enforcement and Counterterrorism

Internationally, there is no enforcement mechanism for violations of international criminal law other than treaties and other voluntary agreements between nations. Nevertheless, many nations have become members of the **International Criminal Police Organization**, more commonly known as **INTERPOL**. An international association of more than 140 nations that agree to share intelligence and provide assistance in the effort to suppress international crime, INTERPOL is based in Ste. Cloud, France, and each member nation has a bureau that serves as a liaison. INTERPOL is more of an investigative consortium than a law enforcement agency. Its value lies in the cooperative sharing of information between members, as well as the coordination of counterterrorist and criminal investigations. Similarly, the **European Police Office (EUROPOL)**, is a cooperative investigative consortium of members of the European Union.

CHAPTER PERSPECTIVE 11.3

The Capture of Mir Aimal Kansi

On the morning of January 25, 1993, a man armed with an AK-47 assault rifle began firing on employees of the Central Intelligence Agency who were waiting in their cars to enter the CIA's headquarters in Langley, Virginia. Two people were killed and three were wounded.

The person responsible was Mir Aimal Kansi, a Pakistani who had been a resident of the United States since 1991. After the shootings, he immediately fled for sanctuary in Pakistan and Afghanistan. He apparently traveled between the two countries, though he found refuge among relatives and local Pakistanis in Quetta.

The United States posted a $2 million reward for Kansi's capture and distributed wanted posters throughout the region. His photograph was distributed on thousands of matchbooks, printed in newspapers, and placed on posters. The hunt was successful, because still-unidentified individuals contacted U.S. authorities in Pakistan and arranged Kansi's capture. In June 1997, Kansi was arrested by a paramilitary FBI team and, with the permission of the Pakistani government, flown to the United States to stand trial in a Virginia state court.

At his trial, prosecutors argued that Kansi had committed the attack in retaliation for U.S. bombings of Iraq. He was convicted of murder on November 10, 1997. Kansi was executed by lethal injection at the Greensville Correctional Center in Virginia on November 14, 2002.

The Police and Terrorist Environments

Before the terrorist environments became so prevalent, the primary mission of police agencies was to serve as law enforcement. Today, the mission has refocused onto internal security. In these cases, the police become responsible for day-to-day civil protection operations. Internal security missions require law enforcement units to be stationed at strategic locations and to perform security-focused (rather than crime-focused) patrols. These responsibilities are sometimes threat-specific and mirror the terrorist environment of the times. Other security duties include promoting airport security, securing borders, tracking illegal immigrants, looking out for fugitives and other suspects, and conducting surveillance of groups and people who fit terrorist profiles. The police are also the first to stabilize the immediate vicinity around attack sites and are responsible for maintaining long-term order in cities suffering under terrorist campaigns. Thus, the role of law enforcement agencies varies in scale and mission depending on the characteristics of the environment, and can include the following.

Police Repression

Ideally, the security role of the police will be carried out professionally, within the context of constitutional constraints, and with respect for human rights. However, the reality is that many police agencies around the world are highly aggressive and sometimes abusive—particularly those in authoritarian and weakly democratic countries. They are less concerned with human rights than with maintaining order. The consequences are that police agencies in authoritarian environments perform a very different mission than the professionalized police forces in most stable democracies. For example, police in Brazil and Colombia have been implicated in practicing social cleansing against undesirables. Social cleansing involves intimidating a range of people defined as undesirable, including political dissidents, supporters of political dissidents, morals criminals (such as prostitutes and drug users), and marginal demographic groups (such as homeless children). Deaths and physical abuse have been documented during social cleansing campaigns.

Case: The United Kingdom

In the United Kingdom, where factions of the Irish Republican Army were highly active in London and other cities, the British police were considered the front line against IRA terrorism.

They usually displayed a high degree of professionalism without resorting to repressive tactics and consequently enjoyed widespread popular support. For example, London's Metropolitan Police (also known as The Met) became experts in counterterrorist operations when the Irish Republican Army waged a terrorist campaign during the 1970s. The Met's criminal investigations bureau generally used high-quality detective work rather than authoritarian techniques to investigate terrorist incidents. The British criminal justice system also generally protected the rights of the accused during trials of IRA suspects. However, in the rush to stop the IRA's terrorist campaign (especially during the 1970s), miscarriages of criminal justice did occur. Examples of these miscarriages include the so-called **Guildford Four** (made famous by the American film *In the Name of the Father*) and **Birmingham Six** Cases.

Case: Terrorist Profiling in the United States

Before September 11, 2001, the American approach to domestic counterterrorism was a law enforcement approach. After the attacks, this shifted to a more security-focused approach. The FBI and other agencies created a **terrorist profile** that was similar to standard **criminal profiles** used in law enforcement investigations. Criminal profiles are descriptive composites that include the personal characteristics of suspects, such as their height, weight, race, gender, hair color, eye color, and clothing. Suspects who match these criminal profiles can be detained for questioning. The composite of the new terrorist profile included the following characteristics: Middle Eastern heritage, temporary visa status, Muslim faith, male gender, and young adult age. Based on these criteria—and during a serious security crisis—the FBI and Immigration and Naturalization Service administratively detained hundreds of men. Material witness warrants were used from the outset to detain many of those detained for questioning, and new guidelines were promulgated that permitted field offices to conduct surveillance of religious institutions, Web sites, libraries, and organizations without an a priori (before the fact) finding of criminal suspicion.

These detentions and guidelines were criticized. Critics argued that the detentions were improper because the vast majority of the detainees had not been charged with violating the law. Critics of the surveillance guidelines contended that it gave too much power to the state to investigate innocent civilians. Many also maintained that there was a danger that these investigations could become discriminatory **racial profiling**, that is, detaining people because of their ethno-national or racial heritage. Nevertheless, the new security policies continued to use administrative detentions and enhanced surveillance as counterterrorist methods.

Domestic Laws and Counterterrorism

An important challenge for lawmakers in democracies is balancing the need for counterterrorist legislation against the protection of constitutional rights. In severely strained terrorist environments, it is not uncommon for nations—including democracies—to pass authoritarian laws that promote social order at the expense of human rights. Policy makers usually justify these measures by using a balancing argument, in which the greater good is held to outweigh the suspension of civil liberties. As illustrated in the following examples, severe threats to the state are sometimes counteracted by severe laws.

The United Kingdom Revisited: Suspending Civil Liberties in Northern Ireland

After the British army was deployed to Northern Ireland, it became severely pressed by an IRA terrorist campaign. The **Northern Ireland Act**, passed in 1993, created conditions of quasi-martial law. The act suspended several civil liberties. It empowered the British military to engage in warrantless searches of civilian homes, temporarily detain people without charge, and question suspects. The military could also intern (remove from society) suspected terrorists and turn over for prosecution those for whom enough evidence had been seized. Nearly a quarter of a million warrantless searches were conducted by the army, which resulted in the seizure of thousands of arms and the internment or imprisonment of hundreds of suspects.

Another program aimed at prosecuting and imprisoning Irish terrorists was implemented during the 1980s. This was the **supergrass** program, which was a policy of convincing Provos and members of the **Irish National Liberation Army (INLA)** to defect from their movements and inform on their former comrades. Many decided to participate. British-led authorities were thus able to successfully prosecute and imprison a number of Provos and INLA members. However, many of these dissidents were released in the late 1980s when cases taken up on appeal successfully challenged the admissibility of supergrass testimony. In the end, supergrass disrupted Irish militant groups during the 1980s (particularly the INLA) but did not have a long-term impact on Northern Ireland's terrorist environment.

Qualified Amnesty in Italy

Italy suffered thousands of terrorist attacks during the heyday of the Red Brigade terrorist campaign in the 1970s and 1980s. As part of its effort to combat the Red Brigade, the so-called **repentance laws** offered Red Brigade members qualified amnesty for demonstrations of repentance for their crimes. Repentance could be established by cooperating within a sliding scale of collaboration. Thus, those who collaborated most generously had a proportionally large amount of time removed from their sentences, whereas those whose information was less useful had less time removed. A significant number of the roughly 2,000 imprisoned Red Brigade terrorists accepted repentance reductions in their sentences.

The USA PATRIOT Act in the United States

In the aftermath of the September 11 attacks on the United States, Congress quickly passed legislation with the intent to address the new security threat. On October 26, 2001, President George W. Bush signed this legislation into law. It was labeled the Uniting and Strengthening America by Providing Appropriate Tools Required to Intercept and Obstruct Terrorism Act, also known as the USA PATRIOT Act. Its provisions include the following:

◆ Revising the standards for government surveillance, including federal law enforcement access to private records
◆ Enhancing electronic surveillance authority, such as tapping into e-mail, electronic address books, and computers
◆ Using "roving wiretaps" to permit surveillance of an individual's telephone conversations on any phone anywhere in the country
◆ Requiring banks to identify sources of money deposited in some private accounts and requiring foreign banks to report on suspicious transactions
◆ Using nationwide search warrants
◆ Deporting immigrants who raise money for terrorist organizations
◆ Detaining immigrants without charge for up to one week on suspicion of supporting terrorism

Debate about these and other provisions came from across the ideological spectrum. Civil liberties watchdog organizations questioned whether the provisions would erode constitutional protections. At the same time, conservatives questioned the possibility of government intrusion into individuals' privacy. To address some of these concerns, lawmakers included a sunset provision mandating that the USA PATRIOT Act's major provisions will automatically expire unless periodically extended. Lawmakers also required the Department of Justice to submit reports on the impact of the act on civil liberties. In early 2005, the House of Representatives and U.S. Department of Justice advocated restricting the act's ability to access certain personal records without a warrant.[11] When the act was renewed in 2006, it thus incorporated compromise provisions that included restrictions on federal agents' access to library records.

The treatment of terrorist suspects is an ongoing debate. Chapter Perspective 11.4 discusses the controversial issue of the use of physical coercion and torture.

CHAPTER PERSPECTIVE 11.4

The Torture Debate

From October through December 2003, Iraqi detainees held at the U.S.-controlled Abu Ghraib prison near Baghdad were abused by American guards.[a] This abuse included sexual degradation, intimidation with dogs, stripping prisoners naked, forcing them into "human pyramids," and making them stand in extended poses in so-called stress positions.[b] The U.S. Congress and global community became aware of these practices in April 2004 when graphic photographs were published in the media, posted on the Internet, and eventually shown to Congress. Criminal courts martial were convened, and several guards were convicted and sentenced to prison.[c]

Unfortunately for the United States, not only was its image tarnished, but further revelations about additional incidents raised serious questions about these and other practices. For example, in March 2005, U.S. Army and Navy investigators alleged that 26 prisoners in American custody had possibly been the victims of homicide.[d]

A debate about the definition and propriety of torture ensued.

Torture is officially eschewed by the United States, both morally and as a legitimate interrogation technique. Morally, such practices are officially deemed inhumane and unacceptable. As an interrogation method, American officials have long argued that torture produces bad intelligence because victims are likely to admit whatever the interrogator wishes to hear. However, during the war on terrorism, a fresh debate began about how to define torture and whether physical and psychological stress methods that fall outside of this definition are acceptable.

Assuming that the application of coercion is justifiable to some degree to break the resistance of a suspect, the question becomes whether physical and extreme psychological coercion are also justifiable.[e] For instance, do the following techniques constitute torture?

♦ Water boarding, in which prisoners believe that they will drown.
♦ Sexual degradation, when prisoners are humiliated by stripping them or forcing them to perform sex acts.
♦ Stress positions, whereby prisoners are forced to pose in painful positions for extended periods.
♦ Creating a chronic state of fear.
♦ Environmental stress, accomplished by adjusting a detention cell's temperature.
♦ Sleep deprivation.

When images such as those from Abu Ghraib became public, the political consequences were serious. Nevertheless, policy makers continued their debate on which practices constitute torture, and whether some circumstances warrant the imposition of as much stress as possible on suspects—up to the brink of torture.[f]

Notes

a. See Johanna McGeary, "The Scandal's Growing Stain," *Time,* May 17, 2004.
b. See Peter Grier and Faye Bowers, "Abu Ghraib Picture Begins to Fill In," *The Christian Science Monitor,* August 24, 2004.
c. See Faye Bowers, "In US Stand on Torture, More Trials to Come," *The Christian Science Monitor,* January 18, 2005.
d. Douglas Jehl and Eric Schmitt, "U.S. Military Says 26 Deaths May Be Homicide," *The New York Times,* March 16, 2005.
e. See Viveca Novak, "Impure Tactics," *Time*, February 21, 2005.
f. See Sonni Efron, "Torture Becomes a Matter of Definition," *Los Angeles Times,* January 23, 2005.

International Law: Legalistic Responses by the World Community

International law is based on tradition, custom, and formal agreements between nations. It is essentially a cooperative concept, because there is no international enforcement mechanism comparable to domestic courts, law enforcement agencies, or crimes codes. All of these institutions exist in some form at the international level, but it should be remembered that nations voluntarily recognize their authority. They do this through formal agreements. Bilateral and multilateral agreements are used to create an environment conducive to maintaining legal order.

Photo 11.4 Torture or not? An Iraqi detainee is forced by American captors to stand in a stress position in Abu Ghraib prison in Baghdad.

Nations enter into treaties to create predictability and consistency in international relations. When threats to international order arise—such as hijackings, kidnappings, and havens for wanted extremists—the international community often enters into multilateral agreements to manage the threat. The following examples illustrate the nature of multilateral counterterrorist agreements.

International Conventions on Hijacking Offenses

In response to the spate of airline hijackings that occurred in the late 1960s and early 1970s, the world community enacted a number of international treaties to promote cooperation in combating international terrorism directed against international travel services. These included the following:

Tokyo Convention on Offences and Certain Other Acts Committed on Board Aircraft. Enacted in 1963 as the first airline crimes' treaty, it required all signatories to "make every effort to restore control of the aircraft to its lawful commander and to ensure the prompt onward passage or return of the hijacked aircraft together with its passengers, crew, and cargo."[12]

Hague Convention of 1970. This required signatories to extradite "hijackers to their country of origin or to prosecute them under the judicial code of the recipient state."[13]

Montreal Convention of 1971. This extended international law to cover "sabotage and attacks on airports and grounded aircraft, and laid down the principle that all such offenses must be subject to severe penalties."[14]

Protecting Diplomats

In reply to the spate of attacks on embassies and assaults on diplomats in the late 1960s and early 1970s, several international treaties were enacted to promote cooperation in combating international terrorism against diplomatic missions. These included the following:

Convention to Prevent and Punish Acts of Terrorism Taking the Form of Crimes Against Persons and Related Extortion That Are of International Significance. A treaty among members of the Organization of American States, this "sought to define attacks against internationally protected persons as common crimes, regardless of motives."[15]

Prevention and Punishment of Crimes Against Internationally Protected Persons, Including Diplomatic Agents. This sought to establish a common international framework for suppressing extremist attacks against those who are protected by internationally recognized status.[16]

Extradition Treaties

Nations frequently enter into treaties that allow law enforcement agencies to share intelligence and operational information that can be used to track and capture terrorists. An example is extradition treaties, which require parties to bind over terrorist suspects at the request of fellow signatories. Strong extradition treaties and other criminal cooperation agreements are powerful tools in the war against terrorism and, when properly implemented, can be quite effective. However, these treaties are collaborative and not easily enforceable when one party declines to bind over a suspect or is otherwise uncooperative. In this event, there is little recourse other than to try to convince the offending party to comply with the terms of the treaty.

International Courts and Tribunals

The United Nations has established several institutions to address the problems of terrorism, genocide, torture, and international crime and to bring the perpetrators of crimes against humanity to justice. They are international courts, and their impact can be significant when nations agree to recognize their authority. Examples of UN authority include the following institutions:

International Court of Justice. The International Court of Justice is the principal judicial arm of the United Nations.

International Criminal Court. The International Criminal Court (ICC) was established to prosecute crimes against humanity, such as genocide.

International Criminal Tribunal for the Former Yugoslavia. The International Criminal Tribunal for the Former Yugoslavia (ICTY) has investigated allegations of war crimes and genocide arising out of the wars that broke out after the fragmentation of Yugoslavia during the 1990s.

International Criminal Court for Rwanda. The International Criminal Court for Rwanda (ICTR) has investigated allegations of war crimes and genocide that resulted from the breakdown of order in Rwanda during the 1990s.

Table 11.6 summarizes the activity profile of legalistic counterterrorist options.

TABLE 11.6   COUNTERTERRORIST OPTIONS: LEGALISTIC RESPONSES
The purpose of legalistic responses is to provide protection to the general public, protect the interests of the state, and criminalize the behavior of the terrorists.

Counterterrorist Option	Activity Profile		
	Rationale	Practical Objectives	Typical Resources Used
Law enforcement	• Enhancement of security apparatus • Demilitarization of counterterrorist campaign	• Day-to-day counterterrorist operations • Bringing terrorists into the criminal justice system	• Police personnel • Specialized personnel
Domestic laws	Criminalization of terrorist behavior	• Enhancement of criminal penalties for terrorist behavior • Bringing terrorists into the criminal justice system	• Criminal justice system • Legislative involvement
International law	International consensus and cooperation	Coalitional response to terrorism	• International organizations • State resources

Chapter Summary

This chapter discussed options for responding to terrorism within the context of several categories and subcategories. The decision-making process for selecting counterterrorist options is predicated on several key factors:

◆ The characteristics of the terrorist movement
◆ The nature of the overall terrorist environment
◆ The political goals of the counterterrorist actor

When assessing the practical utility of resorting to the use of force, it is important to remember that many of these responses are inconsistently effective against determined terrorists in the long term. Successes have been won against domestic terrorists—especially when governments have been unconstrained in their use of force and coercion—but this is not a universal outcome. Internationally, short-term successes have resulted in the resolution of specific terrorist incidents. Long-term successes, however, have sometimes been difficult to achieve. Nevertheless, the use of force has produced some success in disrupting terrorist groups and reducing the intensity of terrorist environments.

Repressive operations other than war include a number of options. Intelligence collection and analysis are extremely useful in building activity profiles of terrorist groups and in understanding the dynamics of terrorist environments. Intelligence is also useful for generalized prediction but less so in predicting the precise location and timing of specific terrorist attacks. Regarding enhanced security, because target hardening usually involves establishing fixed barriers, surveillance technologies, and security posts, determined terrorists can design methods to circumvent these precautions. Nevertheless, there is an increased potential for failure from the terrorists' perspective, and it should be presumed that enhanced security deters less determined and less resourceful terrorists.

Conciliatory responses have achieved both short-term and long-term success in ending terrorist environments. There have also been a number of failed conciliatory operations. Diplomatic options have enjoyed marked success in some cases but have been frustrated by entrenched hostilities and uncooperative parties in others. Social reforms have enjoyed long-term success when reforms are gradually accepted as legitimate by target populations. Concessionary options are risky because of the perception of appeasement of the terrorists, but are nonetheless sometimes successful.

Legalistic responses are in many ways the front line for counterterrorist policies. Law enforcement agencies are usually the first responders to incidents, and they are responsible for ongoing civil security and investigations. Problems arise when repression or miscarriages of justice discredit police agencies. Nevertheless, security-oriented police duties have successfully resolved or controlled terrorist environments. Domestic laws are adaptations of legal systems to domestic terrorist crises. Some of these adaptations—both authoritarian and democratic—have been quite successful. International laws and institutions have likewise enjoyed some success, but because they are inherently cooperative, parties to treaties and other agreements must comply with their terms. Otherwise, international laws and institutions have very little enforcement authority.

DISCUSSION BOX

This chapter's Discussion Box is intended to stimulate critical debate about the purpose of elite antiterrorism units.

The Utility of Elite Counterterrorist Units

Elite military and police counterterrorism units have been mustered into the security establishments of many nations. Many include highly trained professionals who can operate in a number of environments under extremely hazardous conditions. Their missions include hostage rescues, punitive strikes, abductions, and reconnaissance operations.

When elite units perform well, hostages are rescued, crises are resolved, and terrorist environments are disrupted. Sometimes, however, they find themselves involved in ambiguous political situations or tenuous operational conditions. Special operations are simply often high-risk, high-gain situations.

Nevertheless, proponents of elite counterterrorist units argue that conventional forces are not trained or configured to fight shadow wars, that only special operations forces can do so effectively. Critics of these units argue that conventional forces can accomplish the same objectives and goals and that, aside from the very good special operations units, other elite units have not proven themselves particularly effective.

Historical examples suggest that deploying special operations forces is a high-risk and high-gain option.

Discussion Questions

1. How necessary are elite counterterrorism units? Why?

2. How effective do you think these elite units are?

3. What other counterterrorist options do you think can be effective without resorting to the deployment of elite units?

4. Which counterterrorist options work most efficiently in conjunction with elite units? Which options work least efficiently?

5. In the long term, what impact will elite units have in the war against international terrorism?

Key Terms and Concepts

The following topics were discussed in this chapter and can be found in the glossary:

11th Parachute Division (Paras)
1st Navy Parachute Infantry Regiment
75th Ranger Regiment
antiterrorism
Anti-Terrorism and Effective Death Penalty Act
Bureau for the Protection of the Constitution
Central Intelligence Agency (CIA)
coercive covert operations
concessionary options
conciliatory responses
counterterrorism
covert operation
criminal profile
cyberwar
decommissioning
Defense Intelligence Agency (DIA)
Delta Force (1st Special Forces Operational Detachment–Delta)
diplomacy
Diplomatic Security Service
disinformation
economic sanctions
Emergency Search Team

enhanced security
European Police Office (EUROPOL)
Executive Order 12333
extradition treaty
Federal Bureau of Investigation (FBI)
Force 777
French Navy Special\ Assault Units
GEO (Grupo Especial de Operaciones)
GIGN (Groupe d'Intervention Gendarmerie Nationale)
Golani Brigade
Good Friday Agreement
Green Berets (Special Forces Groups)
Green Police
GSG-9 (Grenzschutzgruppe 9)
Hague Convention of 1970
Hostage Rescue Team
human intelligence
intelligence
intelligence community
International Court of Justice
International Criminal Court (ICC)
International Criminal

Police Organization (INTERPOL)
international law
Kansi, Mir Aimal
law enforcement
legalistic responses
MI-5
MI-6
Military Intelligence Service
Montreal Convention\ of 1971
Mossad
National Security Agency (NSA)
nonmilitary repressive responses
Northern Ireland Act
Operation Eagle Claw
Operation El Dorado Canyon
Operation Enduring Freedom
Operation Infinite Justice
Operation Peace for Galilee
Parachute *Sayaret*
paramilitary repressive options

peace process
Police Border Guards
preemptive strike
punitive strike
racial profiling
repentance laws
repressive responses
Rewards for Justice
    Program
Royal Marine
    Commandos

Sayaret
Sayaret Matkal
Sea Air Land Forces (SEALs)
signal intelligence
sky marshals
social reform
Special Air Service (SAS)
Special Boat Service (SBS)
Special Operations
    Command
special operations forces

supergrass
suppression campaigns
terrorist profiles
Tokyo Convention on
    Offences and Certain
    Other Acts Committed on
    Board Aircraft
Weinrich, Johannes
Wrath of God
YAMAM
YAMAS

Terrorism on the Web

Log on to the Web-based student study site at **www.sagepub.com/martinessstudy** for additional Web sources and study resources.

Web Exercise

Using this chapter's recommended Web sites, conduct an online investigation of counterterrorism.

1. Compare and contrast the services described online by the referenced organizations. How do their services differ?
2. How important are the services provided by organizations that monitor terrorist behavior and fugitives? Why?
3. In what ways do you think the Internet will contribute to the counterterrorist effort in the future?

For an online search of counterterrorism, readers should activate the search engine on their Web browser and enter the following keywords:

"Terrorism and Counterterrorism"
"Antiterrorism"

Recommended Readings

The following publications provide information about counterterrorist units and intelligence agencies.

Andrew, Christopher. *Her Majesty's Secret Service: The Making of the British Intelligence Community*. New York: Penguin Books, 1987.

Bamford, James. *Body of Secrets: Anatomy of the Ultra-Secret National Security Agency, From the Cold War Through the Dawn of a New Century*. New York: Doubleday, 2001.

Coulson, Danny O., and Elaine Shannon. *No Heroes: Inside the FBI's Secret Counter-Terror Force*. New York: Pocket Books, 1999.

Crank, John P., and Patricia E. Gregor. *Counter-Terrorism After 9/11: Justice, Security and Ethics Reconsidered*. New York, Anderson, 2005.

Harclerode, Peter. *Secret Soldiers: Special Forces in the War Against Terrorism*. London: Cassel, 2000.

Howard, Russell D., and Reid L. Sawyer. *Defeating Terrorism: Shaping the New Security Environment*. New York: McGraw-Hill, 2004.

Thomas, Gordon. *Gideon's Spies: The Secret History of the Mossad*. New York: St. Martin's, 1999.

12

Future Trends and Projections

Previous chapters provided a great deal of information about the causes of terrorism, the motives behind political violence, terrorist environments, and counterterrorist responses. Many examples of postwar terrorist movements and environments were presented to illustrate theoretical concepts and trends. The discussion in this chapter synthesizes many of these concepts and trends, and projects emerging trends likely to characterize terrorist environments in the near future.

One concept must be understood at the outset: Projecting future trends is not synonymous with predicting specific events, and the two should be differentiated as follows:

♦ Projections involve theoretical constructs of trends based on available data.
♦ Predictions are practical applications of data (that is, intelligence) to anticipate specific behaviors by extremists.

One must develop a logical longitudinal framework to evaluate the future of political violence. A longitudinal framework uses past history, trends, and cycles to project future trends. It allows scholars, students, and practitioners to "stand back" from immediate crises and contemporary terrorist environments to try to construct a reasonable picture of the near future. These projections must be made with the understanding that the more near-term a forecast is, the more likely it is to be realistic. Conversely, far-term forecasts are less likely to be realistic because projections must be consistently updated using contemporary data.

The discussion in this chapter will review the future of political violence from the following perspectives:

♦ What does the future hold?
♦ Revisiting the war on terrorism

Photo 12.1 A U.S. Marine fights to pacify an insurgent stronghold in the city of Fallujah, Iraq.

What Does the Future Hold?

David C. Rapoport designed a theory that holds that modern terrorism has progressed through three waves that lasted for roughly 40 years each, and that we now live in a fourth wave. His four waves are as follows:[1]

- ◆ The anarchist wave: 1880s to the end of World War I
- ◆ The anticolonial wave: end of World War I until the late 1960s
- ◆ The New Left wave: late 1960s to the near present
- ◆ The religious wave: about 1980 until the present

If Rapoport's theory is correct, the current terrorist environment will be characterized by the New Terrorism for the immediate future. That said, it can also be argued that the sources of extremist behavior in the modern era will generally remain unchanged in the near future. The modern era of terrorism is likely to continue to occur for the following reasons:

- ◆ People who have been relegated to the social and political margins—or who believe that they have been so relegated—often form factions that resort to violence.
- ◆ Movements and nations sometimes adopt religious or ethno-national supremacist doctrines that they use to justify aggressive political behavior.
- ◆ Many states continue to value the "utility" of domestic and foreign terrorism.

These factors have precipitated new trends in terrorist behavior that began to spread during the 1990s and early 2000s. These include the following:

- ◆ Increasing use of communications and information technologies by extremists
- ◆ Adaptations of cell-based organizational and operational strategies by global revolutionary movements
- ◆ Continued use of relatively low-tech tactics such as suicide bombers
- ◆ Efforts to construct or obtain relatively high-tech weapons of mass destruction or, alternatively, to convert existing technologies into high-yield weapons

Whither the Old Terrorism?

Traditional terrorism was characterized by the following commonalities:

- ◆ Leftist ethno-nationalist motives
- ◆ Leftist ideological motives
- ◆ Surgical and symbolic selection of targets
- ◆ Deliberate media manipulation and publicized incidents
- ◆ Relatively low casualty rates
- ◆ Identifiable organizational profile
- ◆ Hierarchical organizational profile
- ◆ Full-time professional cadres

Although this model did not become a relic in the 1990s, it did stop being the primary model in the 2000s. There is today in the West no single binding or common ideological foundation for political violence—no modern equivalent to revolutionary Marxism. Religious extremism is self-isolating, and there is no longer an international solidarity movement in the West for ethno-nationalist or religious violence. It remains to be seen what the long-term impact will be on actual religious terrorist violence, but it is unlikely that there will be an ideological glue to bind extremist sentiment in the West in the near future.

Many of today's terrorists have begun to promote vaguely articulated objectives and motives. They have also become cell-based stateless revolutionaries, unlike their predecessors, who tended to organize themselves hierarchically and had state sponsors. Even though the number of international incidents has declined during the New Terrorism, casualty rates have been

TABLE 12.1   SUPPLANTING THE OLD WITH THE NEW

The "old" terrorism was, in many ways, symmetrical and predictable. It was not characterized by terrorist environments exhibiting massive casualty rates or indiscriminate attacks. Its organizational profile was also characterized by traditional organizational configurations.

Terrorist Environment	Activity Profile				
	Target Selection	Casualty Rates	Organizational Profile	Tactical/ Weapons Selection	Typical Motives
The "old" terrorism	Surgical and symbolic	Low and selective	Hierarchical and identifiable	Conventional and low to medium yield	Leftist and ethnocentric
New Terrorism	Indiscriminate and symbolic	High and indiscriminate	Cellular	Unconventional and high yield	Sectarian

higher. This is not to say that the old terrorism has disappeared or will do so in the near future. It will continue alongside a growing—and more aggressive—threat from the New Terrorism.

Table 12.1 compares several key characteristics of the old and new terrorisms.

New Threats

The sources of extremist behavior have not changed in the recent past and will not change in the near future. Intense expressions of intolerance and resentment are likely to continue to motivate many extremists to blame entire systems, societies, and groups of people for their problems. Under these conditions, many people will continue to see political violence as a justifiable option. The likelihood of political violence will remain high as long as intolerance and blaming are motivated by passionate feelings of national identity, racial supremacy, religious dogma, or ideological beliefs. Both state and dissident terrorists will continue to exploit these tendencies for their own purposes.

Terrorist Environments in the 21st Century

State-Sponsored Terrorism

State-sponsored terrorism is not likely to soon disappear. It will remain a feature of future terrorist environments. Authoritarian states will continue to use terrorism as domestic policy, and aggressive states will continue to foment acts of international terrorism as foreign policy.

Domestically, political repression is—and long has been—a common practice of regimes more interested in protecting the authority of the state than human rights. The old instruments of repression will continue to be common tools of authoritarian government. These include security institutions such as police, military, and paramilitary forces. One should expect the pervasiveness of security institutions to be augmented by the continued improvement of surveillance and communications technologies. One should also expect that some of the world's more ruthless regimes will occasionally deploy weapons of mass destruction against dissident ethno-national groups, as the regime of Saddam Hussein did against Iraqi Kurds.

Internationally, aggressive regimes in the postwar era frequently supported sympathetic proxies to indirectly confront their adversaries. Doing so was often a safe and low-cost alternative to overt conflict. It is reasonable to assume that some regimes will continue the practice in the near future, especially in regions where highly active proxies have the opportunity to severely press their sponsoring regime's rivals. For example, in early 2002, a ship bearing Iranian arms was intercepted by Israeli security forces before it could off-load the arms. The weapons, which were bound for Palestinian nationalist fighters, were almost certainly part of a Tehran proxy operation.

Dissident Terrorism

Patterns of dissident terrorism during the 1990s and 2000s were decreasingly ideological and increasingly cultural. Ethno-nationalist terrorism continued to occur on a sometimes grand scale, and religious terrorism spread among radical Islamic groups. In addition, stateless international terrorism began to emerge as the predominant model in the global arena. These trends are likely to continue as vestiges of the East-West ideological competition give way to patterns of religious extremism and seemingly interminable communal conflicts. This clash of civilizations scenario has been extensively debated since it was theorized by Samuel Huntington.[2]

Religious Terrorism

Terrorism motivated by religion has become a global problem. Religious terrorism spread during the 1980s and grew to challenge international and domestic political stability during the 1990s and early 2000s. The frequency of, and casualties from, sectarian attacks grew quickly during this period. Religious terrorists also became adept at recruiting new members and organizing themselves as semiautonomous cells across national boundaries.

The trends that developed toward the end of the 20th century suggest that religious violence will continue to be a central aspect of terrorism in the 21st century. Internationally, religious terrorists of many nationalities have consistently attacked targets that symbolize enemy interests. Unlike the relatively surgical strikes of secular leftists in previous years, religious terrorists have proven particularly homicidal. Al Qaeda, for example, was responsible in April 2002 for the explosion of a fuel tanker truck at a synagogue in Tunisia that killed 17 people, 12 of them German tourists. This kind of lethality has become a central element in international religious terrorism. Domestically, terrorist movements in Pakistan, Israel, Malaysia, the Philippines, and elsewhere have actively sought to overthrow or destabilize their governments and damage symbols representing foreign interests. For example, Western civilians were taken as hostages by radical Islamic, antigovernment, Abu Sayyaf terrorists in the Philippines during 2001 and 2002.

Ideological Terrorism

Few ideologically motivated insurgencies survived into the 2000s, because most ideological terrorist environments and wars of national liberation were resolved during the 1980s and 1990s. Nevertheless, a few Marxist insurgencies persisted, and Western democracies witnessed the growth of neofascist and anarchistic movements. Violence from neofascists tended to consist of relatively low-intensity hate crimes, mob brawls, and occasional low-yield bombings. Violence from the new anarchists usually involved brawls with the police at international conferences or occasional confrontations with racist skinheads. Right-wing movements and groups are likely to persist and grow in the near term. With the growth of aboveground neofascist parties in Europe, they could become a vanguard or armed wing of a renewed fascist movement.

The World in Conflict: Future Sources of Terrorism

During the 20th century, seemingly unmanageable regional and internal conflicts raged for years, and often decades, before the warring parties made peace with each other. Enormous carnage occurred during communal conflicts, and millions of people died. Terrorism on a massive scale was not uncommon. It was also during this period that high-profile acts of international terrorism became a familiar feature of the international political environment.

In the 1990s and 2000s, contending nations and communities continued to engage in significant violence. Terrorism and violent repression were used by many adversaries, who claimed them as necessary methodologies to resolve the problems of their political environments. This trend is unlikely to abate in the near future, and it is reasonable to project that some conflicts may increase in intensity.

The following cases are projected sources and targets of terrorist violence in the near future.

The Middle East

Political, religious, and ethno-nationalist conflicts have been endemic to the Middle East for decades, and will continue for the near future. Many of these conflicts are long-standing and seemingly intractable, at least for the short term. Fundamentalist movements have also captured the imaginations of the many young Arabs disenchanted with the perceived failures of Arab nationalism and socialism. Large numbers of foreign fighters found their call to action after the U.S.-led invasion of Iraq and rallied to the side of the Iraqi insurgents. As a result, many international targets were attacked throughout the Middle East, a trend that will likely continue. For example, in August 2003, Iraqi insurgents bombed the United Nations headquarters in Iraq, killing UN High Commissioner for Human Rights Sergio Vieira de Mello. Religious tensions between Shi'a and Sunni Muslims also increased after sectarian violence erupted in Iraq. At the same time, Al Qaeda morphed into a model for other Islamists, a phenomenon that is likely to continue to be replicated by other movements and cells. In a sense, the organization of Al Qaeda has become an ideology of Al Qaeda.

Palestine and Israel

Palestinian nationalism entered its second generation during the intifada. The intifada also became a war fought by young gunmen, terrorists, and suicide bombers who grew up amid the turmoil of the 1960s, 1970s, and 1980s. Black September, the Palestine Liberation Organization's (PLO) expulsion from Lebanon, the airline hijackings, and many other incidents occurred when members of the new generation were children and teenagers. The new generation unhesitatingly began a new round of violence against Israel, one that was to prove more lethal and more pervasive.

Latin America

Most of the communal and ideological conflicts in Central and South America were resolved by the late 1990s. Violent leftist dissidents and repressive rightist regimes were no longer prominent features of the Latin American political environment. Some conflicts remained unresolved in the early 2000s, however. The worst scenario for near-term terrorism is in Colombia, where a weak central government has long exercised minimal control over a society beset by leftist rebels, rightist paramilitaries, and unchecked **narcotraficantes**. Aside from this and a few pockets of political violence, near-term indigenous terrorist violence (as opposed to international threats) will likely be sporadic and less intense than between the end of World War II and the early 1990s.

Europe

International terrorists have long used Europe as a battleground, as proven by the attacks in Madrid on March 11, 2004, and in London on July 7, 2005. This is unlikely to change, particularly in light of the discovery of Islamist cells and arrests of suspected Al Qaeda operatives in several western European countries.

Domestically, there is a possibility that followers of neofascist movements may target immigrant workers or other people they define as undesirables. There are also pockets of ethno-nationalist conflict that linger in Spain and the United Kingdom, though the likelihood that these conflicts will escalate almost zero. The Balkans remains the most unstable region in Europe, and sporadic violence is likely to occur there from time to time. However, the large-scale communal conflicts and genocidal behavior that resulted from the breakup of Yugoslavia have been suppressed by NATO and UN intervention. Russia's conflict in Chechnya will continue to be a source of terrorism, as well as a possible source of radical Islamist activism.

Africa

Ethno-national communal conflict has long been a recurring feature in Africa. In East Africa, periodic outbreaks of violence have claimed hundreds of thousands of lives. Similar outbreaks have occurred in West Africa. Another pattern of conflict in sub-Saharan Africa has been the use of state-sponsored domestic terrorism by authoritarian regimes to suppress dissident ethno-nationalist and political sentiment. Some internal conflicts—such as in Sudan, Somalia, Liberia, and Sierra Leone—are likely to flare up periodically in the near future. These conflicts have been characterized by many examples of terrorism on a large scale. In North Africa, radical Islamist movements have proven themselves motivated to commit acts of terrorism, and they are very capable of doing so. Algerian, Moroccan, and Egyptian members of Al Qaeda and other fundamentalist movements have attacked foreign interests, religious sites, and government officials as part of a vaguely defined international jihad.

Asia

Established patterns of ethno-national conflict and domestic terrorism suggest that some Asian nations and regions will continue to experience outbreaks of political violence. These outbreaks are likely to range from small-scale attacks to large-scale conflicts. Ethno-national groups continue to wage war in Sri Lanka, Kashmir, the Caucasus, and elsewhere. Terrorism is an accepted method of armed conflict among many of these groups. Ideological rebellions are less common than during the postwar era, but pockets of Marxist rebellion are still to be found, in Nepal and the Philippines, for example. These leftist remnants have occasionally used terrorist tactics rather effectively. Radical Islamic movements, inspired in part by Al Qaeda internationalism, have appeared in several Asian countries, including Indonesia, the Philippines, Malaysia, India, Pakistan, and the Central Asian republics. Some of these movements have demonstrated their willingness to engage in terrorism.

Case: The United States and the West

During the Cold War, the United States and its Western allies were frequent targets of leftist terrorism. In the modern era, the United States and the West have become targets of the New Terrorism. International terrorists still associate them with international exploitation, but add other dimensions to the new environment, such as fundamentalist religion, anti-Semitism, and a willingness to use new technologies and weapons of mass destruction. Symbolic targets—embassies, military installations, religious sites, tourists, and business visitors to foreign countries—have been attacked worldwide. Table 12.2 summarizes several examples of ongoing conflict likely to result in terrorist violence in the near future.

High-Tech Terrorism

New technologies allow terrorists to communicate efficiently, broaden their message, and wield unconventional weapons in unexpected ways—all central characteristics of asymmetrical warfare. Because of incremental improvements in communications and computer technologies, it is reasonable to conclude that the trend among terrorists and their supporters will be to use them extensively. This is also true of the increasing availability of weapons components for weapons of mass destruction, as well as the continuing softening of terrorists' reluctance to use them.

TABLE 12.2   A WORLD STILL IN CONFLICT: PROJECTED SOURCES OF POLITICAL VIOLENCE

Regional and domestic conflicts are certain to engender terrorist movements in the near future. Some of these conflicts are long-term disputes that have been ongoing for decades and that are often characterized by international spillovers.

Conflict	Activity Profile		
	Opposing Parties	**Contending Issues**	**Duration**
Palestinians/Israelis	Palestinian nationalists, Palestinian fundamentalists, Israelis	Palestinian state, Israeli security	Decades, from late 1940s and Israeli Independence
Northern Ireland	Catholic Unionists, Protestant Loyalists, British administration	Union with Irish Republic, loyalty to United Kingdom	Decades, from late 1960s and first Provo campaign
India/Pakistan	Kashmiri jihadis, India, Pakistan	Status of Kashmir, jihadi terrorism, border disputes	Decades, from late 1940s and end of British Empire
Colombia	Marxist rebels, narcotraficantes, paramilitaries, Colombian government	Social revolution, drug trade, state authority	Decades, from 1960s
Stateless New Terrorism	Internationalist terrorists, targeted interests	Vague goals, international stability	New Model, from 1990s

Information Technologies

The Internet provides opportunities for commercial, private, and political interests to spread their message and communicate with outsiders. Its use by extremists has already become a common feature of the modern era. Information technologies are being invented and refined constantly and will continue to be central to the New Terrorism. These technologies will facilitate networking between groups and cells and will permit propaganda to be spread widely and efficiently. The Internet and e-mail are certain to be used to send instructions about overall goals, specific tactics, new bomb-making techniques, and other facets of the terrorist trade. Both overt and covert information networks permit widely dispersed cells to exist and communicate covertly.

Information and computer technologies, of course, can also be used offensively. It is quite possible, even probably, that extremists will adopt cyberwar techniques to try to destroy information and communications systems. Cyberterrorism may very well become a central facet of the terrorist environment in the near future.

Weapons of Mass Destruction

The scenario of terrorists acquiring **weapons of mass destruction** is no longer the stuff of novels and films. The possibility that terrorists may use such weapons on a previously unimaginable scale is in fact very real indeed.

Terrorists are not without incentives to construct and use weapons of mass destruction in lieu of conventional weapons. The psychological and economic impact of such devices, for example, can easily outweigh the destructive effect of the initial attack (which might be relatively small).

Experts have argued that terrorists are making concerted efforts to acquire the requisite components to develop weapons of mass destruction and that their doing so is probably just a matter of time. Terrorists who are motivated by race or religion (or both) are likely to have little compunction about using chemical, biological, or radiological weapons against what they define as subhumans or nonbelievers.

As an alternative to producing weapons of mass destruction, terrorists have demonstrated their ability to convert available tech-

Photo 12.2 Anthrax-laced letters sent to Capitol Hill offices. Several letters were mailed during the immediate aftermath of the September 11 attacks.

nologies into high-yield weapons. The destructive and psychological consequences of such potential have not been lost on modern terrorists. Passenger airliners, to take the most obvious example, were used effectively indeed as ballistic missiles in the September 11 attacks. In another case, a precursor to 9/11, elite French GIGN counterterrorist police thwarted Algerian terrorists from using an airliner as a missile over Paris in December 1994.

Case: Exotic Technologies

Some technologies can theoretically be converted into weapons by terrorists who have a high degree of scientific knowledge and training. These exotic technologies include the following:

Electromagnetic Pulse Technologies. **Electromagnetic pulse (EMP) technologies** use an electromagnetic burst from a generator that can disable electronic components such as microchips. If used on a sizable scale, EMP could destroy large quantities of military or financial information. High-energy radio frequency devices are another type of generator that can disable electronic components.

Plastics. Weapons constructed from plastics and other materials such as ceramics could possibly thwart detection from metal detectors. Handguns, rifles, and bullets can be constructed of these materials.

Liquid Metal Embrittlement. Some chemicals can theoretically weaken metals when applied. They can embrittle, or make rigid, various metals. Were motivated extremists to obtain **liquid metal embrittlement** technology, they would find it relatively easy to transport and could conceivably apply it to commercial vehicles and aircraft.

Tech-Terror: Feasibility and Likelihood

The feasibility of terrorists' acquiring and using emerging technologies must be calculated by addressing the availability of these technologies. This has occurred in part because the scientific knowledge needed for assembling these weapons is available from a number of sources, including the Internet. Some weapons assembly requires expertise, but not necessarily extensive scientific training. Radiological weapons, for example, are relatively unsophisticated devices and require only toxic radioactive materials and a dispersion device. Other devices, such as nuclear weapons components, have so far been exceedingly difficult to assemble or steal. Nevertheless, the feasibility of obtaining weapons of mass destruction has increased not only because of the dissemination of technical know-how, but also because terrorists do not necessarily need to acquire new or exotic technologies. Older technologies and materials—such as pesticides, carbon monoxide, and ammonium nitrate and fuel oil (ANFO)—can be used to construct high-yield weapons. Aerosols and other devices can also be used as relatively unsophisticated delivery systems.

The likelihood that new technologies will be acquired and used is moot: Many have already been acquired and used. For example, apparently apolitical and anarchistic hackers—some of them teenagers—have vandalized information and communications systems, demonstrating that cyberwar is no longer an abstract concept. There is little reason to presume that this trend will diminish and many good reasons to presume that it will increase. The increasing availability of new technologies, when combined with the motivations and morality of the New Terrorism, suggests very strongly that technology will be an increasingly potent weapon in terrorist arsenals.

Soft Targets and Terrorist Symbolism

Terrorists throughout the postwar era tended to select targets that were both symbolic and soft. **Soft targets** include civilians and passive military figures, both unlikely to offer resistance until after the terrorists have inflicted casualties or other destruction. This tactic was sometimes quite effective in the short term and occasionally forced targeted interests to grant concessions. Those who practice New Terrorism have regularly selected soft targets that symbolize enemy interests. These targets are chosen in part because of their symbolic value but also because they are likely to result in significant casualties. Suicide bombers have become particularly adept at maximizing casualties. Thus, regardless of terrorist motives or environments, it is highly likely that violent extremists will continue to attack passive symbolic targets.

The Future of International Terrorism in the United States

With the turn of the 21st century, it has become very clear that the near future of international terrorism in the United States will be considerably threatening. Trends suggest that the United States will be a preferred target for international terrorists both domestically and abroad. This is not a new phenomenon. What is new is that the American homeland is vulnerable to attack for the first time in its history. International terrorism on U.S. soil is also not new. Many of these incidents were discussed in other chapters. The asymmetrical nature of the New Terrorism and the destructive magnitude of newly obtainable weapons and technology, however, are unlike any previous threats.

The Future of the Violent Left in the United States

Single-issue extremism is certain to be a feature of the radical left. Radical environmentalists have attracted a small but loyal constituency. New movements have also shown themselves adept at attracting new followers. For example, a nascent anarchist movement has taken root in the United States and other western democracies. This movement is loosely rooted in an antiglobalist ideology that opposes alleged exploitation by prosperous nations of poorer nations in the new global economy. It remains to be seen whether violent tendencies will develop within this trend. Sporadic incidents from single-issue terrorists are likely to occur from time to time.

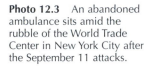
Photo 12.3   An abandoned ambulance sits amid the rubble of the World Trade Center in New York City after the September 11 attacks.

Photo 12.4   The war on terrorism continues.

The Future of the Violent Right in the United States

The future of the politically violent right lies in a possible weak network–weak sponsor scenario. In advocating leaderless resistance, for example, the violent Patriot and neo-Nazi right proved that they had learned from the 1980s and terrorist groups such as the Order. Conspiracies uncovered by law enforcement authorities beginning in the mid-1990s thus exhibited a covert and cell-based organizational philosophy. Threats are also possible from religious extremists who might reinvigorate the violent moralist movement. An extremist pool also lingers within the remnant Patriot movement, racial supremacist communities, and newer movements such as the neo-Confederate movement. Antigovernment and racial conspiracy theories continue to proliferate. Publications such as *The Turner Diaries*[3] and *The Myth of the Six Million*[4] continue to spread racial and anti-Semitic extremism. Promulgating these theories keeps reactionary tendencies alive on the right.

Chapter Perspective 12.1 describes **Holocaust** denial, a conspiracy belief that claims to debunk the Nazi Holocaust. It is a classic example of reactionary sentiment and the conspiracy mythology that continues characterize racial supremacist extremism.

CHAPTER PERSPECTIVE 12.1

Holocaust Denial[a]

Among the many conspiracy theories found on the far and fringe right wing and among Middle Eastern extremists is the argument that the Nazi Holocaust, the genocide of European Jews, never occurred. The underlying belief is that the Holocaust is a hoax, that the Nazis never systematically murdered any ethnic group, and that death camps such as Auschwitz were merely detention or work facilities. Holocaust denial has become a fundamental tenet for many tendencies on the reactionary right and has been repeated by fairly high-profile individuals. Publications and Web sites have been dedicated to this theory, though it is illegal in some European countries.

In the United States, this suggestion was promulgated by two right-wing organizations—the Liberty Lobby and the Institute for Historical Review—beginning in the late 1970s and early 1980s. Neo-Nazi groups, Ku Klux Klan factions, and Patriot survivalists have all been known to cite the so-called hoax of the Holocaust to justify their anti-Semitic conspiracy theories. Although Holocaust denial is mostly a white racial supremacist phenomenon in the United States, it has been endorsed by other domestic groups such as the Nation of Islam, which has historically promoted African American racial supremacy.

The theory has found an international audience among Western neofascists, anti-Zionists, and extremist groups in the Middle East. Many of these groups and movements have adopted overtly anti-Semitic rhetoric. Holocaust denial should be considered a facet of broader anti-Semitic tendencies, such as the promulgation of *The Protocols of the Learned Elders of Zion*, a forgery discussed in Chapter 7.

Note

a. For an excellent discussion of Holocaust denial, see Deborah E. Lipstadt, *Denying the Holocaust: The Growing Assault on Truth and Memory* (New York: Plume, 1994).

Revisiting the War on Terrorism

Counterterrorist experts in the modern terrorist era will be required to concentrate on achieving several traditional objectives. These can realistically, and in a practical sense, only minimize the terrorist threats of the near future, not eliminate them. Objectives include the following:

◆ Disrupting and preventing terrorist conspiracies from operationalizing their plans
◆ Deterring would-be terrorist cadres from crossing the line between extremist activism and political violence
◆ Implementing formal and informal international treaties, laws, and task forces to create a cooperative counterterrorist environment
◆ Minimizing physical destruction and human casualties

It is clear that no single model or method for controlling terrorism will apply across timelines or environments. Because of this reality, projecting counterterrorist models must include a longitudinal framework based on theory and practical necessity. The models used in the near future will continue to reflect the same categories of responses of the recent past, such as the use of force, operations other than war, and legalistic measures. To be practicable, however, these models will have to be updated and adapted to emerging threats continually. If they are, perhaps terrorism will be controlled to some degree—by keeping dissident terrorists off balance and state terrorists isolated, thereby preventing them from having an unobstructed hand in planning and carrying out attacks or other political violence.

Government Responses

Assuming that policy makers grasp the limitations of relying exclusively on coercive methods, it is likely that they will develop alternative measures. Operations other than war include conciliatory options, which may provide long-term solutions to future extremism. Conciliatory options of the past—such as peace processes, negotiations, and social reforms—did have some success in resolving both immediate and long-standing terrorist crises. If skillfully applied, future adaptations of such options could help avoid political violence. In the past, these options were usually undertaken with the presumption that some degree of coercion would be kept available should the conciliatory options fail. This is a pragmatic consideration that is likely to continue.

Societal Responses

Because extremism has historically originated primarily from domestic conflict (though sometimes from national traumas such as invasion), it is important that future efforts to counter extremist ideas incorporate societal and cultural responses. It is also important that new societal and cultural norms reflect demographic changes and political shifts. As discussed in previous chapters, dissent can certainly be repressed, but repression is rarely a long-term solution absent social reforms and political inclusion. This in turn is an ideal more likely to succeed in democratic systems than under authoritarian regimes.

Societal responses must, of course, be adapted to the idiosyncrasies of each nation and region. This is difficult in many cases—and seemingly nothing more than impossibly idealistic in others—because many if not most regimes and contending groups have little interest in reducing social tensions. More often than not, they simply try to manipulate such tensions to their benefit.

Countering Extremism

Extremist ideologies and beliefs are the fertile soil for politically violent behavior. Ethnocentrism, nationalism, ideological intolerance, racism, and religious fanaticism are core motivations for terrorism. History has shown that coercive measures used to counter these tendencies are often only marginally successful. The reason is uncomplicated: A great deal of extremist behavior is rooted in passionate ideas, long histories of conflict, and codes of self-sacrifice (explored in Chapter 3). It is difficult to forcibly reverse these tendencies, and though coercion can eliminate cadres and destroy extremist organizations, sheer repression is a risky long-term solution.

New Fronts in a New War

One projection for near-term counterterrorism stands out: Models must be flexible enough to respond to new environments and must avoid stubborn reliance on methods that "fight the last war." This reality is particularly pertinent to the war against terrorism. It, unlike previous wars, was declared against behavior as much as against terrorist groups and revolutionary cadres. The fronts in the new war are amorphous and include the following:

Covert shadow wars are fought outside public scrutiny using unconventional methods. **Shadow wars** require deploying military, paramilitary, and coercive covert assets to far regions of the world.

Domestic (homeland) security measures are required to harden targets, deter attacks, and thwart conspiracies. Internal security requires the extensive use of nonmilitary security personnel, such as customs officials, law enforcement agencies, and immigration authorities.

Counterterrorist financial operations are directed against bank accounts, private foundations, businesses, and other potential sources of revenue for terrorist networks. Intelligence agencies can certainly hack into financial databases, but a broad-based coalition of government and private financial institutions is necessary for this task.

Global surveillance of communications technologies requires tracking telephones, cell phones, browser traffic, and e-mail. Agencies specializing in electronic surveillance, such as the American National Security Agency, are the most capable institutions to carry out this mission.

Identifying and disrupting transnational terrorist cells and support networks requires international cooperation to track extremist operatives and "connect the dots" on a global scale. Primary responsibility for this task lies with intelligence communities and law enforcement agencies.

The new fronts in the new war clearly highlight the need to continuously upgrade physical, organizational, and operational counterterrorist measures; flexibility and creativity are essential. Failure in any regard will hinder adaptation to the terrorist environment of the 21st century. The inability to control and redress long-standing bureaucratic and international rivalries, to take one example, could be disastrous.

Continued Utility of Force

Violent coercion will continue to be a viable counterterrorist option. Terrorist cells, especially in disputed regions where they enjoy popular support, cannot be dismantled solely with law enforcement, intelligence, or nonmilitary assets. Situations sometimes require a warlike response with military assets, from small special operations units to large deployments of significant air, naval, and ground forces. Traditional military force has a demonstrated record against terrorist efforts that is relevant to coercive counterterrorist policies in the near future.

History has shown, of course, that military and paramilitary operations are not always successful. As discussed in previous chapters, some of these operations have ended in disaster. Others have been marginally successful. It is therefore likely that future uses of force will likewise fail on occasion. Nevertheless, the past utility of this option, and its symbolic value, are certain to encourage its continued use. Absent a viable threat of force, states are highly unlikely to dissuade committed revolutionaries or aggressive states from acts of political violence.

Countering Terrorist Financial Operations

Mohammed Atta, the leader of the September 11 cell, was closely affiliated with Al Qaeda members in several countries. He apparently received wire transfers of money in Florida from operatives in Egypt and was in close contact with Syrians who managed financial resources in Hamburg, Germany.[5] Estimates suggest that the total cost of the attack was $300,000, a small sum for the destruction and disruption that resulted.[6] Very few if any of these funds came from state sponsors. They instead came from private accounts run by Al Qaeda operatives. It is very clear from the Atta example that Al Qaeda—and logically, other practitioners of the New Terrorism—have successfully established themselves as stateless revolutionaries. This profile is unlikely to change in the near future.

Finding Hidden Fortunes

Terrorists and their extremist supporters have amassed sizable funds from a variety of sources, including the following:

♦ Transnational crime—the trafficking and smuggling activities of Colombian and Sri Lankan groups, for example
♦ Personal fortunes—such as Osama bin Laden's personal financial resources
♦ Extortion—for example, the criminal profits of Abu Sayyaf
♦ Private charities and foundations—such as those that support Hezbollah and Hamas[7]

Large portions of these assets were deposited in anonymous bank accounts, enabling funds to be electronically transferred between banking institutions and other accounts internationally in minutes. During the months following the September 11 attacks, government agencies from a number of countries made a concerted effort to identify and trace the terrorists' banking accounts. Law enforcement and security agencies also began to closely scrutinize the activities of private charities and foundations in an attempt to determine whether they were front groups secretly funneling money to supporters of terrorist organizations.

One problem encountered on a global scale involved the tradition and policy of **customer anonymity** found in some banking systems, such as those of the Cayman Islands and Switzerland. Many mainstream executives and policy makers were very hesitant to endorse an abrogation of the sanctity of customer anonymity. Their rationale was straightforward: Individual privacy and liberty could be jeopardized if security officers were permitted to peruse the details of hundreds of thousands of bank accounts looking for a few terrorist accounts that might or might not exist. From a practical business perspective, customers were likely to reconsider doing business with financial institutions that could no longer guarantee their anonymity.

A Resilient Adversary

Terrorists have adapted quickly to the new focus on financial counterterrorist measures. Implementing a process that apparently began during the global crackdown after September 11, 2001, terrorists and their supporters removed assets from financial institutions and began investing in valuable commodities such as gold, diamonds, and other precious metals and gems.[8] From the counterterrorist's perspective, this tactic might cripple the global effort to electronically monitor, track, and disrupt terrorist finances. From the terrorist's, the chief encumbrance is the literal burden of transporting precious commodities. Such an encumbrance, however, is acceptable in light of the difficulty for counterterrorist agents in identifying and interdicting couriers or to locate and raid repositories. It is thus likely that this adaptation will continue to be made as circumstances require, perhaps making it virtually impossible to trace skillfully hidden terrorist assets.

Another adaptation is an ancient practice known as **hawala**, a transnational system of brokers who know and trust each other. Persons wishing to transfer money approach hawala brokers and, for a fee, ask the broker to transfer money to another person. Using the name and location of the recipient, the initial broker will contact a broker in the recipient's country. The recipient contacts the local broker, who delivers the money. To prevent fraud, the sending broker gives the recipient broker a code number (such as the string of numbers on a $20 bill). The recipient (who receives the code from the sender) must give the broker this code number before picking up the funds. No records are kept of the transaction, thus ensuring anonymity. It is a useful system because it relies on honor to succeed and money is never physically moved.

Counterterrorist Surveillance Technologies

It is technologically feasible to access virtually every private electronic transaction, including telephone records and conversations, computer transactions and communications (such as e-mail), and credit card records. Digital fingerprinting and facial imaging permit security agencies to access records virtually instantaneously. Because these technologies are inherently intrusive, they have been questioned by political leaders and civil libertarian organizations. Nevertheless, surveillance technologies are considered invaluable counterterrorist instruments.

Case: Carnivore (DCS-1000)

In July 2000, it was widely reported that the FBI had a surveillance system that could monitor Internet communications. Called **Carnivore**, the system was said to be able to read Internet traffic moving through cooperating Internet service providers. All that was required was for Carnivore to be installed on an Internet provider's network at the provider's facilities. Under law, the FBI could not use Carnivore without a specific court order under specific guidelines, much like other criminal surveillance orders.

The FBI received a great deal of negative publicity, especially after it was reported that the agency had evaded demands for documents under a Freedom of Information Act (FOIA) request filed by the Electronic Privacy Information Center (EPIC), a privacy rights group.[9] Concern was also raised by critics when it was reported in November 2000 that Carnivore had been successfully tested and exceeded expectations. This report was not entirely accurate. In fact, Carnivore did not operate properly when it was used in March 2000 to monitor a criminal suspect's e-mail; it had inadvertently intercepted the e-mail of innocent Internet users. This glitch embarrassed the Department of Justice (DOJ) and angered the DOJ's Office of Intelligence Policy and Review.

By early 2001, the FBI gave Carnivore a less ominous sounding new name, redesignating the system DCS-1000. Despite the political row, which continued well into 2002, in part because of the continued FOIA litigation, Carnivore was cited as a potentially powerful tool in the new war on terrorism. The use of DCS-1000 after 2003 was apparently reduced markedly, allegedly because Internet surveillance was outsourced to private companies.

Case: Echelon

The U.S. National Security Agency (NSA) manages a satellite surveillance network called **Echelon**, apparently in cooperation with NSA counterparts in Australia, Canada, Great Britain, and New Zealand. Its purpose is to monitor voice and data communications.

A kind of global wiretap that filters through communications using antennae, satellite, and other technologies, Echelon can reportedly intercept Internet transfers, telephone conversations, and data transmissions. It is not publicly known how much traffic can be intercepted or how it is done, but the network is apparently very capable.

The Case for International Cooperation

Cooperation between nations has always been essential to counterterrorist operations. International treaties, laws, and informal agreements were enacted during the postwar era to create a semblance of formality and consistency to global counterterrorist efforts. However, cooperation at the operational level was not always consistent or mutually beneficial. With the advent of the New Terrorism and international counterterrorist warfare, international cooperation at the operational level has become a priority for policy makers. A good example is the new frontline missions of intelligence and criminal justice agencies.

Intelligence and Law Enforcement

The world's intelligence communities and criminal justice systems have always been important counterterrorist instruments. After the September 11 attacks, these institutions were tasked with increased responsibilities—largely because of their demonstrated ability to incapacitate and punish terrorists. These institutions, perhaps more than military institutions, are also able to apply steady and long-term pressure on terrorist networks. They are in many cases more adept than military assets at keeping terrorists on the run over time. This is not to say that they are a panacea for future terrorism, but international cooperation between intelligence and law enforcement agencies does provide the means to track operatives, identify networks, and interdict other assets on a global scale.

International law enforcement cooperation in particular provides worldwide access to extensive criminal justice systems and their well-established **terminal institutions** (such as prisons). Such institutions—under the jurisdiction of criminal justice and military justice systems—can house individual terrorists after they have been captured, prosecuted, and convicted. These institutions can also effectively end terrorist careers. When faced with the prospect of lifelong incarceration, terrorists are also likely to be willing, for example, to exchange favors for intelligence. In a cooperative environment, of course, such data may be or would be shared among allied governments.

Reorganization of Homeland Security and Intelligence in the United States

Strong proposals were made to revamp the American homeland security community within nine months of the September 11 attacks. These came in reaction and response to both the apparent failure of the pre–September 11 domestic security community to adapt to the new terrorist environment, and the highly publicized operational problems.

Before September 11, the United States had relied on administratively separated federal law enforcement and services agencies to provide homeland security. These agencies are defined for our purposes as follows:

Law Enforcement Agencies. These are bureaus within large cabinet agencies charged with enforcing federal criminal laws. The FBI, the Drug Enforcement Administration; and the Bureau of Alcohol, Tobacco, and Firearms are examples. Traditionally, these agencies had investigated security threats the same way they investigated crimes—by working cases and making arrests.

Services Agencies. These agencies regulate and manage services for the general population. Service agencies include large cabinet agencies, regulatory agencies, and independent agencies. The departments of Health and Human Services, Energy, and Defense and the Central Intelligence Agency are examples. These agencies had had a variety of missions, including regulating immigration, inspecting nuclear facilities, and responding to emergencies.

Among these agencies, the FBI was one of the few that performed a quasi-security mission, explicitly adopting as one of its primary missions the protection of the United States from foreign intelligence and terrorist threats. An ideal policy framework would have required the FBI and CIA to coordinate and share counterterrorist intelligence in a spirit of absolute cooperation. In theory, the FBI would focus on investigating possible domestic security threats, and the CIA would pass along foreign intelligence that might affect domestic security. Table 12.3 summarizes the pre–September 11 security duties of several U.S. federal agencies.

TABLE 12.3	FEDERAL AGENCIES AND HOMELAND SECURITY: BEFORE THE POST–SEPTEMBER 11 ORGANIZATIONAL CRISIS

Before the organizational crisis, homeland security was the responsibility of a number of federal agencies. These were not centrally coordinated, and they answered to different authorities. Cooperation was theoretically ensured by liaison protocols, special task forces, and oversight. In reality, there was a great deal of functional overlap and bureaucratic turf issues.

Agency	Activity Profile		
	Parent Organization	Mission	Enforcement Authority
Central Intelligence Agency	Independent agency	Collection and analysis of foreign intelligence	No domestic authority
Coast Guard	Department of Transportation	Protection of U.S. waterways	Domestic law enforcement authority
Customs	Department of the Treasury	Examination of people and goods entering the United States	Domestic inspection, entry, and law enforcement authority
Federal Bureau of Investigation	Department of Justice	Investigating and monitoring criminal and national security threats	Domestic law enforcement authority
Federal Emergency Management Agency	Independent agency	Responding to natural and human disasters	Coordination of domestic emergency responses
Immigration and Naturalization Service	Department of Justice	Managing the entry and naturalization of foreign nationals	Domestic inspection, monitoring, and law enforcement authority
Secret Service	Department of the Treasury	Establishing security protocols for president, vice president, and special events	Domestic protection of president and vice president and special law enforcement authority (including counterfeiting)

One problem that became quite clear in the year following September 11 was that the old organizational model did not adapt well to the new security crisis. This failure proved damaged both its operations and its image and was politically embarrassing. A series of revelations and allegations called into question previous assertions by the FBI and CIA that neither had had prior intelligence about the September 11 attacks. It was discovered, for example, that

- The FBI had been aware for years that foreign nationals were enrolling in flight schools
- The CIA had compiled intelligence data about some members of the Al Qaeda cell that carried out the attacks

Policy makers and elected leaders wanted to know why neither the FBI nor the CIA had connected the dots to create a single intelligence profile. Serious interagency and internal problems became publicly apparent when a cycle of recriminations, press leaks, and congressional interventions damaged the united front image that the White House had projected. Policy makers determined that problems in the security community included the following:

- Long-standing interagency rivalries
- Entrenched and cumbersome bureaucratic cultures and procedures
- No central coordination of programs
- Fragmentation of counterterrorist operations
- Poor coordination of counterterrorist intelligence collection and analysis
- Disconnect between field offices and Washington headquarters
- Turf-based conflict between the FBI and CIA

To remedy the highly publicized political and operational disarray of the domestic security effort, President Bush in June 2002 initiated a process that completely reorganized the American security community. The Department of Homeland Security Act of 2002 was signed into law on November 25, 2002. It created a large cabinet-level **Department of Homeland Security** that, because of the operational fragmentation of intelligence and security, absorbed the functions of several federal agencies.[10] Homeland Security thus became the third largest federal agency, trailing only the Department of Veterans Affairs and the Department of Defense in size.

Table 12.4 summarizes the security duties of several U.S. federal agencies immediately after the creation of the new department. Figure 12.1 depicts the Homeland Security organizational chart.

Subsequent commission reports led to sweeping changes in the U.S. intelligence community. These reports included the following:

- In July 2004, the National Commission on Terrorist Attacks on the United States, also known as the 9/11 Commission, issued a detailed report on the September 11, 2001 attacks.[11]
- In March 2005, the Commission on the Intelligence Capabilities of the United States Regarding Weapons of Mass Destruction issued a detailed report on intelligence failures regarding the possession and proliferation of weapons of mass destruction.[12]

TABLE 12.4   FEDERAL AGENCIES AND HOMELAND SECURITY: AFTER THE
ORGANIZATIONAL CRISIS

In the wake of the post–September 11 organizational crisis, the Bush administration subsumed the homeland security duties of several federal agencies under the jurisdiction of a new Department of Homeland Security. The goal was to coordinate operations and to end to overlapping duties.

   The following table is a good snapshot of a nation's reorganization of national security in response to a significant shift in a terrorist environment. It is also an example of how two agencies that arguably precipitated the organizational crisis—the FBI and CIA—were able to maintain their independence.

| Agency | Activity Profile | | |
	New Parent Organization	New Directorate	New Directorate's Duties
Central Intelligence Agency	No change: independent agency	No change	No change: collection and analysis of foreign intelligence
Coast Guard	Department of Homeland Security	Border and Transportation Security	Coordination of all national entry points
Customs	Department of Homeland Security	Border and Transportation Security	Coordination of all national entry points
Federal Bureau of Investigation	No change: Department of Justice	No change	No change: investigating and monitoring criminal and national security threats
Federal Emergency Management Agency	Department of Homeland Security	Emergency Preparedness and Response	Coordination of national responses to terrorist incidents
Immigration and Naturalization Service (some functions)	Department of Homeland Security	Border and Transportation Security	Coordination of all national entry points
Secret Service	Department of Homeland Security	Secret Service	Establishing security protocols for president and special events

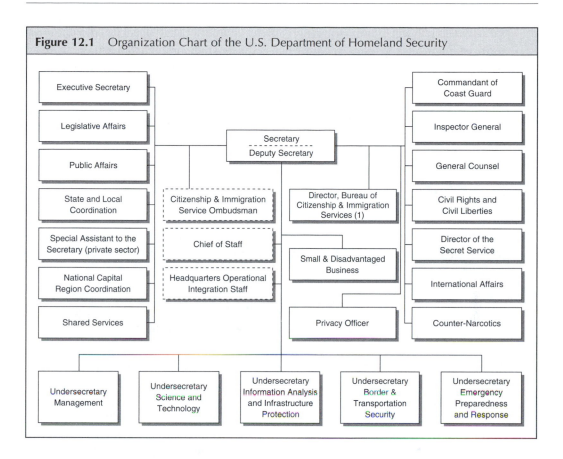

Figure 12.1 Organization Chart of the U.S. Department of Homeland Security

Clearly, the attacks of September 11 were the catalyst for a broad reconfiguration of the American security culture.

Chapter Summary

This chapter explored trends that suggest the near-term future of terrorism. An underlying theme throughout this discussion was that the near future will reflect the emerging profile of the New Terrorism. Traditional terrorism is certainly still a factor, but is no longer an exclusive or predominant model. These trends can be analyzed within the context of previous terrorist environments, with the caveat that these will continue to adapt to emerging political environments. The near future of terrorism will be shaped by ongoing regional conflicts, new technologies, and renewed attacks against symbolic soft targets.

Counterterrorism in the post–September 11 era will have to adapt to new fronts. These fronts require creative use of overt and covert operations, homeland security measures, intelligence, and cooperation among counterterrorist agencies. Disruption of terrorist financial operations will prove one of the most important and the most challenging priorities in the new war. Most threats in the near future for the United States will come from international religious extremists, domestic rightists, and single-issue leftists. Because of revelations about bureaucratic inefficiency in the aftermath of the September 11 attacks, the United States implemented a restructuring of its homeland security community.

DISCUSSION BOX

This chapter's Discussion Box is intended to stimulate critical debate about the possible use, by democracies and authoritarian regimes, of antiterrorist technologies to engage in surveillance.

Toward Big Brother?

Electronic surveillance has become a controversial practice in the United States and elsewhere. The fear is that civil liberties can be jeopardized by unregulated interception of telephone conversations, e-mail, and fax transmissions. Detractors argue that government use of these technologies can conceivably move well beyond legitimate application against threats from crime, espionage, and terrorism. Absent strict protocols to rein in these technologies, a worst-case scenario envisions state intrusions into the everyday activities of innocent civilians. Should this happen, critics foresee a time when privacy, liberty, and personal security become values of the past.

Discussion Questions

1. How serious is the threat from abuses in the use of new technologies?
2. How should new technologies be regulated? Can they be regulated?
3. Is it sometimes necessary to sacrifice a few freedoms to protect national security and to ensure the long-term viability of civil liberty?
4. Should the same protocols be used for domestic electronic surveillance and foreign surveillance? Why?
5. What is the likelihood that new surveillance technologies will be used as tools of repression by authoritarian regimes in the near future?

Key Terms and Concepts

The following topics were discussed in this chapter and can be found in the glossary:

Carnivore (DCS-1000)
    customer anonymity
cyberterrorism
Department of Homeland
    Security

Echelon
electromagnetic pulse
    (EMP) technologies
hawala
liquid metal embrittlement

shadow wars
soft targets
terminal institution
weapons of mass
    destruction

Terrorism on the Web

Log on to the Web-based student study site at **www.sagepub.com/martinessstudy** for additional Web sources and study resources.

Recommended Readings

The following publications are an eclectic assortment of recommendations that provide classic—and arguably timeless—insight into the nature of dissident resistance, ideologies of liberty, state manipulation, and revolution.

Hamilton, Alexander, John Jay, and James Madison. *The Federalist: A Commentary on the Constitution of the United States*. New York: Modern Library, 1937.

Koestler, Arthur. *Darkness at Noon*. New York: Macmillan, 1963.

Mill, John Stuart. *On Liberty*, ed. David Spitz. New York: W.W. Norton, 1975.

Moore, Barrington, Jr. *Social Origins of Dictatorship and Democracy: Lord and Peasant in the Making of the Modern World*. Boston, MA: Beacon, 1966.

Orwell, George. *Animal Farm*. New York: Harcourt Brace Jovanovich, 1946.

Appendix A

Map References

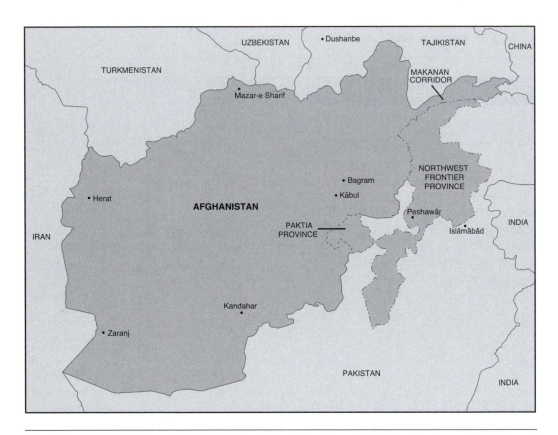

Afghanistan. After the invasion of the country by the Soviet Union, a jihad was waged to drive out the Soviet Army. Muslims from throughout the world joined the fight, forming prototypical revolutionary movements that culminated in the creation of organizations such as Al Qaeda. Al Qaeda remained active in the region bordered by Pakistan's Northwest Frontier province and along the border in Afghanistan's Paktia region.

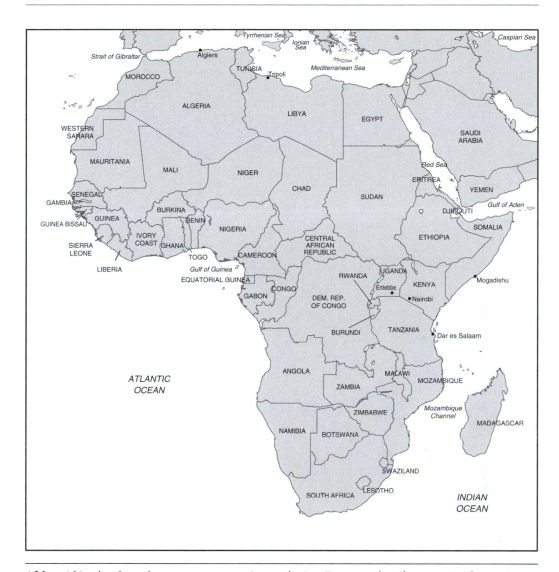

Africa. Africa has been home to most terrorist typologies. For example, Libya practiced terrorism as foreign policy. Religious dissident terrorism occurred in Algeria. Communal terrorism broke out in Rwanda. Communal and paramilitary terrorist violence became common in Somalia, Liberia, and Sierra Leone after the complete breakdown of government authority. International terrorist attacks occurred in Kenya and Tanzania.

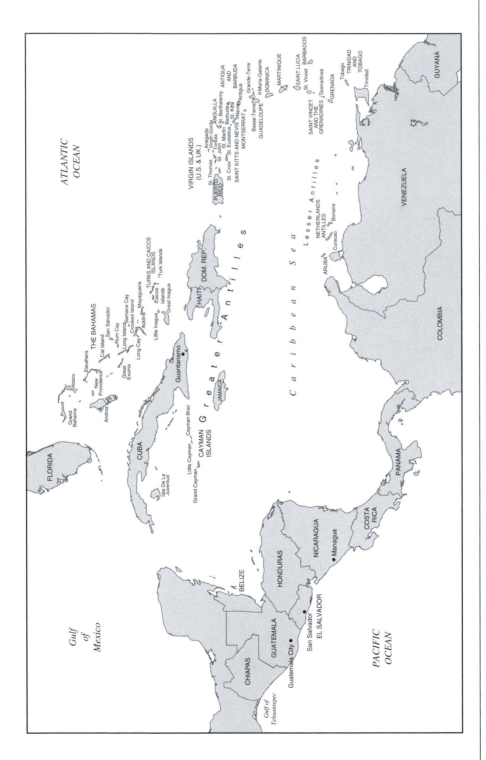

Central America and the Caribbean. Most terrorist violence originated in Cold War proxy conflicts and was waged by rebels, paramilitaries, and state security forces. Civil wars in El Salvador and Guatemala were markedly brutal, as was the contra insurgency in Nicaragua. The Zapatista insurgency in Mexico championed Chiapas Indians. Cuba was an active partner with the Soviet Union, fomenting and participating in conflicts in Africa and Latin America. The American base at Guantanamo Bay, Cuba, became the primary destination for suspected members of Islamic terrorist groups.

Europe. Terrorist violence in Europe originated from ideological, ethno-nationalist, and Middle Eastern sources. The Balkans, Northern Ireland, and the Basque region of Spain have been sources of ethno-nationalist terrorism. Ideological terrorism peaked in Germany and Italy during the 1970s, while Middle Eastern spillovers continued into the 21st century.

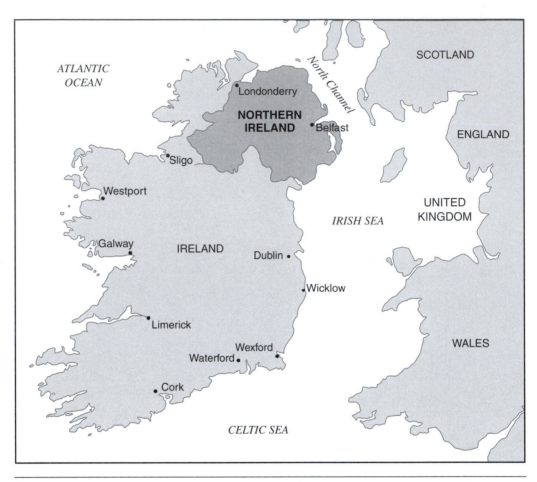

Ireland. When the Irish Free State was declared in 1921, six northern counties remained under British Rule. The modern politics of these counties, known as Northern Ireland, were marked by violent conflict between Protestant unionists and Catholic nationalists. This conflict was termed the Troubles.

Israel, the West Bank, and Gaza. Although a geographically small region, the area has been a center for communal and religious conflict that has frequently spilled over into the international domain and brought nations to the brink of war.

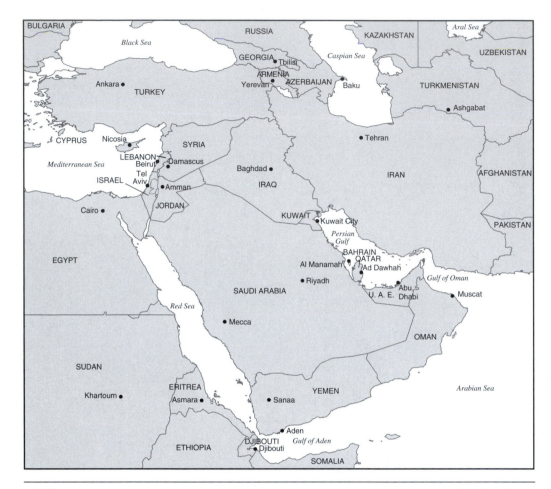

Middle East. Religious and ethno-nationalist terrorism have plagued the region for decades in innumerable conflicts.

Iraq. The regime of Saddam Hussein used terror as a regular instrument of state policy. During the Anfal campaign against ethnic Kurds in northern Iraq, the Iraqi army used chemical weapons against civilians. Iraq also invaded Iran and Kuwait and provided safe haven for secular terrorist operatives. However, the 2003 invasion of Iraq discovered no arsenals of weapons of mass destruction and no links to international Islamist terrorist networks such as Al Qaeda.

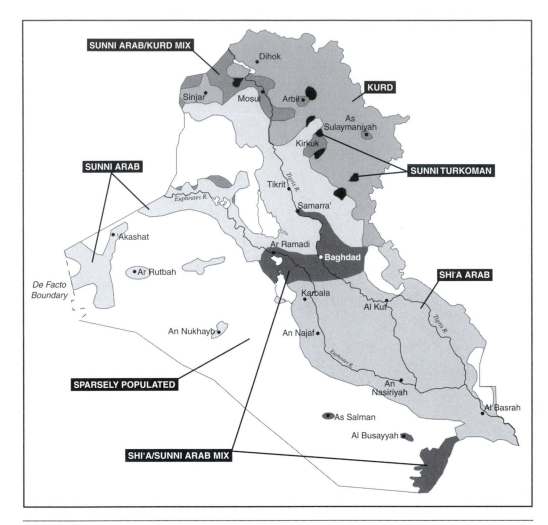

Distribution of Ethno-religious Groups in Iraq. Iraq is a diverse nation of several ethnicities and faiths. Although the predominant groups are Shi'a and Sunni Arabs, there are also sizable populations of Kurdish, Turkoman, Assyrian, and Christian Iraqis.

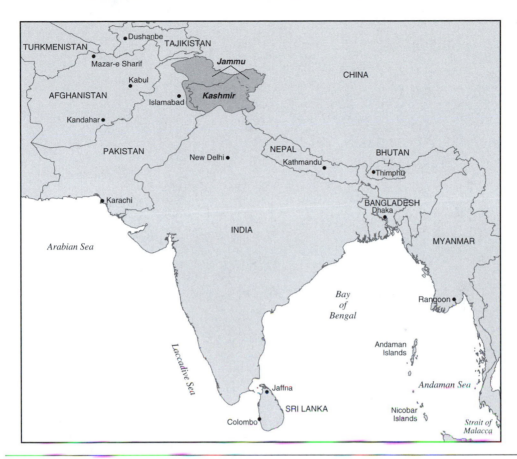

Indian Subcontinent. Several areas of confrontation resulted in terrorist violence. Conflicts in Sri Lanka, Kashmir, and Afghanistan have been particularly active, with ethnic and religious violence taking many thousands of lives.

Southeast Asia. Terrorism has typically originated from ethno-nationalist, ideological, state, and religious sources. Significant cases of dissident and state conflict occurred during and after the Cold War era. Religious terrorism became prominent in the Philippines and Indonesia in the aftermath of the September 11, 2001, attacks. The killing fields of Cambodia are a special case of exceptional state-sponsored violence.

South America. This continent has been a source for many case studies on terrorist environments. Cases include urban guerrillas in Montevideo and Buenos Aires; state repression in Chile, Argentina, and Uruguay; dissident terrorism in Peru; and criminal dissident terrorism in Colombia. The tri-border region became a center of suspected radical Islamist activism.

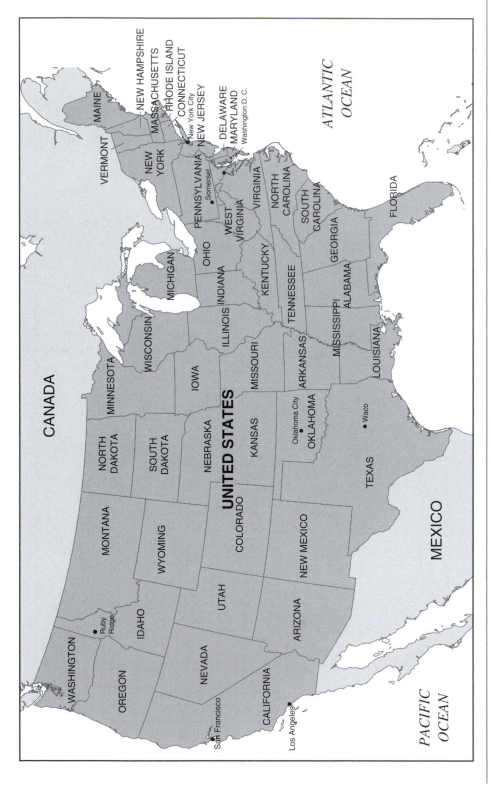

United States. Most terrorism in the United States has been characterized by low-intensity violence in urban areas, with the exceptions of the Oklahoma City bombing and the September 11, 2001, attacks.

Appendix B

National Intelligence Estimate: The Terrorist Threat to the US Homeland

In July 2007, the White House released a new National Intelligence Estimate (NIE), entitled The Terrorist Threat to the US Homeland. The report, published by the National Intelligence Council, was highly critical of central aspects of the Bush Administration's counterterrorist strategy. In particlular, the NIE concluded that the strategy for eliminating Al Qaeda's leadership on Pakistan had failed. It also concluded that Al Qaeda had been significantly strengthened during the two years prior to the NIE.

**National
Intelligence
Estimate**

NIC

**The Terrorist Threat to
the US Homeland**

July 2007

The Terrorist Threat to the US Homeland

July 2007

Office of the Director of National Intelligence

The Director of National Intelligence serves as the head of the Intelligence Community (IC), overseeing and directing the implementation of the National Intelligence Program and acting as the principal advisor to the President, the National Security Council, and the Homeland Security Council for intelligence matters.

The Office of the Director of National Intelligence is charged with:

- Integrating the domestic and foreign dimensions of US intelligence so that there are no gaps in our understanding of threats to our national security;
- Bringing more depth and accuracy to intelligence analysis; and
- Ensuring that US intelligence resources generate future capabilities as well as present results.

National Intelligence Council

Since its formation in 1973, the National Intelligence Council (NIC) has served as a bridge between the intelligence and policy communities, a source of deep substantive expertise on critical national security issues, and as a focal point for Intelligence Community collaboration. The NIC's key goal is to provide policymakers with the best, unvarnished, and unbiased information. Its primary functions are to:

- Support the DNI in his role as Principal Intelligence Advisor to the President and other senior policymakers.
- Lead the Intelligence Community's effort to produce National Intelligence Estimates (NIEs) and other NIC products that address key national security concerns.
- Provide a focal point for policymakers, warfighters, and Congressional leaders to task the Intelligence Community for answers to important questions.
- Reach out to nongovernment experts in academia and the private sector—and use alternative analyses and new analytic tools—to broaden and deepen the Intelligence Community's perspective.

NIEs are the DNI's most authoritative written judgments concerning national security issues. They contain the coordinated judgments of the Intelligence Community regarding the likely course of future events.

National Intelligence Estimates and the NIE Process

National Intelligence Estimates (NIEs) are the Intelligence Community's (IC) most authoritative written judgments on national security issues and designed to help US civilian and military leaders develop policies to protect US national security interests.

NIEs usually provide information on the current state of play but are primarily "estimative"—that is, they make judgments about the likely course of future events and identify the implications for US policy.

The NIEs are typically requested by senior civilian and military policymakers, Congressional leaders and at times are initiated by the National Intelligence Council (NIC). Before a NIE is drafted, the relevant National Intelligence Officer is responsible for producing a concept paper, or terms of reference (TOR), and circulates it throughout the Intelligence Community for comment. The TOR defines the key estimative questions, determines drafting responsibilities, and sets the drafting and publication schedule. One or more IC analysts are usually assigned to produce the initial text. The NIC then meets to critique the draft before it is circulated to the broader IC. Representatives from the relevant IC agencies meet to hone and coordinate line-by-line the full text of the NIE. Working with their Agencies, representatives also assign the level of confidence they have in key judgments. IC representatives discuss the quality of sources with collectors, and the National Clandestine Service vets the sources used to ensure the draft does not include any that have been recalled or otherwise seriously questioned.

All NIEs are reviewed by National Intelligence Board, which is chaired by the DNI and is composed of the heads of relevant IC agencies. Once approved by the NIB, NIEs are briefed to the President and senior policymakers. The whole process of producing NIEs normally takes at least several months.

The NIC has undertaken a number of steps to improve the NIE process under the DNI. These steps are in accordance with the goals and recommendations set out in the Senate Select Committee on Intelligence and WMD Commission reports and the 2004 Intelligence Reform and Prevention of Terrorism Act. Most notably, over the last two years the IC has:

- ♦ *Created new procedures to integrate formal reviews of source reporting and technical judgments.* The Director of CIA, as the National HUMINT Manager, as well as the Directors of NSA, NGA, and DIA and the Assistant Secretary/INR are now required to submit formal assessments that highlight the strengths, weaknesses, and overall credibility of their sources used in developing the critical judgments of the NIE.
- ♦ *Applied more rigorous standards.* A textbox is incorporated into all NIEs that explains what we mean by such terms as "we judge" and that clarifies the difference between judgments of likelihood and confidence levels. We have made a concerted effort to not only highlight differences among agencies but to explain the reasons for such differences and to display them prominently in the Key Judgments.

The US Homeland Threat Estimate:
How It Was Produced

The Estimate, *Terrorist Threats to the US Homeland,* followed the standard process for producing National Intelligence Estimates (NIEs), including a thorough review of sourcing, in-depth community coordination, the use of alternative analysis, and review by outside experts. Starting in October 2006, the NIC organized a series of roundtables with IC experts to scope out terms of reference (TOR) for the Estimate. Drafters from throughout the Community contributed to the draft. In May, a draft was submitted to IC officers in advance of a series of coordination meetings that spanned several days. The National Clandestine Service, FBI, and other IC collection officers reviewed the text for the reliability and proper use of the sourcing. As part of the normal coordination process, analysts had the opportunity—and were encouraged—to register "dissents" and provide alternative analysis. Reactions by the two outside experts who read the final product were highlighted in the text. The National Intelligence Board, composed of the heads of the 16 IC agencies and chaired by the ODNI, reviewed and approved the Estimate on 21 June. As with other NIEs, it is being distributed to senior Administration officials and Members of Congress.

What We Mean When We Say: An Explanation of Estimative Language

When we use words such as "we judge" or "we assess"—terms we use synonymously—as well as "we estimate," "likely" or "indicate," we are trying to convey an analytical assessment or judgment. These assessments, which are based on incomplete or at times fragmentary information are not a fact, proof, or knowledge. Some analytical judgments are based directly on collected information; others rest on previous judgments, which serve as building blocks. In either type of judgment, we do not have "evidence" that shows something to be a fact or that definitively links two items or issues.

Intelligence judgments pertaining to likelihood are intended to reflect the Community's sense of the probability of a development or event. Assigning precise numerical ratings to such judgments would imply more rigor than we intend. The chart below provides a rough idea of the relationship of terms to each other.

Remote	Unlikely	Even chance	Probably, Likely	Almost certainly

We do not intend the term "unlikely" to imply an event will not happen. We use "probably" and "likely" to indicate there is a greater than even chance. We use

(Continued)

words such as "we cannot dismiss," "we cannot rule out," and "we cannot dis-count" to reflect an unlikely—or even remote—event whose consequences are such it warrants mentioning. Words such as "may be" and "suggest" are used to reflect situations in which we are unable to assess the likelihood generally because relevant information is nonexistent, sketchy, or fragmented.

In addition to using words within a judgment to convey degrees of likelihood, we also ascribe "high," "moderate," or "low" confidence levels based on the scope and quality of information supporting our judgments.

♦ "High confidence" generally indicates our judgments are based on high-quality information and/or the nature of the issue makes it possible to render a solid judgment.

♦ "Moderate confidence" generally means the information is interpreted in various ways, we have alternative views, or the information is credible and plausible but not corroborated sufficiently to warrant a higher level of confidence.

♦ "Low confidence" generally means the information is scant, questionable, or very fragmented and it is difficult to make solid analytic inferences, or we have significant concerns or problems with the sources.

KEY JUDGMENTS

We judge the US Homeland will face a persistent and evolving terrorist threat over the next three years. The main threat comes from Islamic terrorist groups and cells, especially al-Qa'ida, driven by their undiminished intent to attack the Homeland and a continued effort by these terrorist groups to adapt and improve their capabilities.

We assess that greatly increased worldwide counterterrorism efforts over the past five years have constrained the ability of al-Qa'ida to attack the US Homeland again and have led terrorist groups to perceive the Homeland as a harder target to strike than on 9/11. These measures have helped disrupt known plots against the United States since 9/11.

♦ We are concerned, however, that this level of international cooperation may wane as 9/11 becomes a more distant memory and perceptions of the threat diverge.

Al-Qa'ida is and will remain the most serious terrorist threat to the Homeland, as its central leadership continues to plan high-impact plots, while pushing others in extremist Sunni communities to mimic its efforts and to supplement its capabilities. We assess the group has protected or regenerated key elements of its Homeland attack capability, including: a safehaven in the Pakistan Federally Administered Tribal Areas (FATA), operational lieutenants, and its top leadership. Although we have discovered only a handful of individuals in the United States with ties to al-Qa'ida senior leadership since 9/11, we judge that al-Qa'ida will intensify its efforts to put operatives here.

♦ As a result, we judge that the United States currently is in a heightened threat environment.

We assess that al-Qa'ida will continue to enhance its capabilities to attack the Homeland through greater cooperation with regional terrorist groups. Of note, we assess that al-Qa'ida will probably seek to leverage the contacts and capabilities of al-Qa'ida in Iraq (AQI), its most visible and capable affiliate and the only one known to have expressed a desire to attack the Homeland. In addition, we assess that its association with AQI helps al-Qa'ida to energize the broader Sunni extremist community, raise resources, and to recruit and indoctrinate operatives, including for Homeland attacks.

We assess that al-Qa'ida's Homeland plotting is likely to continue to focus on prominent political, economic, and infrastructure targets with the goal of producing mass casualties, visually dramatic destruction, significant economic aftershocks, and/or fear among the US population. The group is proficient with conventional small arms and improvised explosive devices, and is innovative in creating new capabilities and overcoming security obstacles.

♦ We assess that al-Qa'ida will continue to try to acquire and employ chemical, biological, radiological, or nuclear material in attacks and would not hesitate to use them if it develops what it deems is sufficient capability.

We assess Lebanese Hizballah, which has conducted anti-US attacks outside the United States in the past, may be more likely to consider attacking the Homeland over the next three years if it perceives the United States as posing a direct threat to the group or Iran.

We assess that the spread of radical—especially Salafi—Internet sites, increasingly aggressive anti-US rhetoric and actions, and the growing number of radical, self-generating cells in Western countries indicate that the radical and violent segment of the West's Muslim population is expanding, including in the United States. The arrest and prosecution by US law enforcement of a small number of violent Islamic extremists inside the United States— who are becoming more connected ideologically, virtually, and/or in a physical sense to the global extremist movement—points to the possibility that others may become sufficiently radicalized that they will view the use of violence here as legitimate. We assess that this internal Muslim terrorist threat is not likely to be as severe as it is in Europe, however.

We assess that other, non-Muslim terrorist groups—often referred to as "single-issue" groups by the FBI—probably will conduct attacks over the next three years given their violent histories, but we assess this violence is likely to be on a small scale.

We assess that globalization trends and recent technological advances will continue to enable even small numbers of alienated people to find and connect with one another, justify and intensify their anger, and mobilize resources to attack—all without requiring a centralized terrorist organization, training camp, or leader.

♦ The ability to detect broader and more diverse terrorist plotting in this environment will challenge current US defensive efforts and the tools we use to detect and disrupt plots. It will also require greater understanding of how suspect activities at the local level relate to strategic threat information and how best to identify indicators of terrorist activity in the midst of legitimate interactions.

Photo Credits

Part I

Part I Photo, page 1: U.S. Department of Defense.

Chapter 1

Photo 1.1, page 2: www.rewardsforjustice.net
Photo 1.2, page 4: Hulton Archive/Getty Images.
Photo 1.3, page 6: Hulton Archive/Getty Images.
Photo 1.4, page 6: U.S. Department of State.
Photo 1.5, page 7: U.S. Department of Defense.

Chapter 2

Photo 2.1, page 30: Hulton Archive/Getty Images.
Photo 2.2, page 32: Hulton Archive/Getty Images.
Photo 2.3, page 34: Hulton Archive/Getty Images.
Photo 2.4, page 36: Hulton Archive/Getty Images.
Photo 2.5, page 38: U.S. Department of Justice.

Chapter 3

Photo 3.1, page 46: Corbis.
Photo 3.2, page 54: Hulton Archive/Getty Images.
Photo 3.3, page 58: Hulton Archive/Getty Images.

Part II

Part II Photo, page 65: U.S. Department of Justice.

Chapter 4

Photo 4.1, page 67: U.S. Diplomatic Security Service.
Photo 4.2, page 74: Hulton Archive/Getty Images.
Photo 4.3, page 75: Hulton Archive/Getty Images.
Photo 4.4, page 77: Time & Life Pictures/Getty Images.

Chapter 5

Photo 5.1, page 92: Library of Congress.
Photo 5.2, page 94: Hulton Archive/Getty Images.
Photo 5.3, page 96: Hulton Archive/Getty Images.
Photo 5.4, page 101: Hulton Archive/Getty Images.
Photo 5.5, page 103: Hulton Archive/Getty Images.

Chapter 6

Photo 6.1, page 112: Hulton Archive/Getty Images.
Photo 6.2, page 115: Hulton Archive/Getty Images.

Photo 6.3, page 126: Hulton Archive/Getty Images.
Photo 6.4, page 127: U.S. Diplomatic Security Service.
Photo 6.5, page 129: Hulton Archive/Getty Images.

Chapter 7

Photo 7.1, page 142: U.S. Department of State.
Photo 7.2, page 146: U.S. Department of State.
Photo 7.3, page 147: Hulton Archive/Getty Images.
Photo 7.4, page 150: Hulton Archive/Getty Images.

Chapter 8

Photo 8.1, page 154: Hulton Archive/Getty Images.
Photo 8.2, page 165: Hulton Archives/Getty Images.
Photo 8.3, page 173: Federal Emergency
Management Agency (FEMA).
Photo 8.4, page 173: Federal Bureau of Investigation.

Part III

Part III Photo, page 179: U.S. Department of Defense.

Chapter 9

Photo 9.1, page 182: Hulton Archive/Getty Images.
Photo 9.2, page 189: U.S. Department of State.
Photo 9.3, page 191: U.S. Department of State.

Chapter 10

Photo 10.1, page 197: Hulton Archive/Getty Images.
Photo 10.2, page 214: U.S. Department of State.
Photo 10.3, page 216: Hulton Archive/Getty Images.
Photo 10.4, page 216: U.S. Department of State.

Chapter 11

Photo 11.1, page 226: Hulton Archive/Getty Images.
Photo 11.2, page 227: U.S. Department of Defense.
Photo 11.3, page 236: Hulton Archive/Getty Images.
Photo 11.4, page 243: www.antiwar.com.

Chapter 12

Photo 12.1, page 249: U.S. Department of Defense.
Photo 12.2, page 256: Federal Bureau of Investigation.
Photo 12.3, page 258: New York City Web site.
Photo 12.4, page 258: U.S. Department of Defense.

Glossary

The glossary summarizes terms used in this textbook. Readers should refer to the glossary to refresh their knowledge of discussions and case studies explored in chapters, tables, and chapter perspectives.

1st Navy Parachute Infantry Regiment An elite French unit that is similar to the British Special Air Service and the American Delta Force. It deploys small intelligence and special operations squads that are trained to operate in desert, urban, and tropical environments. They are part of the core of French counterterrorist special operations forces.

11th Parachute Division (Paras) An elite French army unit that can deploy as a quick-reaction force to deal with terrorist threats.

75th Ranger Regiment An elite combat unit of the U.S. Army that can deploy large formations for counterterrorist missions.

Abbas, Abu The leader of the Palestine Liberation Front.

Absolute deprivation A sociological term that indicates the lack of basic human needs for survival.

Abu Hafs Al-Masri Brigades An Al Qaeda-affiliated group which claimed responsibility for several significant terrorist attacks. These included the August 2003 bombing of the United Nations headquarters in Baghdad, Iraq, the November 2003 bombings of two synagogues in Turkey, and the March 11, 2004 attack on commuter trains in Madrid, Spain (so-called Operation Death Trains).

Abu Nidal Organization (ANO) The designation given to Abu Nidal's movement.

Abu Sayyaf A Muslim insurgency on the island of Basilan in the Philippines; the group has ideological and other links to Al Qaeda. Founded by Abdurajak Janjalani, who was killed by Filipino police in 1998.

Achille Lauro A cruise ship that was hijacked by members of the Palestine Liberation Front. During the incident, the terrorists murdered a wheelchair-bound Jewish American.

Act of political will The notion that one can force change by an absolute commitment to a cause. All that is required is complete and uncompromising dedication to achieving one's goals.

Afghan Arabs A term given to foreign volunteers, mostly Arabs, who fought as mujahideen during the war against the occupation of Afghanistan by the Soviet Army.

African National Congress (ANC) The principal anti-apartheid movement in predemocracy South Africa.

Air France Flight 8969 An airliner hijacked in December 1994 by the Algerian Armed Islamic Group. After the hijackers killed three passengers, the plane was permitted to depart the Algerian airfield and made a refueling stop in Marseilles. Intending to fly to Paris, the hijackers demanded three times the amount of fuel needed to make the journey. The reason for this demand was that they planned to blow up the aircraft over Paris or possibly crash it into the Eiffel Tower. French commandos disguised as caterers stormed the plane in Marseilles, thus bringing the incident to an end.

AK-47 A durable assault rifle designed by Mikhail Kalashnikov in the Soviet Union in 1947. It became a common weapon among conventional and irregular forces around the world.

Al-Aqsa Martyr Brigades A Palestinian nationalist movement affiliated with the Palestine Liberation Organization. Noted for its use of suicide bombers, it has committed terrorist violence against Israelis.

Albigensian Crusade A Christian Crusade in southern France during the 13th century. Legend holds that concerns were raised about loyal and innocent Catholics who were being killed along with members of the enemy Cathar sect. The pope's representative, Arnaud Amaury, allegedly replied, "Kill them all, God will know his own."

Al Jazeera An independent news service based in the Persian Gulf state of Qatar.

al-Megrahi, Abdel Basset An alleged agent of Libya's security service who was convicted and sentenced to life imprisonment by a Scottish court sitting near the Hague, Netherlands, for his participation in the bombing of Pan Am Flight 103 over Lockerbie, Scotland.

Al Qaeda An international network of Islamic mujahideen organized by Osama bin Laden in the aftermath of the anti-Soviet jihad in Afghanistan. Responsible for many acts of international and domestic terrorism.

Al Qaeda Organization for Holy War in Iraq A movement led by Abu Musab al-Zarqawi that waged holy war in Iraq against foreign interests, Shi'a organizations, and "apostate" Iraqis.

al-Sabbah, Hasan ibn The founder and leader of the Assassin movement during the 11th century.

Amal A pro-Syrian Lebanese Shi'a paramilitary movement.

Amir, Yigal A Jewish extremist who assassinated Israeli Prime Minister Yitzhak Rabin on November 4, 1995.

Ammonium nitrate and fuel oil explosives (ANFO) A powerful explosive compound made from ingredients easily obtained from fertilizers and common gasoline.

Amnesty International An international watchdog organization that monitors human rights issues around the world, focusing on the status of political prisoners.

Anarchism A political ideology developed during the 19th century that championed the working class and opposed central control by governments.

Anfal campaign A genocidal campaign waged by the Iraqi army in 1988 against its Kurdish population. Mustard gas and nerve agents were used against civilians.

Anglo-Israelism A mystical ideology developed by Richard Brothers during the 18th century. It posited that the Lost Ten Tribes of ancient Israel migrated through Europe and settled in the British Isles. Thus, the British were God's chosen people. The theory was adapted to the American context as the Christian Identity cult.

Animal Liberation Front An American-based single-issue movement that protests animal abuse. Responsible for committing acts of violence such as arson and vandalism.

Anthrax A disease afflicting farm animals that can also be contracted by humans. A possible ingredient for biological weapons.

Antistate terrorism Dissident terrorism directed against a particular government or group of governments.

Antiterrorism Official measures that seek to deter or prevent terrorist attacks. These measures include target hardening and enhanced security.

Apartheid The former policy of racial separation and white supremacy in South Africa.

Arafat, Yasir The founder and leader of the Palestine Liberation Organization until his death on November 11, 2004.

Armed Forces for National Liberation (Fuerzas Armadas de Liberación Nacional, or FALN) A Puerto Rican independencista terrorist group active during the 1970s and 1980s. Responsible for more bombings than any other single terrorist group in American history.

Armed Islamic Group (GIA) An Algerian Islamic resistance movement responsible for terrorist violence in Algeria and France.

Armed propaganda The use of symbolic violence to spread propaganda about an extremist movement.

Armed Resistance Unit A cover name used by the leftist May 19 Communist Organization in the United States.

Armed Revolutionary Nuclei An Italian neofascist terrorist group. Its most notorious act was the bombing of the main train station in Bologna, which killed 85 and injured 180.

Armenian Revolutionary Army One of several Armenian terrorist groups active in the postwar era that targeted Turkish interests to avenge the Turkish genocide of Armenians.

Armenian Secret Army for the Liberation of Armenia (ASALA) One of several

Armenian terrorist groups active in the postwar era that targeted Turkish interests to avenge the Turkish genocide of Armenians.

Army of God A shadowy and violent Christian fundamentalist movement in the United States that has attacked moralistic targets, such as abortion providers.

Aryan Nations A racial supremacist hate group founded in the mid-1970s by Richard Butler. Originally based in Idaho, the group is organized around Christian Identity mysticism.

Aryan Republican Army (ARA) A neo-Nazi terrorist group that operated in the midwestern United States from 1994 to 1996. Inspired by the example of the Irish Republican Army, the ARA robbed 22 banks in seven states before they were captured. Their goal had been to finance racial supremacist causes and to hasten the overthrow of what they called the Zionist Occupation Government. Some members also considered themselves Christian Identity fundamentalist Phineas Priests.

Asahara, Shoko The spiritual leader of the Aum Shinriko cult.

Ásatrú A mystical belief in the ancient Norse gods' pantheon. Some Ásatrú believers are racial supremacists.

Askaris Government-supported death squads in South Africa that assassinated members of the African National Congress and their supporters prior to the end of apartheid.

Assassination The act of killing a symbolic victim in a sudden and premeditated attack. Many assassinations are politically motivated.

Assault rifles Automatic military weapons that use rifle ammunition.

Asymmetrical warfare A term used to describe tactics, organizational configurations, and methods of conflict that do not use previously accepted or predictable rules of engagement.

Augustine A Christian philosopher who developed the concept of the just war.

Aum Shinrikyō A cult based in Japan and led by Shoko Asahara. Responsible for releasing nerve gas into the Tokyo subway system, injuring 5,000 people.

Authoritarian regimes Governments that practice strict control over public and political institutions and emphasize public order. The media and other public information outlets are regulated and censored

by the government. Some authoritarian regimes have democratic institutions.

Auto-genocide Self-genocide. When members of the same ethnic or religious group commit genocide against fellow members.

Axis of Evil In January 2002, U.S. President Bush identified Iraq, Iran, and North Korea as the axis of evil. In that speech, he promised that the United States "will not permit the world's most dangerous regimes to threaten us with the world's most destructive weapons."

Ayyash, Yehiya A Palestinian terrorist bomb maker affiliated with Hamas. He was responsible for scores of Israeli casualties and was eventually assassinated by a remotely controlled explosive device hidden in his cell phone. Also known as the Engineer.

Baader-Meinhof Gang A leftist terrorist movement active in West Germany during the 1970s and 1980s. Also referred to as the Red Army Faction.

Ba'ath Party A pan-Arab nationalist party.

Bakunin, Mikhail An early philosophical proponent of anarchism in Russia.

Basque Fatherland and Liberty (Euskadi Ta Azkatasuna, or ETA) Founded in 1959 to promote the independence of the Basque region in northern Spain. Although ETA adopted terrorism as a tactic in reply to the Franco government's violent repression of Basque nationalism, its campaign of violence escalated in the post-Franco era. At least six ETA factions and subfactions have been formed.

Beka'a Valley A valley in eastern Lebanon that became a center for political extremism and a safe haven for terrorists.

Berri, Nabih A leader of the Shi'a Amal paramilitary, prominent during the Lebanese Civil War during the 1970s and 1980s, as well as during the TWA Flight 847 incident.

Bhagwan Shree Rajneesh The leader of a Hindu-oriented cult based in the Pacific Northwest of the United States. The group was responsible for poisoning the salad bars of 10 Oregon restaurants with *Salmonella* bacteria in 1984. More than 700 people fell ill.

Bin Laden, Osama Founder and leader of the Islamic terrorist Al Qaeda network.

Biological agents A term used to refer to potential ingredients in biological weapons.

Biometric technology Digital technologies that allow digital photographs of faces to be matched against wanted suspects. Biometrics was used at American football's 2001 Super Bowl championship, when cameras scanned the faces of sports fans as they entered the stadium and matched their digital images against those of criminal fugitives and terrorists. The game became derisively known as the "Snooperbowl."

Birmingham Six Six men who were wrongfully convicted of a 1974 bombing of two pubs in Birmingham, England. They were released in 1991 after an appellate court ruled that the police had used fabricated evidence.

Black Hebrew Israelites An American racial supremacist cult that promotes African racial supremacy. Branches are also known as the Tribe of Judah, Nation of Israel, and Temple of Love. One branch based in Miami and led by Hulon Mitchell (also known as Yahweh Ben Yahweh) was responsible for at least 14 murders during the 1980s.

Black Hundreds An anti-Semitic movement in czarist Russia that was responsible for pogroms and other violence against Russian Jews during the late 19th and early 20th centuries.

Black Liberation Army (BLA) An African American terrorist group active during the 1970s. The BLA tended to target police officers and banks.

Black Panther Party for Self-Defense An African American nationalist organization founded in 1966 in Oakland, California. The Black Panthers eventually became a national movement. It was not a terrorist movement, but some members eventually engaged in terrorist violence.

Black Power An African American nationalist ideology developed during the 1960s that stressed self-help, political empowerment, cultural chauvinism, and self-defense.

Black Pride A movement in the United States that emphasized Afrocentric political agendas; experiments in economic development of African American communities; and cultural chauvinism that was expressed in music, art, and dress.

Black September A campaign waged by the Jordanian army in September 1970 to suppress what was perceived as a threat to Jordanian sovereignty from Palestinian fighters and leaders based in Jordan.

Black Widows The term given by the Russian media and authorities to Chechen women who participate in terrorist attacks against Russian interests. Many Black Widows engage in suicide operations, and such women either volunteer, are manipulated, or are coerced to enlist. They are allegedly the relatives of Chechen men who have been killed in the conflict.

Blacklisting A policy of prohibiting political activists from obtaining employment in certain industries.

Blood Libel, The A medieval European Christian slander that declared that the blood of Christian children was ritualistically used during Jewish holidays. The slander continues to be promulgated in the 21st century.

Bloody Sunday An incident on January 30, 1972, in Londonderry, Northern Ireland, when British paratroopers fired on demonstrators, killing 13 people.

Boland Amendment A bill passed by Congress in December 1982 that forbade the expenditure of U.S. funds to overthrow the Sandinista government.

Booth, John Wilkes A Confederate sympathizer who assassinated President Abraham Lincoln after the conclusion of the American Civil War.

Botulinum toxins (botulism) A rather common form of food poisoning. It is a bacterium rather than a virus or fungus and can be deadly if inhaled or ingested even in small quantities.

Bourgeoisie A term frequently used by Marxists to describe the middle class.

Brothers, Richard The English founder of Anglo-Israelism during the 18th century.

Bubonic plague Known as the Black Death in medieval Europe, this disease is spread by bacteria-infected fleas that infect hosts when bitten. The disease is highly infectious and often fatal.

Bureau for the Protection of the Constitution A government agency in Germany responsible for domestic security and intelligence.

Burke, Edmund The British father of conservatism. An 18th-century intellectual who denounced the excesses of the French Revolution and other challenges to the traditional European order.

Butler, Richard The founder and leader of the Aryan Nations organization. Also a promulgator of Christian Identity mysticism.

Cadre Politically indoctrinated and motivated activists. Frequently the core of a revolutionary movement.

Carlos "The Jackal" The *nom de guerre* for Ilich Ramirez Sanchez, a Venezuelan revolutionary who became an international terrorist. He acted primarily on behalf of the Popular Front for the Liberation of Palestine.

Carmichael, Stokely A leader of the Student Non-Violent Coordinating Committee in the American South during the 1960s. One of the first advocates of Black Power.

Carnivore A surveillance technology developed for use by the Federal Bureau of Investigation (FBI) that could reportedly monitor Internet communications. Under law, the FBI could not use Carnivore without a specific court order under specific guidelines, much like other criminal surveillance orders. The FBI eventually redesignated the system DCS-1000.

Castro, Fidel The leader of the Cuban Revolution.

Caudillo A term given to a strong leader in Spain. Francisco Franco was the caudillo during his regime.

Cell-based terrorist environment A terrorist environment in which terrorists have organized themselves into small semi-autonomous units rather than in traditionally hierarchical configurations. Difficult to combat because there is no central organization.

Cells Autonomous groups of terrorists who may be loosely affiliated with a larger movement but who are largely independent of hierarchical control.

Central Intelligence Agency (CIA) The principal intelligence agency in the United States. The theoretical coordinator of American foreign intelligence collection.

Chemical agents Chemicals that can potentially be converted into weapons. Some chemical agents, such as pesticides, are commercially available. Other chemical agents can be manufactured by terrorists using commonly available instruction guides.

Chesimard, JoAnne *See* Assata Shakur.

Child soldiers Children who have been pressed into military service.

Chlorine gas A chemical agent that destroys the cells that line the respiratory tract.

Christian Identity The American adaptation of Anglo-Israelism. A racial supremacist mystical belief that holds that Aryans are the chosen people of God, the United States is the Aryan Promised Land, nonwhites are soulless beasts, and Jews are biologically descended from the devil.

Christian Right A mostly Protestant fundamentalist movement in the United States that links strict evangelical Christian values to political agendas.

Cinque *See* DeFreeze, Donald.

Classical ideological continuum Symbolic political designations derived from the French Revolution. The concepts of left, center, and right have become part of modern political culture.

Codes of self-sacrifice Philosophical, ideological, or religious doctrines that create a warrior ethic in followers of the doctrine. Codes of self-sacrifice instill a sense of a "higher calling" that allows for the adoption of a superior morality. Followers consider acts of violence carried out in the name of the code completely justifiable.

Coercive covert operations Covert counterterrorist measures that seek to resolve terrorist crises by disrupting and destroying terrorist groups.

Collateral damage A term used to describe unintended casualties. Usually applied to civilians who have been mistakenly killed.

Collective nonviolence An activist philosophy of the civil rights movement in the American South that advocated peaceful civil disobedience.

Communal terrorism Group-against-group terrorism, in which rival demographic groups engage in political violence against each other.

Composite-4 (C-4) A powerful military-grade plastic explosive.

Concessionary options Conciliatory counterterrorist measures that seek to resolve terrorist crises by acceding to the terrorists' demands.

Conciliatory responses Counterterrorist measures that seek to resolve terrorist crises by addressing the underlying conditions that cause extremist violence.

Conservatism A political ideology that seeks to preserve traditional values.

Contagion effect Copycat terrorism in which terrorists imitate each other's behavior and tactics. This theory is still debated.

Contras Rightist Nicaraguan counter-revolutionaries trained and supported by the United States during the 1980s.

Counterculture A youth-centered movement in the United States and other Western countries during the 1960s and 1970s. It questioned status quo social and political values.

Counterterrorism Proactive policies that specifically seek to eliminate terrorist environments and groups. There are a number of possible categories of counterterrorist response, including counterterrorist laws, which specifically criminalize terrorist behavior and supportive operations.

Convention to Prevent and Punish Acts of Terrorism Taking the Form of Crimes Against Persons and Related Extortion That Are of International Significance A treaty among members of the Organization of American States that defined attacks against internationally protected persons as common crimes. Its goal was to establish common ground for recognizing the absolute inviolability of diplomatic missions.

Covert operations Counterterrorist measures that seek to resolve terrorist crises by secretly disrupting or destroying terrorist groups, movements, and support networks.

Crazy states States whose behavior is not rational, in which the people live at the whim of the regime or a dominant group. Some crazy states have little or no central authority and are ravaged by warlords or militias. Other crazy states have capricious, impulsive, and violent regimes in power that act out with impunity.

Creativity A mystical belief practiced by the racial supremacist World Church of the Creator in the United States. Creativity is premised on a rejection of the White race's reliance on Christianity, which is held to have been created by the Jews as a conspiracy to enslave Whites. According to Creativity, the White race itself should be worshipped.

Criminal profiles Descriptive profiles of criminal suspects developed by law enforcement agencies to assist in the apprehension of the suspects.

Crucifixion A form of public execution during the time of the Roman Empire It involved affixing condemned persons to a cross or other wooden platform. The condemned were either nailed through the wrist or hand, or tied upon the platform; they died by suffocation as their bodies sagged.

Crusades A series of Christian military campaigns during the Middle Ages instigated by the Pope and Western Christian rulers. Most of these campaigns were invasions of Muslim territories, although the Crusaders also attacked Orthodox Christians, conducted pogroms against Jews, and suppressed "heresies."

Cult of personality The glorification of a single strong national leader and political regime.

Customer anonymity A policy adopted in some countries that allows national banks to guarantee customer privacy.

Cyberterrorism The use of technology by terrorists to disrupt information systems.

Cyberwar The targeting of terrorists' electronic activities by counterterrorist agencies. Bank accounts, personal records, and other data stored in digital databases can theoretically be intercepted and compromised.

Days of Rage Four days of rioting and vandalism committed by the Weathermen in Chicago in October 1969.

DCS-1000 The redesignated name given to Carnivore, the FBI's Internet surveillance technology.

Death Night The term given to an incident that occurred in West Germany on October 18, 1977, when three imprisoned leaders of the Red Army Faction committed suicide with weapons that were smuggled into a high-security prison. Many Germans have never believed the West German government's official explanation for the events of Death Night, and some have suggested that the government was responsible for the deaths.

Death squads Rightist paramilitaries and groups of people who have committed numerous human rights violations. Many death squads in Latin America and elsewhere have been supported by the government and the upper classes.

Decommissioning The process of disarmament by the Irish Republican Army that was set as a condition for the Good Friday Agreement peace accords.

Defense Intelligence Agency (DIA) The central agency for military intelligence of the U.S. armed forces.

DeFreeze, Donald The leader of the Symbionese Liberation Army, a California-based American terrorist group active during the 1970s. DeFreeze adopted the nom de guerre Cinque.

Delta Force (1st Special Forces Operational Detachment–Delta) A secretive American special operations unit that operates in small covert teams. Its mission is similar to those of the British Special Air Service and French 1st Navy Parachute Infantry Regiment. Delta Force operations probably include abductions, reconnaissance, and punitive operations.

Democratic Front for the Liberation of Palestine (DFLP) A faction of the Palestine Liberation Organization, which split from the Popular Front for the Liberation of Palestine in 1969 and further split into two factions in 1991. It is a Marxist organization that believes in ultimate victory through mass revolution, and it has committed small bombings and assaults against Israel, including border raids.

Department of Homeland Security A new department of the U.S. government created to coordinate homeland security in the aftermath of the September 11, 2001, homeland attacks.

Dictatorship of the proletariat The Marxist belief that the communist revolution will result in the establishment of a working class-centered government.

DINA The Chilean secret service, allegedly responsible for human rights violations and domestic state terrorism during the 1970s and 1980s.

Diplomacy Conciliatory counterterrorist measures that seek to resolve terrorist crises by negotiating with terrorists or their supporters.

Diplomatic Security Service A security bureau within the U.S. Department of State that protects diplomats and other officials.

Direct action A philosophy of direct confrontation adopted by the Students for a Democratic Society and other members of the American New Left movement.

Direct Action (Action Directe) An anti-American and anti-NATO terrorist group in France that was founded in 1979. Direct Action operated mainly in Paris and committed approximately 50 terrorist attacks. It acted in solidarity with Germany's Red Army Faction, Italy's Red Brigades, Belgium's Combatant Communist Cells, and Palestinian extremists.

Directorate for Inter-Services Intelligence (ISI) The chief Pakistani security service.

Dirty bomb A highly toxic bomb which contains conventional bomb components and toxic substances such as radioactive materials or toxic chemicals. The conventional bomb sends out a cloud of radioactive or chemical toxins.

Dirty War A term given to a campaign of state-sponsored terror waged in Argentina during the 1970s. Tens of thousands of people were tortured, disappeared, or killed.

Disinformation Counterterrorist measures that seek to resolve terrorist crises by disseminating damaging information, thus perhaps causing internal dissension and distrust among the terrorists and their supporters.

Dissident terrorism Bottom-up terrorism perpetrated by individuals, groups, or movements in opposition to an existing political or social order.

Drug cartel A criminal cartel that is formed to regulate prices and output of illicit drugs. Many Colombian and Mexican traditional organized crime groups have been drug cartels.

Dynamite A commercially available high explosive that has nitroglycerin as its principal chemical ingredient.

Earth Liberation Front A single-issue movement that protests environmental degradation and pollution. A splinter group from the environmentalist group Earthfirst!, the ELF is potentially more radical than the Animal Liberation Front.

Echelon A satellite surveillance network maintained by the U.S. National Security Agency. It is a kind of global wiretap that filters through communications using antennae, satellite, and other technologies. Internet transfers, telephone conversations, and data transmissions are the types of communications that can reportedly be intercepted.

Economic sanctions Counterterrorist measures that seek to influence the behavior of terrorist states by pressuring their national economies.

Einsatzgruppen Special Nazi SS units that implemented genocidal Nazi policies in Eastern Europe.

Electromagnetic pulse (EMP) technologies Technologies using an electromagnetic burst from a generator that can disable electronic components such as microchips. If used on a sizable scale, EMP could destroy large quantities of military or financial information.

Electronic trigger Remotely controlled bombs are commonly employed by terrorists. The trigger is activated by a remote electronic or radio signal.

Emergency Search Team A paramilitary unit within the U.S. Department of Energy that has counterterrorist capabilities.

End justifies the means, The A concept wherein the desired goal is so just that the methods used to obtain the goal are acceptable regardless of their immediate consequences.

Engels, Friedrich Karl Marx's colleague and compatriot during the genesis of what was to become communist ideology.

Engineer, The The designation given to the Hamas bomb maker Yehiya Ayyash.

Enhanced security Counterterrorist measures that "harden" targets to deter or otherwise reduce the severity of terrorist attacks.

Ensslin, Gudrun A founder and leader of the Red Army Faction/Baader-Meinhof Gang in West Germany. She committed suicide in prison in October 1977 along with Andreas Baader and Jan-Carl Raspe. The incident became known as Death Night.

Episode-specific sponsorship State-sponsored terrorism limited to a single episode or campaign.

Establishment, The A designation coined by the New Left in the United States during the 1960s. It referred to mainstream American political and social institutions.

ETA (Euskadi Ta Azkatasuna) *See* Basque Fatherland and Liberty.

Ethnic cleansing A term created by Serb nationalists during the wars following the breakup of Yugoslavia. It described the suppression and removal of non-Serbs from regions claimed for Serb settlement. A euphemism for genocide.

Ethno-nationalist communal terrorism Violence involving conflict between populations that have distinct histories, customs, ethnic traits, religious traditions, or other cultural idiosyncrasies.

Euphemistic language Code words used by all participants in a terrorist environment to describe other participants and their behavior.

European Police Office (EUROPOL) A cooperative investigative consortium of members of the European Union. It has a mission similar to INTERPOL.

Extradition treaties International agreements to turn over criminal fugitives to the law enforcement agencies of fellow signatories.

Extremism Political opinions that are intolerant toward opposing interests and divergent opinions. Extremism forms the ideological foundation for political violence. Radical and reactionary extremists often rationalize and justify acts of violence committed on behalf of their cause.

Extremism in defense of liberty is no vice An uncompromising belief in the absolute righteousness of a cause. A moralistic concept that clearly defines good and evil. The statement was made by Senator Barry Goldwater during the 1964 presidential election in the United States.

Extraordinary renditions A practice initially sanctioned during the Reagan administration in about 1987 as a measure to capture drug traffickers, terrorists, and other wanted persons. It involves an uncomplicated procedure: Find the suspect anywhere in the world, seize them, transport them to the United States, and force their appearance before a federal court.

Falange The strongest right-wing party during the Spanish Civil War and during Franco's regime. It represented the Spanish variant of fascism.

Fallout Dangerous radioactive debris emitted into the atmosphere by a nuclear explosion that falls to earth as toxic material.

FALN *See* Armed Forces for National Liberation (Fuerzas Armadas de Liberación Nacional)

Far left The extremist, but not necessarily violent, left wing. Usually strongly influenced by Marxist ideology. Radical in political orientation.

Far right The extremist, but not necessarily violent, right wing. Reactionary in political orientation.

Fascism An ideology developed during the mid-20th century that emphasized strong state-centered authority, extreme law and order, militarism, and

nationalism. Variants of fascism were applied during the 1930s in Italy, Germany, and Spain, as well as in Latin America during the postwar era.

Fatah, al- The largest and most influential organization within the Palestine Liberation Organization. It formed the political foundation for PLO leader Yasir Arafat.

Federal Bureau of Investigation (FBI) An investigative bureau within the U.S. Department of Justice, it is the largest federal law enforcement agency. among its duties are domestic counterterrorism and intelligence collection.

Fhima, Lamen Khalifa An alleged agent of Libya's security service who was acquitted by a Scottish court sitting near the Hague, Netherlands, of charges that he participated in the bombing of Pan Am Flight 103 over Lockerbie, Scotland.

Force 17 An elite security unit within Fatah, founded in 1970. It has engaged in paramilitary and terrorist attacks and has served as Yasir Arafat's guard force.

Force 777 An elite Egyptian counterterrorist unit that has had mixed success in resolving terrorist crises.

Four Olds, The During the Great Proletarian Cultural Revolution in China, Maoists waged an ideological struggle to eliminate what they termed *the Four Olds*: old ideas, old culture, old customs, and old habits.

Fourteen Words A rallying slogan for the racial supremacist right wing in the United States. Originally coined by a convicted member of the terrorist group the Order, the Fourteen Words are: "We must secure the existence of our people and a future for White children."

Franco, Francisco The leader of the right-wing rebellion in Spain against the leftist republic during the Civil War of the 1930s. He became the dictator of postwar Spain.

Free press A media environment in which few official restrictions are placed on reporting the news. The free press relies on ethical and professional standards of behavior to regulate reporting practices.

Freedom Birds A term given to female combatants in the Liberation Tigers of Tamil Eelam.

Freedom fighter One who fights on behalf of an oppressed group. A very contextual term.

French Navy Special Assault Units Elite units of the French Navy that are trained for operations against seaborne targets, coastlines, and harbors. Their mission is similar to those of the British Special Boat Service and the U.S. Navy SEALs.

Fringe left The revolutionary left. Often violent.

Fringe right The revolutionary right. Often violent.

Fund for the Martyrs An Iranian fund established for the benefit of Palestinian victims of the intifada against Israel.

Furrow, Buford Oneal A former member of Aryan Nations who went on a shooting spree in Los Angeles, California, in August 1999. His spree included an attack at a Jewish community center in which five people were wounded, and the murder of an Asian mail carrier.

Gasoline bomb A simple explosive consisting of a gasoline-filled container with a detonator. Perhaps the most common gasoline bomb is the Molotov Cocktail.

Gender communal terrorism A conceptual designation describing political violence directed against the women of an ethno-national, religious, or political group.

Genocidal state terrorism State-initiated genocide. The state either involves itself directly in the genocidal campaign or deploys proxies to carry out the genocide.

Genocide The suppression of a targeted demographic group with the goal of repressing or eliminating its cultural or physical distinctiveness. The group is usually an ethno-national, religious, or ideological group.

GIGN (Groupe d'Intervention Gendarmerie Nationale) An elite paramilitary unit recruited from the French gendarmerie, or military police. GIGN is a counterterrorist unit with international operational duties.

Goddess of Democracy A statue erected by Chinese dissidents during the Tiananmen Square protests against dictatorship. Similar in design to the Statue of Liberty in the United States, it was crushed by a Chinese tank during the government's suppression of the protests.

Golani Brigade An elite unit within the Israel Defense Force that normally operates in conventionally organized military units. It has been deployed frequently for suppression campaigns against Hezbollah in South Lebanon.

Goldstein, Baruch A lone wolf terrorist who fired on worshippers inside the Ibrahim Mosque at the Cave of the Patriarch's holy site in the city of Hebron, Israel. According to official government estimates, he killed 29 people and wounded another 125. In reprisal for the Hebron massacre, the Palestinian Islamic fundamentalist movement Hamas launched a bombing campaign that included the first wave of human suicide bombers.

Good Friday Agreement A significant outcome of the Northern Ireland peace process. The agreement was overwhelmingly approved by voters on April 10, 1998, in the Irish Republic and Northern Ireland. It signaled the mutual acceptance of a Northern Ireland assembly and the disarmament, or decommissioning, of all paramilitaries. Also known as the Belfast Agreement.

Grail A designation for the Soviet-made SA-7 surface-to-air missile.

Great Proletarian Cultural Revolution A period from 1965 to 1969 in China during which the Communist Party instigated a mass movement to mobilize the young postrevolution generation. Its purpose was to eliminate so-called revisionist tendencies in society and to create a newly indoctrinated revolutionary generation.

Greater Jihad In Muslim belief, an individual struggle to do what is right in accordance with God's wishes. All people of faith are required to do what is right and good.

Green Berets (Special Forces Groups) A special operations unit of the U.S. Army. Green Berets usually operate in units called A Teams, comprising specialists whose skills include languages, intelligence, medicine, and demolitions. The traditional mission of the A Team is force multiplication.

Green Police The popular name for Israel's Police Border Guards.

Grey Wolves A rightist ultranationalist movement in Turkey that promotes the establishment of a Greater Turkey called Turan, which would unite all Turkish people in a single nation. Responsible for numerous acts of terrorism.

GSG-9 (Grenzschutzgruppe 9) An elite German paramilitary unit that was organized after the disastrous failed attempt to rescue Israeli hostages taken by Black September at the 1972 Munich Olympics. It is a paramilitary force that has been used domestically and internationally as a counterterrorist and hostage rescue unit.

Guerrilla A term first used during Spanish resistance against French occupation troops during the Napoleonic Wars. It refers to irregular hit-and-run tactics.

Guevara, Ernesto "Che" An Argentine revolutionary and intellectual who was instrumental in the success of the Cuban Revolution. Eventually killed in Bolivia, he developed his own philosophy of a continent-wide revolution in Latin America.

Guildford Four Four people who were wrongfully convicted of an October 1974 bombing in Guildford, England. They served 15 years in prison before being released in 1989 when their convictions were overturned on appeal.

Hague Convention of 1970 A treaty requiring the extradition or prosecution of hijackers.

Hague Conventions A series of international agreements that tried to establish rules for conflict.

Hale, Matthew Leader of the World Church of the Creator in the United States until his conviction in 2004 on charges of conspiring to assassinate a federal judge in Illinois.

Hamas (Islamic Resistance Movement) A Palestinian Islamic movement that waged a protracted terrorist campaign against Israel.

Hate crimes Crimes motivated by hatred against protected groups of people. They are prosecuted as aggravated offenses rather than as acts of terrorism.

Hawala An ancient transnational trust-based system used to transfer money via brokers.

Heinzen, Karl A 19th-century revolutionary theorist in Germany who advocated revolutionary terrorism and the use of explosives. Heinzen supported the acquisition of new weapons technologies to utterly destroy the enemies of the people. According to Heinzen, these weapons should include poison gas and new high-yield explosives.

Hezbollah A Lebanese Shi'a movement that promotes Islamic revolution. It was prominent in the resistance against the Israeli presence in South Lebanon and frequently engaged in terrorism.

Hitler, Adolf The leader of the National Socialist German Workers, or Nazi, Party in Germany during the 1930s and 1940s.

Ho Chi Minh The communist leader of Vietnam in the resistance against the French colonial presence. He later became the ruler of North Vietnam and continued his war by fighting the South Vietnamese government and the American military presence in the South.

Holy Spirit Mobile Force A cultic insurgency in Uganda inspired and led by Alice Lakwena. In late 1987, she led thousands of her followers against the Ugandan army. To protect themselves from death, the fighters anointed themselves with holy oil, which they believed would ward off bullets. Thousands of Lakwena's followers were slaughtered in 1987 in the face of automatic weapons and artillery fire.

Hostage Rescue Team (HRT) A paramilitary group organized under the authority of the U.S. Federal Bureau of Investigation. The HRT is typical of American paramilitary units in that it operates under the administrative supervision of federal agencies that perform traditional law enforcement work.

House Un-American Activities Committee A congressional committee created in the aftermath of a Red Scare during the 1930s to investigate threats to American security.

Human intelligence Intelligence that has been collected by human operatives rather than through technological resources. Also referred to as HUMINT.

Human Rights Watch An international watchdog organization that monitors human rights issues around the world.

Hussein, Saddam The ruler of Iraq who became a prominent figure in state sponsorship of international terrorism and the development of weapons of mass destruction during the 1980s and 1990s. He also attempted to annex Kuwait during the early 1990s, causing the Gulf War.

Ideological communal terrorism Communal violence in the postwar era that usually reflected the global ideological rivalry between the United States and the Soviet Union.

Ideologies Systems of political belief.

Imperialism A term used to describe the doctrine of national expansion and exploitation.

Improvised explosive devices (IEDs) So-called roadside bombs that were constructed and deployed by Iraqi insurgents during the U.S.-led occupation.

Information is power A political and popular concept that the control of the dissemination of information, especially through media outlets, enhances the power of the controlling interest.

Inkatha Freedom Party A Zulu-based movement in South Africa.

Intelligence The collection of data for the purpose of creating an informational database about terrorist movements and predicting their behavior.

Intelligence community The greater network of intelligence agencies. In the United States, the Central Intelligence Agency is the theoretical coordinator of intelligence collection.

International Court of Justice The principal judicial arm of the United Nations. Its 15 judges are elected from among member states, and each sits for a nine-year term. The court hears disputes between nations and gives advisory opinions to recognized international organizations.

International Criminal Court (ICC) A court established to prosecute crimes against humanity, such as genocide. Its motivating principle is to promote human rights and justice. In practice, this has meant that the ICC has issued arrest warrants for the prosecution of war criminals.

International Criminal Court for Rwanda (ICTY) A tribunal that has investigated allegations of war crimes and genocide resulting from the breakdown of order in Rwanda during the 1990s. The indictments against suspected war criminals detail what can only be described as genocide on a massive scale.

International Criminal Police Organization (INTERPOL) An international network of law enforcement that cooperates in the investigation of crimes. It is based in Ste. Cloud, France.

International Criminal Tribunal for the Former Yugoslavia A tribunal that has investigated allegations of war crimes and genocide arising out of the wars that broke out after the fragmentation of Yugoslavia during the 1990s. Several alleged war criminals, including former Yugoslavian president Slobodan Milosevic, have been brought before the court. Others remain at large but under indictment.

International mujahideen Islamic revolutionaries who have adopted a pan-Islamic ideology.

International terrorism Terrorism that is directed against targets symbolizing international interests. These attacks can occur against domestic targets that have international symbolism or against targets in the international arena.

Intifada The protracted Palestinian uprising against Israel. Literally, shaking off.

Iran-Contra scandal Lt. Col. Oliver North's efforts to circumvent the Boland Amendment were revealed by the press when a covert American cargo plane was shot down inside Nicaragua and an American mercenary was captured.

Irish National Liberation Army (INLA) The INLA grew out of the split in the IRA during the 1970s. The group adopted Marxist theory as its guiding ideology and fought to reunite Northern Ireland with Ireland. The INLA considered itself to be fighting in unity with other terrorist groups that championed oppressed groups around the world. Its heyday was during the 1970s and mid-1980s.

Islamic jihad A label adopted by some Islamic terrorists seeking to establish an Islamic state. Groups known as Islamic Jihad exist in Lebanon, Palestine, and Egypt.

It became necessary to destroy the town to save it An extremist goal to destroy an existing order without developing a clear vision for the aftermath. A moralistic concept to justify terrorist behavior. The statement was allegedly made by an American officer during the war in Vietnam.

Izzedine al-Qassam Brigade A militant movement within the overarching Hamas movement of Palestinian Islamic revolutionaries.

Jamahiriya Security Organization (JSO) The Libyan state security agency during the reign of Muammar el-Qaddafi. Apparently responsible for promoting Libya's policy of state-sponsored terrorism.

Jewish Defense League A militant Jewish organization founded in the United States by Rabbi Meir Kahane.

Jihad A central tenet in Islam that literally means a sacred "struggle" or "effort." Although Islamic extremists have interpreted jihad to mean waging holy war, it is not synonymous with the Christian concept of a crusade.

Jihadi One who wages jihad.

Joint operations State-sponsored terrorism in which state personnel participate in the terrorist enterprise.

Jones, Jim Founder and leader of the American apocalyptic cult People's Temple of the Disciples of Christ. Founded in California, the group relocated to Guyana in the late 1970s. More than 900 followers committed suicide in November 1978 by drinking flavored sugar water laced with poison.

Journalistic self-regulation The theoretical practice of ethical reporting among members of the press.

Jus ad bellum Correct conditions for waging war. An element of the Just War Doctrine.

Jus in bello One's correct behavior while waging war. An element of the Just War Doctrine.

Just War doctrine A moral and ethical doctrine that raises the questions of whether one can ethically attack an opponent, how one can justifiably defend oneself, and what types of force are morally acceptable in either context. The doctrine also addresses who can morally be defined as an enemy and what kinds of targets may be morally attacked.

Justice Commandos of the Armenian Genocide One of several Armenian terrorist groups active in the postwar era that targeted Turkish interests to avenge the Turkish genocide of Armenians.

Kach (Kahane Chai) Militant movements in Israel that carried on after Rabbi Meir Kahane was assassinated. They advocate the expulsion of Arabs from territories claimed as historically Jewish land. *Kach* means "only thus." *Kahane Chai* means "Kahane lives."

Kaczynski, Theodore "Ted" The Unabomber in the United States, who sent bombs hidden in letters and packages to protest technological society. A case study in politically motivated lone-wolf attacks. Also a case study of a medium degree of criminal skill.

Kahane, Rabbi Meir Founder and leader of the Jewish Defense League.

Kalashnikov, Mikhail The inventor of the famous AK-47 assault rifle.

Kansi, Mir Amal A terrorist who used an AK-47 assault rifle against employees of the Central Intelligence Agency who were waiting in their cars to enter the CIA's headquarters in Langley, Virginia. Two people were killed, and three were wounded. He was later captured in Pakistan, was sent to the United States for prosecution, and was convicted of murder.

Kerner Commission A term given to the presidentially appointed National Advisory Commission on Civil Disorders, which released a report in 1968 on civil disturbances in the United States.

Khaled, Leila A Palestinian nationalist who successfully hijacked one airliner and failed in an attempt to hijack another. She acted on behalf of the Popular Front for the Liberation of Palestine.

Khmer Rouge A Cambodian Marxist insurgency that seized power in 1975. During their reign, between 1 and 2 million Cambodians died, many of them in the infamous Killing Fields.

Khomeini, Ayatollah Ruhollah An Iranian religious leader who led the Iranian Revolution, which eventually ousted Shah Mohammad Reza Pahlavi.

Kidnapping/hostage-taking A method of propaganda by the deed in which symbolic individuals or small groups are taken captive as a way to publicize the terrorists' cause.

Kill one man, terrorize a thousand A paraphrasing of a quotation by the Chinese military philosopher Wu Ch'i. Variously ascribed to the Chinese military philosopher Sun Tzu and Chinese communist leader Mao Zedong.

Komiteh Revolutionary tribunals established after the Islamic revolution in Iran.

Kony, Josef Founder and leader of Uganda's cultic Lord's Resistance Army.

Kosovo A region in southern Yugoslavia that many Serbs consider their spiritual homeland. It became an international flash point when ethnic Albanian Kosovars sought to secede from Yugoslavia.

Kosovo Liberation Army (KLA) An ethnic Albanian dissident movement seeking independence for the Kosovo region of the former Yugoslavia. The group has used terrorism against Serb civilians and security forces.

Kropotkin, Petr An early philosophical proponent of anarchism in Russia.

Ku Klux Klan (KKK) A racial supremacist organization founded in 1866 in Pulaski, Tennessee. During its five eras, the KKK was responsible for thousands of acts of terrorism.

Kuclos A symbol adopted by the Ku Klux Klan consisting of a cross and a teardrop-like symbol enclosed by a circle. Literally, Greek for circle.

Kurdistan The regional homeland of the Kurdish people. It is divided between Iran, Iraq, Syria, and Turkey.

Kurds An ethno-national group in the Middle East. Several nationalist movements fought protracted wars on behalf of Kurdish independence.

La Cosa Nostra The American version of the Italian Mafia. Traditionally organized into family networks. Literally, "this thing of ours."

Labeling Attaching euphemistic terms to the participants in a terrorist environment.

Lakwena, Alice Leader of the cultic Holy Spirit Mobile Force in Uganda. She claimed to be inspired by the Christian Holy Spirit and preached that her movement would defeat government forces and purge Uganda of witchcraft and superstition. Her followers were annihilated, and she fled the country.

Laskar Jihad (Militia of the Holy War) An armed Islamic group organized in April 2000 in Indonesia. Under the leadership of Ja'afar Umar Thalib, the group waged a communal holy war in Indonesia, primarily against Christians on Indonesia's Molucca Islands.

Law enforcement The use of law enforcement agencies and criminal investigative techniques in the prosecution of suspected terrorists.

Leaderless resistance A cell-based strategy of the Patriot and neo-Nazi movements in the United States requiring the formation of "phantom cells" to wage war against the government and enemy interests. Dedicated Patriots and neo-Nazis believe that this strategy will prevent infiltration from federal agencies.

Left, center, right Designations on the classical ideological continuum. The left tends to promote social change. The center tends to favor incremental change and the status quo. The right tends to favor traditional values.

Legalistic responses Counterterrorist measures that use the law to criminalize specific acts as terrorist behaviors and that use law enforcement agencies to investigate, arrest, and prosecute suspected terrorists.

Lemkin, Dr. Raphael The creator of the word *genocide,* which first appeared in print in his influential book *Axis Rule in Occupied Europe,* published in 1944.

Lenin, Vladimir Ilich The Russian revolutionary leader and theorist who was the principal leader of the Russian Revolution of 1917. He became the first leader of the Soviet Union and was also the author of several books that were very influential in the international communist movement throughout the 20th century.

Lesser Jihad The defense of Islam against threats to the faith. This includes military defense and is undertaken when the Muslim community is under attack.

Liberalism A political ideology that seeks incremental and democratic change.

Liberation Tigers of Tamil Eelam (LTTE) A nationalist group in Sri Lanka that champions the independence of the Tamil people. Responsible for many acts of terrorism.

Liquid metal embrittlement The process of using chemicals to weaken metals. Such chemicals can embrittle, or make rigid, various metals.

Logistically supportive sponsorship State-sponsored terrorism in which the state provides a great deal of logistical support to the terrorists but stops short of directly participating in the terrorist incident or campaign.

Lone wolf model (lone wolves) A designation describing political violence committed by individuals who are motivated by an ideology but who have no membership in a terrorist organization.

Long hot summer A term used during the 1960s to describe urban racial tensions that sometimes led to rioting.

Long live death! A slogan of the Spanish Falange.

Lord's Resistance Army Josef Kony reorganized Uganda's cultic Holy Spirit Mobile Force into the Lord's Resistance Army. He blended together Christianity,

Islam, and witchcraft into a bizarre mystical foundation for his movement. The group was exceptionally brutal and waged near-genocidal terrorist campaigns—largely against the Acholi people that it claimed to champion.

Luddites A movement of English workers during the early 1800s; the Luddites objected to the social and economic transformations of the Industrial Revolution. They targeted the machinery of the new textile factories, and textile mills and weaving machinery were disrupted and sabotaged. After 17 Luddites were executed in 1813, the movement gradually died out.

Lumpenproletariat Karl Marx's designation of the nonproletarian lower classes. He considered it incapable of leading the revolution against capitalism.

Lynch mobs Lynch mobs were groups of white American vigilantes who murdered their victims by hanging, burning, or shooting them to death. Most victims of lynching were African Americans. Lynchings were sometimes carried out in a festive atmosphere.

M-16 The standard assault rifle for the U.S. military; first introduced in the mid-1960s.

MacDonald, Andrew *See* William Pierce.

Mala in se An act designated as a crime that is fundamentally evil, such as murder and rape.

Mala prohibita An act designated as a crime that is not fundamentally evil, such as prostitution and gambling.

Manifesto of the Communist Party The seminal document of communism, written by Karl Marx and Friedrich Engels.

Manipulation of the media The attempt by state and substate participants in a terrorist environment to control or otherwise affect the reporting of news by the media.

Mao Zedong The leader of the Chinese Revolution. His tactical and strategic doctrine of People's War was practiced by a number of insurgencies in the developing world. Mao's interpretation of Marxism was also very influential among communist revolutionaries.

Marighella, Carlos A Brazilian Marxist revolutionary and theorist who developed an influential theory for waging dissident terrorist warfare in urban environments.

Martyr nation A theoretical construct arguing that an entire ethno-national people is willing to endure any sacrifice to promote its liberation.

Martyrdom Martyrdom is achieved by dying on behalf of a religious faith or for some other greater cause. A common concept among religious movements.

Marx, Karl A mid-19th-century philosopher who, along with Friedrich Engels, developed the ideology of class struggle.

Marxism An ideology that believes in the historical inevitability of class conflict, culminating in the final conflict that will establish the dictatorship of the proletariat.

Mass communications The technological ability to convey information to a large number of people. It includes technologies that allow considerable amounts of information to be communicated through printed material, audio broadcasts, video broadcasts, and expanding technologies such as the Internet.

Mathews, Robert Jay The founder and leader of the American neo-Nazi terrorist group the Order, founded in 1983.

May 19 Communist Organization (M19CO) An American Marxist terrorist group that was active in the late 1970s and early 1980s. It was composed of remnants of the Republic of New Africa (described elsewhere in Glossary), the Black Liberation Army, the Weather Underground, and the Black Panthers. M19CO derived its name from the birthday of Malcolm X and Vietnamese leader Ho Chi Minh.

McCarthy, Senator Joseph A Wisconsin senator who initiated a purge of suspected communists in the United States during the early 1950s.

McVeigh, Timothy A member of the Patriot movement in the United States and probably a racial supremacist. Responsible for the constructing and detonating an ANFO bomb that destroyed the Alfred P. Murrah Federal Building in Oklahoma City, Oklahoma, on April 19, 1995. One hundred sixty-eight people were killed.

Means of production A Marxist concept describing the primary source of economic production and activity during the stages of human social evolution.

Media as a weapon For terrorists and other extremists, information can be wielded as a weapon of war. Because symbolism is at the center of most terrorist incidents, the media are explicitly identified by terrorists as potential supplements to their arsenal.

Media gatekeeping Similar to journalistic self-regulation. The theoretical practice of ethical self-regulation by members of the free press.

Media scooping Being the first to obtain and report exclusive news.

Media spin The media's inclusion of subjective and opinionated interpretations when reporting the facts.

Media-oriented terrorism Acts of terrorism carried out to attract and manipulate media coverage.

Meinhof, Ulrike A founder and leader of the Red Army Faction/Baader-Meinhof Gang in West Germany. Meinhof hanged herself in prison on May 9, 1976.

Meredith, James An American civil rights activist who was ambushed, shot, and wounded in June 1966 as he walked through Mississippi to demonstrate that African Americans could safely go to polling places to register to vote.

MI-5 An intelligence agency in Great Britain responsible for domestic intelligence collection.

MI-6 An intelligence agency in Great Britain responsible for international intelligence collection.

Military Intelligence Service A government agency in Germany responsible for international intelligence collection.

Militias Organized groups of armed citizens who commonly exhibit antigovernment tendencies and subscribe to conspiracy theories. The armed manifestation of the Patriot movement.

Mines Military-grade explosives that are buried in the soil or rigged to be detonated as booby traps. Antipersonnel mines are designed to kill people, and antitank mines are designed to destroy vehicles. Many millions of mines have been manufactured and are available on the international market.

Mini-Manual of the Urban Guerrilla An influential essay written by Carlos Marighella that outlined his theory of urban dissident terrorist warfare.

Mitchell, Hulon, Jr. Also known as Yahweh Ben Yahweh (God, Son of God). Founder and leader of a Miami branch of the Black Hebrew Israelites. He taught that Whites are descendants of the devil and worthy of death. Followers who did not practice complete obedience were dealt with harshly, and some were beaten or beheaded. Mitchell and some of his followers were imprisoned. Mitchell's group was implicated in 14 murders during the 1980s.

Molotov Cocktail A simple gasoline bomb consisting of a gasoline-filled bottle with a rag inserted as a wick.

Molotov, Vyacheslav The Soviet foreign minister during the Second World War, for whom the Molotov Cocktail is named.

Monolithic terrorist environment A terrorist environment in which a single national sponsor supports and directs international terrorism. Combating this environment is relatively uncomplicated, because attention can theoretically be directed against a single adversary.

Montoneros A terrorist movement in Argentina during the 1970s. They espoused radical Catholic principles of justice, Peronist populism, and leftist nationalism. The Montoneros became skillful kidnappers and extorted an estimated $60 million in ransom payments. Shootings, bombings, and assassinations were also pervasive. When the military seized control in March 1976, all political opposition was crushed, including the terrorist campaign.

Montreal Convention of 1971 A treaty that extended international law to terrorist attacks on airports and grounded aircraft.

Moussaoui, Zacarias A suspected member of the cell that carried out the attacks of September 11, 2001. He was being held in custody on other charges at the time of the attacks and was later prosecuted in federal court.

Movimiento Nacional Francisco Franco's movement, which consolidated his power and established the model for Spanish fascism.

Mud People A derogatory term given to non-Aryans by followers of the Christian Identity movement. Mud People are considered nonhuman, soulless beasts who dwelt outside the Garden of Eden.

Mujahideen Individuals who wage war in defense of Islam. Literally, holy warriors.

Multinational corporations Large corporations that conduct business on a global scale. They are usually centered in several countries.

Muslim Brotherhood A transnational Sunni Islamic fundamentalist movement that is very active in several North African and Middle Eastern countries. It has been implicated in terrorist violence committed in Egypt, Syria, and elsewhere.

Mussolini, Benito The Italian dictator who led the first successful fascist seizure of power during the 1920s.

Mustard gas A chemical agent that is a mist rather than a gas. It is a blistering agent that blisters the skin, eyes, and nose, and it can severely damage the lungs if inhaled.

Narco-terrorism Political violence committed by dissident drug traffickers who are primarily concerned with protecting their criminal enterprise. This is in contradistinction to drug-related violence.

Narcotraficantes Latin American drug traffickers.

Narodnaya Volya (People's Will) A 19th-century terrorist group in Russia.

National Alliance An overtly Nazi organization based in West Virginia that was founded in 1970 by William Pierce, a former member of the American Nazi Party.

National Liberation Army (Ejercito de Liberacion Nacional, or ELN) A Marxist insurgency founded in the 1960s that has operated primarily in the countryside of Colombia. Its ideological icons are Fidel Castro and Ernesto "Che" Guevara. The ELN has engaged in bombings, extortion, and kidnappings. Targets have included foreign businesses and oil pipelines. The ELN has also participated in the drug trade.

National Security Agency (NSA) An American intelligence agency charged with signal intelligence collection, code making, and code breaking.

Nationalist dissident terrorism Political violence committed by members of ethno-national groups who seek greater political rights or autonomy.

Nativism American cultural nationalism. A cornerstone of Ku Klux Klan ideology.

Nazi Holocaust The genocide waged against European Jews by Germany before and during World War II. The first significant anti-Semitic racial decree promulgated by the Nazis was the Law for the Protection of German Blood and German Honor, passed in September 1935. In the end, approximately 6 million Jews were murdered.

Nechayev, Sergei An early philosophical proponent of anarchism in Russia. Author of *Revolutionary Catechism*.

Neocolonialism A postwar Marxist concept describing Western economic exploitation of the developing world.

Nerve gas A chemical agent, such as sarin, tabun, and VX, that blocks (or short-circuits) nerve messages in the body. A single drop of a nerve agent, whether inhaled or absorbed through the skin, can shut down the body's neurotransmitters.

Netwar An emerging method of conflict that uses network forms of organization and information-age strategies, doctrines, and technologies. Participants in these networks are dispersed small groups who operate as a "flat" organizational network rather than under chains of command.

New African Freedom Fighters The self-defined military wing of an African American nationalist organization called the Republic of New Africa. Composed of former members of the Black Liberation Army and Black Panthers, the group operated in collaboration with other members of the revolutionary underground. It was eventually broken up in 1985 after members were arrested for conspiring to free an imprisoned comrade, bomb the courthouse, and commit other acts of political violence.

New Left A movement of young leftists during the 1960s who rejected orthodox Marxism and took on the revolutionary theories of Frantz Fanon, Herbert Marcuse, Carlos Marighella, and other new theorists.

New Media The use of existing technologies and alternative broadcasting formats to analyze and disseminate information. These formats include talk-show models, tabloid styles, celebrity status for hosts, and blatant entertainment spins. Strong and opinionated political or social commentary also makes up a significant portion of New Media content.

New Order, The An American neo-Nazi group broken up by federal agents in March 1998 in East St. Louis, Illinois. They had modeled themselves after the Order and were charged with planning to bomb the Anti-Defamation League's New York Headquarters; the headquarters of the Southern Poverty Law Center in Birmingham, Alabama; and the Simon Wiesenthal Center in Los Angeles.

New Terrorism A typology of terrorism characterized by a loose cell-based organizational structure, asymmetrical tactics, the threatened use of weapons of mass destruction, potentially high casualty rates, and usually a religious or mystical motivation.

New World Liberation Front The name of an American terrorist group active during the mid-1970s. Organized as a "reborn" manifestation of the Symbionese Liberation Army by former SLA members and new recruits.

New World Order A conspiracy theory common among American neo-Nazis and members of the Patriot movement. It holds that non-American interests are threatening to take over—or have already taken over—key governmental centers of authority. This takeover is part of an international plot to create a one-world government.

News triage The decision-making process within the media, which decides what news to report and how to report it.

Nidal, Abu A Palestinian nationalist responsible for numerous acts of terrorism. Frequently an agent of state-sponsored terrorism.

Nihilism A 19th-century Russian philosophical movement of young dissenters who believed that only scientific truth could end ignorance. Nihilists had no vision for a future society; they asserted only that the existing society was intolerable. Modern nihilist dissidents exhibit a similar disdain for the existing social order but offer no clear alternative for after its destruction.

Nihilist dissidents Antistate dissidents whose goal is to destroy the existing social order with little consideration given for the aftermath of the revolution. They practice "revolution for revolution's sake."

Nihilist dissident terrorism The practice of political violence with the goal of destroying an existing order, committed with little or no regard for the aftermath of the revolution.

Northern Ireland Act A law passed in Northern Ireland in 1993 that created conditions of quasi-martial law. The Act suspended several civil liberties and empowered the British military to engage in warrantless searches of civilian homes, temporarily detain people without charge, and question suspects. The

military could also intern suspected terrorists and turn over for prosecution those for whom enough evidence had been seized.

Nuclear weapons High-explosive military weapons using weapons-grade plutonium and uranium. Nuclear explosions devastate the area within their blast zone, irradiate an area outside the blast zone, and are capable of sending dangerous radioactive debris into the atmosphere that descends to earth as toxic fallout.

Off the grid A tactic used by hardcore members of the Patriot movement who believe in the New World Order conspiracy theory. Believers typically refuse to use credit cards, drivers' licenses, and Social Security, as a way to lower their visibility from the government, banks, and other potential agents of the New World Order.

Office of Homeland Security An American agency established immediately after the September 11 attacks. Its ostensible purpose was to coordinate domestic security. Because the Office lacked the authority to set policy for other federal agencies, it was replaced by the Department of Homeland Security.

Official Secrets Act An act in Great Britain that permitted the prosecution of individuals for the reporting of information that was deemed to endanger the security of the British government.

Official state terrorism Terrorism undertaken as a matter of official government policy.

Okhrana The secret police of czarist Russia. Responsible for writing the anti-Semitic *Protocols of the Learned Elders of Zion.*

One man willing to throw away his life is enough to terrorize a thousand The symbolic power of a precise application of force by an individual who is willing to sacrifice himself can terrorize many other people. A moralistic concept that illustrates how a weak adversary can influence a strong adversary. The statement was made by the Chinese military philosopher Wu Ch'i.

One person's terrorist is another person's freedom fighter The importance of perspective in the use of violence to achieve political goals. Championed groups view violent rebels as freedom fighters, whereas their adversaries consider them terrorists.

One-Seedline Christian Identity A Christian Identity mystical belief that argues that all humans, regardless of race, are descended from Adam. However, only Aryans (defined as northern Europeans) are the true elect of God. They are the Chosen People whom God has favored and who are destined to rule over the rest of humanity. In the modern era, those who call themselves the Jews are actually descended from a minor Black Sea ethnic group and therefore have no claim to Israel.

Operation Death Trains The synchronized detonation of ten bombs aboard commuter trains in Madrid, Spain on March 11, 2004. The operation, carried out by the Abu Hafs al-Masri Brigades, killed 191 people and wounded more than 1,500. The attack led to the ouster of Spanish Prime Minister Jose Maria Aznar's government, and the withdrawal of 1,300 Spanish troops from Iraq.

Operation Eagle Claw An operation in April 1980 launched by the United States to rescue Americans held hostage by Iran at the U.S. embassy in Tehran. The operation failed on the ground in Iran when a helicopter flew into an airplane, and both exploded. Eight soldiers were killed, and the mission was aborted.

Operation Enduring Freedom The designation given to the counterterrorist war waged in the aftermath of the September 11, 2001, homeland attacks.

Operation Infinite Justice The original designation given to the counterterrorist war waged in the aftermath of the attacks of September 11, 2001.

Operation Iraqi Freedom The designation given to the U.S.-led invasion of Iraq in March 2003.

Operation Peace for Galilee An invasion of Lebanon by the Israeli army in June 1982, with the goal of rooting out PLO bases of operation. It was launched in reply to ongoing PLO attacks from its Lebanese bases.

ORDEN A right-wing Salvadoran paramilitary that engaged in death squad activity.

Order of Assassins A religious movement established in the Middle East during the 11th century. It sought to purge the Islamic faith and resist the Crusader invasions. The Assassins were noted for using stealth to kill their opponents.

Order, The An American neo-Nazi terrorist group founded by Robert Jay Mathews in 1983. Centered in the Pacific Northwest, the Order's methods for fighting its war against what it termed the Zionist Occupation Government were counterfeiting, bank robberies, armored car robberies, and murder. The Order had been suppressed by December 1985.

Osawatomie An underground periodical published by the Weather Underground Organization during the 1970s.

Pahlavi, Shah Mohammad Reza The last Shah of Iran, who was ousted in an Islamic revolution led by the Ayatollah Ruhollah Khomeini.

Palestine Islamic Jihad (PIJ) The PIJ is not a single organization but a loose affiliation of factions. It is an Islamic fundamentalist revolutionary movement that seeks to promote jihad, or holy war, and to form a Palestinian state; it is responsible for assassinations and suicide bombings.

Palestine Liberation Front (PLF) The PLF split from the Popular Front for the Liberation of Palestine–General Command in the mid-1970s and further split into pro-PLO, pro-Syrian, and pro-Libyan factions. The pro-PLO faction was led by Abu Abbas, who committed a number of attacks against Israel.

Palestine Liberation Organization (PLO) An umbrella Palestinian nationalist organization. It comprises numerous activist factions, many of which engage in political violence.

Palmer Raids A series of raids in the United States during the administration of American President Woodrow Wilson, targeting communist and other leftist radical groups.

Palmer, R. Mitchell The attorney general who served during the administration of President Woodrow Wilson.

Pan Am Flight 103 An airliner that exploded over Lockerbie, Scotland, on December 21, 1988. In the explosion, 270 people were killed, including all 259 passengers and crew and 11 persons on the ground. Libya was implicated in the incident.

Pan-Arabism An international Arab nationalist movement that held momentum after World War II.

Parachute Sayaret An elite reconnaissance unit within the Israel Defense Force. It has been deployed in small and large units, relying on high mobility to penetrate deep into hostile territory. They participated in the Entebbe rescue and were used against Hezbollah in South Lebanon.

Paramilitaries A term used to describe rightist irregular units and groups that are frequently supported by governments or progovernment interests. Many paramilitaries have been responsible for human rights violations.

Paramilitary repressive options Counterterrorist measures that seek to resolve terrorist crises by deploying armed nonmilitary personnel. These personnel can include covert operatives, localized militia units, or large armed units.

Party of God Lebanon's Hezbollah.

Peace process An ongoing process of negotiations between warring parties with the goal of addressing their underlying grievances and ending armed conflict.

People's Liberation Army The Chinese communist national army, founded by Mao Zedong.

People's Temple of the Disciples of Christ A cult founded and led by Jim Jones. After relocating to Guyana, the hundreds of members of the group committed ritual suicide when followers shot and killed Congressman Leo Ryan and several others; 914 people died, including 200 children.

People's war A concept in irregular warfare in which the guerrilla fighters and the populace are theoretically indistinguishable.

People's Will *See* Narodnaya Volya

Phalangists A Lebanese Christian paramilitary movement.

Phansi A rope used by the Thuggees of India to ritualistically strangle their victims.

Phantom cells An organizational concept articulated by former Klansman Louis Beam in the early 1990s. Rightist dissidents were encouraged to organize themselves into autonomous subversive cells that would be undetectable by the enemy U.S. government or agents of the New World Order.

Phineas Priesthood A shadowy movement of Christian Identity fundamentalists in the United States who believe that they are called by God to purify their race and Christianity. They are opposed to abortion, homosexuality, interracial mixing, and Whites who "degrade" White racial supremacy. It is a calling for men only, so no women can become Phineas Priests. The name is taken from the Bible at chapter 25, verse 6 of the Book of Numbers, which tells the story of a Hebrew man named Phineas who killed an Israelite man and his Midianite wife in the temple.

Phoenix Program A three-year campaign conducted during the Vietnam War to disrupt and eliminate the administrative effectiveness of the communist Viet Cong.

Phosgene gas A chemical agent that causes the lungs to fill with water, choking the victim.

Pierce, William The founder and leader of the neo-Nazi National Alliance in the United States. Also author of *The Turner Diaries,* under the nom de plume of Andrew MacDonald.

Pipe bombs Devices constructed from common pipes that are filled with explosives (usually gunpowder) and then capped on both ends. Nuts, bolts, screws, nails, and other shrapnel are usually taped or otherwise attached to these devices.

Plastic explosives A malleable explosive compound commonly used by terrorists.

Pogroms Anti-Semitic massacres in Europe that occurred periodically from the time of the First Crusade through the Nazi Holocaust. Usually centered in Central and Eastern Europe.

Pol Pot The principal leader of Cambodia's Khmer Rouge during its insurgency and consolidation of power.

Police Border Guards An elite Israeli paramilitary force that is frequently deployed as a counterterrorist force. Known as the Green Police, it operates in two subgroups: YAMAS is a covert group that has been used extensively during the Palestinian intifada, and YAMAM was specifically created to engage in counterterrorist and hostage rescue operations.

Politically sympathetic sponsorship State-sponsored terrorism that does not progress beyond ideological and moral support.

Pope Urban II The Roman Catholic pope who commissioned the first Christian Crusade with the goal of seizing Jerusalem and other territories from Muslim control.

Popular Front for the Liberation of Palestine (PFLP) The PFLP was founded in 1967 by George Habash. It is a

Marxist organization that advocates a multinational Arab revolution, and it has been responsible for dramatic international terrorist attacks. Its hijacking campaign in 1969 and 1970, its collaboration with West European terrorists, and its mentorship of Carlos the Jackal arguably established the model for modern international terrorism.

Popular Front for the Liberation of Palestine– General Command (PFLP-GC) Ahmid Jibril formed the PFLP-GC in 1968 when he split from the Popular Front for the Liberation of Palestine (PFLP) because he considered the PFLP too involved in politics and not sufficiently committed to the armed struggle against Israel. The PFLP-GC was probably directed by Syria and has been responsible for many cross-border attacks against Israel.

Port Huron Statement A document crafted in the United States in 1962 by members of Students for a Democratic Society. It harshly criticized mainstream American values and called for the establishment of a New Left movement in the United States.

Prairie Fire An underground manifesto published by the Weather Underground Organization during the 1970s.

Precision-guided munitions (PGM) Technologically advanced weapons that can be remotely guided to targets. Some PGMs are referred to as smart bombs.

Preemptive strikes Counterterrorist measures that proactively seek out and attack terrorist centers prior to a terrorist incident.

Pressure trigger Weapons such as mines are detonated when physical pressure is applied to a trigger. A variation on physical pressure triggers is trip-wire booby traps. More sophisticated pressure triggers react to atmospheric (barometric) pressure, such as changes in pressure when an airliner ascends or descends.

Progressive Labor Party A Maoist faction of the Students for a Democratic Society in the United States. Active during the late 1960s and 1970s.

Proletariat A Marxist term for the working class.

Propaganda by the deed The notion that revolutionaries must violently act upon their beliefs to promote the ideals of the revolution. Originally promoted by the anarchists.

Property is theft The anarchist philosopher Pierre-Joseph Proudhon's belief that systems based on the acquisition of private property are inherently exploitative.

Protocols of the Learned Elders of Zion A forgery written under the direction of the czarist secret police in the late 19th century. It purports to be the proceedings of a secret international society of Jewish elders who are plotting to rule the world.

Proudhon, Pierre-Joseph Nineteenth-century philosopher and father of anarchism.

Provisional Irish Republican Army (Provos) A terrorist organization in Northern Ireland that champions the rights of Northern Irish Catholics. The PIRA was formed with the goal of uniting Northern Ireland with the Irish Republic. Also known as the Provos.

Provos The popular name given to the Provisional Irish Republican Army.

Publicize the cause The practice by terrorists of disseminating information about their grievances and championed groups. This can be done through symbolic violence and the manipulation of the media.

Punitive strikes Counterterrorist measures that seek out and attack terrorist centers to damage the terrorist organization. Frequently conducted as retribution for a terrorist incident.

Qaddafi, Muammar el- The ruler of Libya who became a prominent figure in state sponsorship of international terrorism during the 1980s.

Qods (Jerusalem) Force A unit of Iran's Revolutionary Guards Corps that promotes the "liberation" of Jerusalem from non-Muslims.

Racial holy war (Rahowa) A term given by racial supremacists to a future race war that they believe will, inevitably, occur in the United States.

Racial profiling Similar to criminal profiling, but it uses race or ethnicity as the overriding descriptor to assist in the apprehension of suspects. Race is a legitimate element for criminal profiling, but it cannot be the principal element. Unfortunately, incidents of racial profiling have been documented for some time in the United States.

Radiological agents Materials that emit radiation that can harm living organisms when inhaled or otherwise ingested. Nonweapons-grade radiological agents could theoretically be used to construct a toxic dirty bomb.

Rajneeshees Followers of the cult leader Bhagwan Shree Rajneesh. In September 1984, the group poisoned the salad bars of 10 Oregon restaurants with *Salmonella* bacteria. More than 700 people fell ill.

Rape of Nanking During a six-week campaign in 1937 to 1938, the Japanese army killed between 200,000 and 300,000 Chinese in the Chinese capital of Nanking. Many thousands were bayoneted, beheaded, or tortured. An estimated 20,000 to 80,000 Chinese women and girls were raped by Japanese soldiers, and thousands of women were either forced into sexual slavery as comfort women or made to perform in perverse sex shows to entertain Japanese troops.

RDX The central component of most plastic explosives.

Reactionary A term given to far-right and fringe-right political tendencies.

Red Army Faction (RAF) A leftist terrorist movement active in West Germany during the 1970s and 1980s. Also referred to as the Baader-Meinhof Gang.

Red Brigades A leftist terrorist movement active in Italy during the 1970s and 1980s.

Red Guards Groups of young Communist radicals who sought to purge Chinese society during the Great Proletarian Cultural Revolution.

Red Scares Periodic anticommunist security crises in the United States, when national leaders reacted to the perceived threat of communist subversion.

Regicide The killing of kings.

Reign of Terror (Régime de la Terreur) A period during the French Revolution when the new republic violently purged those who were thought a threat to the prevailing ideals of the revolution. Terrorism was considered a necessary and progressive revolutionary tactic.

Relative deprivation theory A sociological term that indicates the lack of human needs vis-à-vis other members of a particular society.

Religious communal terrorism Conflict between religious groups involving terrorist violence.

Repentance laws An offer of qualified amnesty by the Italian government to Red Brigades members, requiring them to demonstrate repentance for their crimes. Repentance was established by cooperating within a sliding scale of collaboration. A significant number of imprisoned Red Brigades terrorists accepted repentance reductions in their sentences.

Repressive responses. Counterterrorist measures that seek to resolve terrorist crises by disrupting or destroying terrorist groups and movements. These responses include military and nonmilitary options.

Republic of New Africa (RNA) An African American nationalist organization. Its goal was to form a separate African American nation from portions of several southern states. Some members opted to engage in political violence under the name of the New African Freedom Fighters.

Resistance A cover name used by the leftist May 19 Communist Organization in the United States.

Revolutionary Armed Forces of Colombia (Fuerzas Armados Revolucionarios de Colombia, or FARC) An enduring Marxist insurgent movement in Colombia that has engaged in guerrilla warfare and terrorism since its inception during the 1960s.

Revolutionary Catechism A revolutionary manifesto written by the Russian anarchist Sergei Nechayev.

Revolutionary dissident terrorism The practice of political violence with the goal of destroying an existing order, committed with a plan for the aftermath of the revolution.

Revolutionary Fighting Group *See* May 19 Communist Organization in the United States.

Revolutionary Guards Corps Iranian Islamic revolutionaries who have been deployed abroad, mainly to Lebanon, to promote Islamic revolution.

Revolutionary Justice Organization An adopted alias of Lebanon's Hezbollah.

Revolutionary Tribunal The revolutionary court established during the French Revolution.

Revolutionary United Front (RUF) A rebel movement that arose in Sierra Leone in 1991. Led by Foday Sankoh, RUF forces were responsible for widespread human rights abuses and atrocities.

Revolutionary Youth Movement II (RYM-II) A faction of Students for a Democratic Society in the United States. RYM II adapted their ideological motivations to the political and social context of the 1960s by tailoring the ideologies of orthodox Marxism to the political environment of the 1960s.

Rewards for Justice Program An international bounty program managed by the U.S. Diplomatic Security Service. The program offered cash rewards for information leading to the arrest of wanted terrorists.

Roadside bomb An improvised explosive device (IED) constructed and deployed by Iraqi insurgents against U.S.-led occupation forces.

Rocket-propelled grenades (RPGs) Handheld military weapons that use a propellant to fire a rocket-like explosive.

Royal Marine Commandos British rapid-reaction troops that deploy in larger numbers than the Special Air Service and Special Boat Service. They are organized around units called commandos that are roughly equivalent to a conventional battalion.

RPG-7 A rocket-propelled grenade weapon manufactured in large quantities by the Soviet Bloc.

Rushdie, Salman The author of *The Satanic Verses*. He was forced into hiding after the publication of the book because of calls for his assassination by Iranian clerics.

SA-7 An infrared-targeted Soviet-made surface-to-air missile, also known as the Grail.

Sam Melville–Jonathan Jackson Unit A group that took credit for bombing the Boston State House in Boston, Massachusetts, in 1975. It was a term used by the leftist United Freedom Front.

Sanchez, Ilich Ramirez *See* Carlos the Jackal

Sandinistas A Marxist movement in Nicaragua that seized power after a successful insurgency against the regime of Anastasio Somoza Debayle. The Sandinista regime became the object of an American-supported insurgency.

Sarin A potent nerve gas. The Aum Supreme Truth cult released sarin gas into the Tokyo subway system in March 1995, killing 12 and injuring thousands.

Satanic Verses, The A book published by Salman Rushdie that was denounced by many Muslims, which led to a bounty being placed by Iranian clerics for Rushdie's assassination.

Sayaret Elite Israeli reconnaissance units that engage in counterterrorist operations. Sayaret have been attached to General Headquarters (Sayaret Matkal) and the Golani Brigade. There is also a Parachute Sayaret.

Sayaret Matkal An elite reconnaissance unit within the Israel Defense Force, attached to the IDF General Headquarters. It is a highly secretive formation that operates in small units and is regularly used for counterterrorist operations.

Sea Air Land Forces (SEALs) Similar to the British Special Boat Service and French Navy special assault units, the primary mission of the U.S. Navy SEALs is to conduct seaborne, riverine, and harbor operations. They have also been used extensively on land.

Sectarian violence Religious communal violence.

Semtex A high-grade and high-yield plastic explosive originally manufactured in Czechoslovakia when it was a member of the Soviet Bloc.

Shadow wars Covert campaigns to suppress terrorism.

Shining Path (Sendero Luminoso) A Marxist insurgent movement in Peru. Founded and led by former philosophy professor Abimael Guzmán, the group regularly engaged in terrorism.

Sicarii The Zealot rebels who opposed Roman rule. Named for the curved dagger, or *sica,* that was a preferred weapon.

Signal intelligence Intelligence that has been collected by technological resources. Also referred to as SIGINT.

Signature method Methods that become closely affiliated with the operational activities of specific extremist groups.

Sinn Féin An aboveground political party in Northern Ireland that champions Catholic rights and union with the Irish Republic.

Skinheads A countercultural youth movement that began in England in the late 1960s. An international racist skinhead movement eventually developed in Europe and the United States. The term *skinhead* refers to the members' practice of shaving their heads.

Sleeper cells A tactic used by international terrorist movements in which operatives theoretically establish residence in another country to await a time when they will be activated by orders to carry out a terrorist attack.

Smallpox A formerly epidemic disease that has been eradicated in nature, but that has been preserved in research laboratories. A possible ingredient for biological weapons.

Social cleansing The practice of eliminating defined undesirables from society. These undesirables can include those who practice defined morals crimes or who engage in denounced political and social behaviors.

Social reform Conciliatory counterterrorist measures that seek to resolve terrorist crises by resolving political and social problems that are the focus of the terrorists' grievances.

Social Revolutionary Party A Russian revolutionary movement during the late 19th and early 20th centuries. The group adopted terrorism as a revolutionary method.

Soft targets Civilian and other undefended targets that are easily victimized by terrorists.

Somoza Debayle, Anastasio The U.S.-supported Nicaraguan dictator overthrown by the Sandinista-led insurgency.

Special Air Service (SAS) A secretive organization in the British Army that has been used repeatedly in counterterrorist operations. Organized at a regimental level, but operating in very small teams, the SAS is similar to the French 1st Navy Parachute Infantry Regiment and the American Delta Force.

Special Boat Service (SBS) A special unit under the command of the British Royal Navy. The SBS specializes in operations against seaborne targets and along coastlines and harbors. They are similar to the French Navy's special assault units and the American SEALs.

Special Operations Command The general headquarters for U.S. Special Operations Forces.

Special operations forces Elite military and paramilitary units deployed by many armed forces. They are highly trained and are capable of operating in large or small formations, both overtly and covertly.

Spillover effect Terrorist violence that occurs beyond the borders of the countries that are the target of such violence.

State assistance for terrorism Tacit state participation in and encouragement of extremist behavior. Its basic characteristic is that the state, through sympathetic proxies and agents, implicitly takes part in repression, violence, and terrorism.

State patronage for terrorism Active state participation in and encouragement of extremist behavior. Its basic characteristic is that the state, through its agencies and personnel, actively takes part in repression, violence, and terrorism.

Stateless revolutionaries International terrorists who are not sponsored by, or based in, a particular country.

State-regulated press State-regulated media exist in environments where the state routinely intervenes in the reporting of information by the press. This can occur in societies that otherwise have a measure of democratic freedoms, as well as in totalitarian societies.

Stinger A technologically advanced handheld antiaircraft missile manufactured by the United States.

Stockholm syndrome A psychological condition in which hostages begin to identify and sympathize with their captors.

Strong multipolar terrorist environment A terrorist environment that presumes that state sponsorship guides terrorist behavior but that several governments support their favored groups. It also presumes that there are few truly autonomous international terrorist movements; they all have a link to a state sponsor.

Structural theory A theory used in many disciplines to identify social conditions (structures) that affect group access to services, equal rights, civil protections, freedom, or other quality-of-life measures.

Struggle meetings Revolutionary rallies held during the Chinese Revolution. Denunciations were often made against those thought to be a threat to the revolution.

Student Soviet, The The section of Paris temporarily ceded to students during the 1968 uprising centered at the Sorbonne.

Students for a Democratic Society (SDS) A leftist student movement founded in 1962 in the United States. It rose to become the preeminent activist organization on American campuses throughout the 1960s.

Its factions—the Progressive Labor Party, Revolutionary Youth Movement II, and the Weathermen—were highly active during the 1960s and early 1970s.

Submachine guns Light automatic weapons that fire pistol ammunition.

Suicide bombers Terrorists trained to detonate explosives that have been strapped to their bodies.

Sun Tzu Chinese military philosopher whose book *The Art of War* has been a significant influence on military theory.

Supergrass A policy in Northern Ireland during the 1980s of convincing Provos and members of the Irish National Liberation Army to defect from their movements and inform on their former comrades.

Suppression campaigns Counterterrorist measures that seek to resolve terrorist crises by waging an ongoing campaign to destroy the terrorists' capacity to strike.

Tamil Tigers *See* Liberation Tigers of Tamil Eelam

Terminal institutions Institutions under the jurisdiction of criminal justice and military justice systems that provide final resolution to individual terrorists' careers after they have been captured, prosecuted, convicted, and imprisoned.

Terrorism Elements from the American definitional model define terrorism as a premeditated and unlawful act in which groups or agents of some principal engage in a threatened or actual use of force or violence against human or property targets. These groups or agents engage in this behavior intending the purposeful intimidation of governments or people to affect policy or behavior with an underlying political objective. There are more than 100 definitions of terrorism.

Terrorist One who practices terrorism. Often a highly contextual term.

Terrorist profiles Descriptive profiles of terrorist suspects developed by law enforcement agencies to assist in the apprehension of terrorist suspects. Similar to criminal profiling.

Third World A postwar term created to describe the developing world.

Thuggees A mystical cult that existed for centuries in India. Members ritualistically murdered travelers to honor the goddess Kali.

TNT A commercially available explosive

Tokyo Convention on Offences and Certain Other Acts Committed on Board Aircraft A treaty enacted in 1963 as the first airline crimes' treaty.

Total war The unrestrained use of force against a broad selection of targets to utterly defeat an enemy.

Totalitarian regimes Governments that practice total control over public and political institutions. The media and other public information outlets are completely controlled.

Troubles, The The term given to sectarian violence between Catholics and Protestants in Northern Ireland.

Tupac Amaru Revolutionary Movement (Movimento Revolucionario Tupac Amaru) A Marxist terrorist movement in Peru, primarily active during the 1980s and early 1990s.

Tupamaros A Marxist urban terrorist movement active during the early 1970s in Uruguay. After a number of dramatic attacks, the Tupamaros were eventually annihilated when the Uruguayan military used authoritarian methods, and the general Uruguayan population rejected the movement.

Turner Diaries, The A short novel written by National Alliance founder William Pierce under the pseudonym Andrew MacDonald. It depicts an Aryan revolution in the United States and is considered, by many neo-Nazis, a blueprint for the eventual racial holy war.

TWA Flight 847 In June 1985, hijackers belonging to Lebanon's Hezbollah hijacked a TWA airliner, taking it on a high-profile and media-intensive odyssey around the Mediterranean.

Two-Seedline Christian Identity A Christian Identity mystical belief that rejects the notion that all humans are descended from Adam (see One-Seedline Christian Identity). According to this belief, Eve bore Abel as Adam's son but bore Cain as the son of the Serpent. Outside of the Garden of Eden lived non-White, soulless beasts who were a separate species from humans. When Cain slew Abel, he was cast out of the Garden to live among the soulless beasts. Those who became the descendants of Cain are the modern Jews. They are biologically descended from the devil and are a demonic people worthy of extermination.

Tyrannicide The assassination of tyrants for the greater good of society.

UNABOM The FBI's case designation for their investigation into bombings perpetrated by Theodore Kaczynski. *Un* was short for university, and *a* referred to airlines.

Unabomber The popular nickname applied to Theodore Kaczynski.

United Freedom Front (UFF) A leftist terrorist group in the United States that was active from the mid-1970s through the mid-1980s.

USS Cole An American destroyer that was severely damaged on October 12, 2000, while berthed in the port of Aden, Yemen. Two suicide bombers detonated a boat bomb next to the *Cole*, killing themselves and 17 crew members and wounding 39 other Navy personnel.

Utopia An ideal society.

Vanguard strategy In Marxist and non-Marxist theory, the strategy of using a well-indoctrinated and motivated elite to lead the working-class revolution. In practice, this strategy was adopted in the postwar era by terrorist organizations and extremist movements.

Viet Cong The name given by the United States and its noncommunist South Vietnamese allies to South Vietnamese communist insurgents.

Viet Minh (Viet Nam Doc Lap Dong Minh Hoi) An organization founded by the Vietnamese leader Ho Chi Minh. The Vietnam Independence Brotherhood League began fighting first against the Japanese conquerors of French Indochina and then against the French colonial forces.

Vigilante state terrorism Unofficial state terrorism in which state personnel engage in nonsanctioned political violence. It can include the use of death squads.

Waffen SS The armed SS of Nazi Germany. Elite military units composed of racially selected Germans and fascist recruits from occupied territories.

Warfare Making war against an enemy. In the modern era, it usually refers to conventional and guerrilla conflicts.

Wars of national liberation A series of wars fought in the developing world in the postwar era. These conflicts frequently pitted indigenous guerrilla fighters against European colonial powers or governments perceived as pro-Western. Insurgents were frequently supported by the Soviet Bloc or China.

Weak multipolar terrorist environment A terrorist environment that presupposes that state sponsorship exists, but that the terrorist groups are more autonomous. Under this scenario, several governments support their favored groups, but many of these groups are relatively independent international terrorist movements.

Weapons of mass destruction High-yield weapons that can potentially cause a large number of casualties when used by terrorists. Examples of these weapons include chemical, biological, radiological, and nuclear weapons. They can also be constructed from less exotic compounds, such as ANFO.

Weather Bureau The designation adopted by the leaders of the Weatherman faction of Students for a Democratic Society.

Weather Collectives Groups of supporters of the Weather Underground Organization.

Weather Underground Organization (WUO) The adopted name of the Weathermen after they moved underground.

Weathermen A militant faction of Students for a Democratic Society that advocated, and engaged in, violent confrontation with the authorities. Some engaged in terrorist violence.

Weinrich, Johannes A former West German terrorist "purchased" by the German government from the government of Yemen. He stood trial in Germany for the 1983 bombing of a French cultural center in Berlin, in which one person was killed and 23 others were wounded. He had also been a very close associate of Carlos the Jackal.

White Aryan Resistance A California-based racial supremacist hate group founded by former Klansman Tom Metzger in the early 1980s.

Wrath of God A counterterrorist unit created by the Israelis to eliminate Palestinian operatives who had participated in the massacre of Israeli athletes at the 1972 Munich Olympics. Also called Mivtzan Elohim.

Wu Ch'i Chinese military philosopher who is usually associated with Sun Tzu.

YAMAM One of two operational subgroups deployed by Israel's Police Border Guards (the other subgroup is YAMAS). It engages in counterterrorist and hostage rescue operations.

YAMAS One of two operational subgroups deployed by Israel's Police Border Guards (the other subgroup is YAMAM). It is a covert unit that has been used to neutralize terrorist cells in conjunction with covert Israel Defense Force operatives.

Year Zero The ideological designation given by the Khmer Rouge to the beginning of their genocidal consolidation of power.

Zapatista National Liberation Front Leftist rebels originally centered in Chiapas, Mexico. During the late 1990s, they engaged in guerrilla fighting that ended when they were integrated into the Mexican political process.

Zealots Hebrew rebels who uncompromisingly opposed Roman rule in ancient Palestine.

Zionism An intellectual movement within the Jewish community describing the conditions for the settlement of Jews in Israel.

Notes

Chapter 1

1. *Webster's New Twentieth-Century Dictionary of the English Language,* unabridged, 2nd ed. (New York: Publishers Guild, 1966).

2. Laird Wilcox, "What Is Extremism? Style and Tactics Matter More Than Goals," in *American Extremists: Militias, Supremacists, Klansmen, Communists, & Others,* eds. John George and Laird Wilcox (Amherst, NY: Prometheus, 1996), 54.

3. For an excellent list of traits of extremists, see Laird Wilcox, "What Is Extremism?" 54.

4. For references to the Cuban perspective, see Martin Kenner and James Petras, *Fidel Castro Speaks* (New York: Grove, 1969).

5. Jeffrey Goldberg, "Inside Jihad U.: The Education of a Holy Warrior," *New York Times Magazine,* June 25, 2000, 35.

6. Walter Laqueur, *The New Terrorism: Fanaticism and the Arms of Mass Destruction* (New York: Oxford University Press, 1999), 5.

7. See Alex P. Schmid and Albert J. Jongman, *Political Terrorism* (Amsterdam: North Holland, 1988). The book reports the results of a survey of 100 experts asked for their definitions of terrorism.

8. David J. Whittaker, ed., *The Terrorism Reader* (New York: Routledge, 2001), 8.

9. Ibid., 1.

10. Office for the Protection of the Constitution, 1.

11. Ibid.

12. Ted Robert Gurr, "Political Terrorism: Historical Antecedents and Contemporary Trends," in *Violence in America: Protest, Rebellion, Reform,* vol. 2 (Newbury Park, CA: Sage, 1989).

13. J. P. Gibbs, "Conceptualization of Terrorism," *American Sociological Review* 54 (1989): 329, quoted in *Hate Crime: International Perspectives on Causes and Control,* ed. Mark S. Hamm (Highland Heights, KY: Academy of Criminal Justice Sciences and Cincinnati, OH: Anderson, 1994), 111.

14. Bruce Hoffman, *Inside Terrorism* (New York: Columbia University Press, 1998), 43.

15. Whittaker, *The Terrorism Reader,* 8.

16. U.S. Departments of the Army and the Air Force, *Military Operations in Low Intensity Conflict,* Field Manual 100–20/Air Force Pamphlet 3–20 (Washington, DC: U.S. Government Printing Office, 1990), 3–1.

17. 18 U.S.C. 3077.

18. Terrorist Research and Analytical Center, National Security Division, Federal Bureau of Investigation, *Terrorism in the United States 1995* (Washington, DC: U.S. Department of Justice, 1996), ii.

19. Office of the Coordinator for Counterterrorism, *Patterns of Global Terrorism 1996,* U.S. Department of State Publication 10433 (Washington, DC: U.S. Depart ment of State, 1997), vi.

20. Steven E. Barkan and Lynne L. Snowden, *Collective Violence* (Boston, MA: Allyn & Bacon, 2001).

21. Hoffman, *Inside Terrorism.*

22. Laqueur, *The New Terrorism.*

23. Rachel Ehrenfeld, *Narco Terrorism* (Basic Books, 1990); Jonathan B. Tucker, ed., *Toxic Terror: Assessing Terrorist Use of Chemical and Biological Weapons* (Cambridge, MA: MIT Press, 2000).

24. Kevin Jack Riley and Bruce Hoffman, *Domestic Terrorism: A National Assessment of State and Local Preparedness* (Santa Monica, CA: RAND, 1995), 2.

25. The quote has been more widely reported since the Vietnam War as "we had to destroy the village to save it." See Don Oberdorfer, *Tet!* (Garden City, NY: Doubleday, 1971), 184-185, 332. See also Neil Sheehan, *A Bright Shining Lie: John Paul Vann and America in Vietnam* (New York: Random House, 1988), 719.

26. In Chinese literature and propaganda, bandit is often a synonym for rebel. Sun Tzu, *The Art of War* (New York: Oxford University Press, 1963), 168.

27. Within the context of the campaign, Lyndon Johnson's anti-Communist credentials were decisively validated during the Gulf of Tonkin crisis in August 1964. The destroyer *U.S.S. Maddox* was reported to have been attacked by North Vietnamese torpedo boats off the coast of North Vietnam. American air strikes were launched in reprisal, and Congress unanimously passed the Gulf of Tonkin Resolution, which supported President Johnson's use of measures to protect U.S. interests in Vietnam and elsewhere. For a discussion of the incident within the context of the politics at the time, see Doris Kearns, *Lyndon Johnson and the American Dream* (New York: Harper & Row, 1976), 195. For an insider's discussion, see Merle Miller, *Lyndon: An Oral Biography* (New York: Ballantine, 1980), 465–477.

28. For a discussion of this subject from the cultural perspectives of Bosnian Muslims and Serbs, see Steven M. Weine, *When History Is a Nightmare: Lives and Memories of Ethnic Cleansing in Bosnia-Herzegovina* (New Brunswick, NJ: Rutgers University Press, 1999).

29. Associated Press reporter Peter Arnett attributed the source of the quotation to a U.S. soldier at the city of Ben Tre in South Vietnam during the Tet offensive in 1968. As reported, it was the statement of an officer higher up in the chain of command—a U.S. army major. Although widely repeated at the time and since the war, only Arnett said he heard the statement at the site, and only Arnett filed the quotation from the Ben Tre visit. The U.S. officer was not identified and has not been identified since the war (see Oberdorfer, *Tet!,* and Sheehan, *A Bright Shining Lie*). The U.S. Department of Defense launched an investigation to find the source of the quote. No officer or enlisted man was identified (see Oberdorfer, *Tet!*).

30. Gabriel Weimann and Conrad Winn, *The Theater of Terror: Mass Media and International Terrorism* (New York: Longman, 1994), 104. The authors mention the following societal participants in media events: the direct victims, the terrorists, the broadcasting audience, journalists, and governments.

31. David L. Paletz and Alex P. Schmid, *Terrorism and the Media* (Newbury Park, CA: Sage, 1992), 179.

32. For a discussion of the ambiguities about defining combatants and noncombatants, see Peter C. Sederberg, *Terrorist*

Myths: Illusion, Rhetoric, and Reality (Englewood Cliffs, NJ: Prentice Hall, 1989), 37–39.

33. For a discussion of the ambiguities about defining indiscriminate force, see ibid., 39–40.

Chapter 2

1. See Walter Laqueur, *The New Terrorism: Fanaticism and the Arms of Mass Destruction* (New York: Oxford University Press, 1999), 90.

2. An extraordinary and classic history of the Roman Empire is found in Edward Gibbon's great work of literature, *The History of the Decline and Fall of the Roman Empire* (New York: AMS Press, 1974, reprinted from J. B. Bury, ed., London: Methuen, 1909).

3. For a discussion of the Roman occupation of Judea and the First Jewish Revolt within the broader context of Roman politics, see Michael Grant, *The Twelve Caesars* (New York: Scribner's, 1975).

4. Burke, a Whig Member of Parliament, was a progressive in his time. He opposed absolutism, poor treatment of the American colonists, and the slave trade. He expressed his opposition to Jacobin extremism in a series of writings, including *Reflections on the French Revolution* and *Letters on a Regicide Peace.*

5. For a classic account of the Terror, see Stanley Loomis, *Paris in the Terror: June 1793–July 1794* (Philadelphia: J. B. Lippincott, 1964).

6. A French physician, Joseph Ignace Guillotin, invented the guillotine. He was loyal to the revolution and a deputy to the States-General. Guillotin encouraged the use of the beheading machine as a painless, humane, and symbolically revolutionary method of execution.

7. Pamala L. Griset and Sue Mahan. *Terrorism in Perspective* (Thousand Oaks, CA: Sage, 2002), 4.

8. Ibid.

9. For a good discussion about the consolidation of the Chinese revolution and the Great Proletarian Cultural Revolution that occurred later, see John King Fairbank, *The Great Chinese Revolution: 1800–1985* (New York: Harper & Row, 1987). For an analysis of the ideology of the Ayatollah Khomeini and the Iranian revolution, see Vanessa Martin, *Creating an Islamic State: Khomeini and the Making of a New Iran* (London: I. B. Tauris, 2000).

10. Karl Marx and Friedrich Engels, "Manifesto of the Communist Party," in *The Marx-Engels Reader,* ed. Robert C. Tucker (New York: W. W. Norton, 1972), 335.

11. Laqueur, *The New Terrorism*, 13.

12. Ibid., 81.

13. Ibid., 82.

14. The politics of the French Revolution are evaluated in Theda Skocpol, *States and Social Revolutions: A Comparative Analysis of France, Russia, and China* (Cambridge: Cambridge University Press, 1990), 174–205.

15. For a comparative discussion of ideological conflict in revolutionary environments, see Skocpol, *States and Social Revolutions*. For seminal discussions that formed the ideological foundation for the U.S. Constitution and Bill of Rights, see Alexander Hamilton, John Jay, and James Madison, *The Federalist: A Commentary on the Constitution of the United States, Being a Collection of Essays Written in Support of the Constitution Agreed upon September 17, 1787* (New York: Modern Library, 2000).

16. Laqueur, *The New Terrorism*, 230.

17. Konrad Kellen, "Ideology and Rebellion: Terrorism in West Germany," in *Origins of Terrorism: Psychologies, Ideologies, Theologies, States of Mind,* ed. Walter Reich (Washington, DC: Woodrow Wilson Center Press, 1998), 57.

18. See Harry B. Ellis, *Ideals and Ideologies: Communism, Socialism, and Capitalism* (New York: World, 1972).

19. Leonard Weinberg, "An Overview of Right-Wing Extremism in the Western World: A Study of Convergence, Linkage, and Identity," in *Nation and Race: The Developing Euro-American Racist Subculture,* eds. Jeffrey Kaplan and Tore Bjørgo (Boston, MA: Northeastern University Press, 1998), 8.

20. Ibid., 10.

21. For a good journalistic chronology of the Spanish Civil War, see R. H. Haigh,, D. S. Morris, and A. R. Peters, *The Guardian Book of the Spanish Civil War* (Aldershot, Hants, UK: Wildwood House), 1987.

22. Mark W. Janis, *An Introduction to International Law,* 3d ed. (New York: Aspen Law & Business, 1999), 171.

23. Jessica Stern, *The Ultimate Terrorists* (Cambridge, MA: Harvard University Press, 1999), 18.

24. See Bernhardt J. Hurwood, *Society and the Assassin: A Background Book on*

Political Murder (New York: Parents' Magazine Press, 1970), 17.

25. David E. Sanger, "In Speech, Bush Focuses on Conflicts Beyond Iraq." *The New York Times,* May 1, 2003.

26. National Commission on Terrorist Attacks Upon the United States, *The 9/11 Commission Report* (New York: W. W. Norton, 2004), 32–33, 305, 311.

27. Carol J. Williams, "Suicide Attacks Rising Rapidly," *Los Angeles Times,* June 2, 2005.

Chapter 3

1. See Clifford E. Simonsen and Jeremy R. Spindlove, *Terrorism Today: The Past, the Players, the Future* (Upper Saddle River, NJ: Prentice Hall, 2000), 22–24.

2. Steven E. Barkan and Lynne L. Snowden, *Collective Violence* (Boston: Allyn & Bacon, 2001), 6.

3. Ibid., 7.

4. Relative deprivation theory was pioneered by James Chowning Davies. See James Chowning Davies, "Toward a Theory of Revolution," *American Sociological Review* 25 (1962): 5–19.

5. Jack A. Goldstone, "Introduction: The Comparative and Historical Study of Revolutions," in *Revolutions: Theoretical, Comparative, and Historical Studies* (San Diego, CA: Harcourt Brace Jovanovich, 1986), 1–17, cited in Barkan and Snowden, *Collective Violence*, 53.

6. See Barkan and Snowden, *Collective Violence*, 55.

7. Ibid., 17.

8. Discussed in ibid., 18.

9. Jerrold M. Post, "Terrorist Psycho-logic: Terrorist Behavior as a Product of Psychological Forces," in *Origins of Terrorism: Psychologies, Ideologies, Theologies, States of Mind,* ed. Walter Reich (Washington, DC: Woodrow Wilson Center Press, 1998), 9.

10. Ibid., 28, reporting findings of the Ministry of the Interior, Federal Republic of Germany, *Analysen Zum Terrorismus 1–4* (Darmstadt: Deutscher Verlag, 1981, 1982, 1983, 1984); H. Jäger, G. Schmidtchen and L. Süllwold, eds., *Analysen Zum Terrorismus 2: Lebenlaufanalysen* (Darmstadt: Deutscher Verlag, 1981); W. von Baeyer-Kaette, D. Classens, H. Feger, and F. Neidhardt, eds., *Analysen Zum Terrorismus 3: Gruppeprozesse* (Darmstadt: Deutscher Verlag, 1982).

11. Ibid., 28.

12. Ibid., 20.

13. Ibid.

14. Ibid., 25–42.

15. Ibid., 31.

16. For a good summary of Mao's political ideology, see Stuart R. Schram, *The Political Thought of Mao Tse-tung* (New York: Praeger, 1974).

17. Quotation in Peter Ford, "Why Do They Hate Us?" *Christian Science Monitor,* September 27, 2001.

18. Michael Scheuer (as Anonymous), *Imperial Hubris: Why the West Is Losing the War on Terror* (Washington, DC: Brassey's, 2004), 1.

19. Dana Priest, "Report Says Iraq Is New Terrorist Training Ground," *Washington Post,* January 14, 2005.

20. For a profile of one foreign volunteer's experience in Iraq, see Megan K. Stack, "Getting an Education in Jihad," *Los Angeles Times,* December 29, 2004.

21. For an analysis of American intervention in Central America and the Caribbean, see Lester D. Langley, *The Banana Wars: United States Intervention in the Caribbean, 1898–1934* (Chicago: Dorsey, 1985).

22. See Jay Mallin, *Terror and Urban Guerrillas: A Study of Tactics and Documents* (Coral Gables, FL: University of Miami Press, 1971).

23. Marighella's name has been alternatively spelled with one l and two ls. Marighella himself alternated between spellings.

24. Carlos Marighella, *Mini-Manual of the Urban Guerrilla,* cited in Mallin, *Terror and Urban Guerrillas,* 70–71.

25. For an interesting attack against Marxist ideology, see Milovan Djilas, *The Unperfect Society: Beyond the New Class* (New York: Harcourt, Brace & World, 1969).

26. Barkan and Snowden, *Collective Violence,* 27.

27. Ibid., 27–28.

Chapter 4

1. The one noncommunal dissident terrorist incident that approximated state terrorism in sheer scale was the September 11, 2001, Al Qaeda attack on the United States.

2. Paul R. Pillar, *Terrorism and U.S. Foreign Policy* (Washington, DC: Brookings Institution Press, 2001), 157–198.

3. Ibid., 157.

4. Ibid., 178.

5. Ibid., 86.

6. See Peter Iadicola and Anson Shupe, *Violence, Inequality, and Human Freedom* (Dix Hills, NY: General Hall, 1998), 276–289.

7. For a good discussion of U.S. policy in Nicaragua, see Thomas W. Walker, ed., *Revolution & Counterrevolution in Nicaragua* (Boulder, CO: Westview, 1991).

8. See John King Fairbank, *The Great Chinese Revolution: 1800–1985* (New York: Harper & Row, 1987), 316–341.

9. Ibid., 317.

10. Yehezkel Dror, *Crazy States: A Counterconventional Strategic Problem* (New York: Kraus, Milwood, 1980).

11. Peter C. Sederberg, *Terrorists Myths: Illusion, Rhetoric, and Reality* (Englewood Cliffs, NJ: Prentice Hall, 1989).

12. Raphael Lemkin, *Axis Rule in Occupied Europe: Laws of Occupation, Analysis of Government, Proposals for Redress* (Washington, DC: Carnegie Endowment for International Peace, 1944). Chapter 9 discussed genocide in detail.

13. Quoted in Gerhard von Glahn, *Law Among Nations: An Introduction to Public International Law,* 5th ed. (New York: Macmillan, 1986), 303–304.

14. Bruce Hoffman, *Inside Terrorism* (New York: Columbia University Press, 1998), 195.

15. Frances FitzGerald, *Fire in the Lake: The Vietnamese and the Americans in Vietnam* (New York: Vintage Books, 1972), 549.

Chapter 5

1. David J. Whittaker, ed., *The Terrorism Reader* (New York: Routledge, 2001), 33.

2. Richard Schultz, "Conceptualizing Political Terrorism," *Journal of International Affairs* 32 (1978): 1, 7–15, quoted in Whittaker, *The Terrorism Reader.*

3. Steven E. Barkan and Lynne L. Snowden, *Collective Violence* (Boston: Allyn & Bacon, 2001), 70.

4. Ted Robert Gurr, "Political Terrorism: Historical Antecedents and Contemporary Trends," in *Violence in America: Protest, Rebellion, Reform,* vol. 2., ed. Ted Robert Gurr (Newbury Park, CA: Sage, 1989), 204.

5. Peter C. Sederberg, *Terrorist Myths: Illusion, Rhetoric, and Reality* (Englewood Cliffs, NJ: Prentice Hall, 1989).

6. Christopher Hewitt, "Public's Perspectives," in *Terrorism and the Media,*

eds. David L. Paletz and Alex P. Schmid (Newbury Park, CA: Sage, 1992), 182.

7. For an excellent discussion of this progression, see Brent L. Smith, *Terrorism in America: Pipe Bombs and Pipe Dreams* (Albany: State University of New York Press, 1994).

8. The Provisional IRA was formed in 1969 when radicals broke from the official IRA, which was more political than military.

9. For an excellent analysis of the IRA and the "troubles" in Northern Ireland, see J. Boywer Bell, *The IRA 1968–2000: Analysis of a Secret Army* (London: Frank Cass, 2000).

10. Whittaker, *The Terrorism Reader,* 126.

11. See Carlos Marighella, "Mini-Manual of the Urban Guerrilla," in *Terror and Urban Guerrillas: A Study of Tactics and Documents,* ed. Jay Mallin (Coral Gables, FL: University of Miami Press, 1971), 110–112.

12. Don Podesta, "The Terrible Toll of Human Hatred," *Washington Post National Weekly Edition,* June 8, 1987, quoted in *Violence and Terrorism,* 3d ed., eds. Bernard Schechterman and Martin Slann (Guilford, CT: Dushkin, 1993), 33.

13. Podesta, "The Terrible Toll."

14. Trent N. Thomas, "Global Assessment of Current and Future Trends in Ethnic and Religious Conflict," in Ethnic Conflict and Regional Instability, eds. Robert L. Pfaltzgraff, Jr., and Richard A. Schultz (Carlisle, PA: Strategic Studies Institute, 1994), 33–41, quoted in *Violence and Terrorism,* 5th ed., eds. Bernard Schechterman and Martin Slann (Guilford, CT: Dushkin, 1999), 24.

15. Richard Clutterbuck, *Terrorism, Drugs and Crime in Europe After 1992* (New York: Routledge, 1990), 73.

16. *Irish News,* November 30, 1993. Reporting on a joint study by the Northern Ireland Economic Research Centre in Belfast and the Economic and Social Research Institute in Dublin.

17. Some northerners are ethnic Arabs, but many are not. Nevertheless, non-Arab Muslims have been heavily influenced by their Arab neighbors and fellow Muslims and hence have developed an arabized culture.

18. The Axis powers were an alliance of Germany, Japan, Italy, and their allies.

19. Podesta, "The Terrible Toll."

20. R. Crosbie-Weston, "Terrorism and the Rule of Law," in *20th Century,* ed. R. W. Cross (London: Purnell Reference Books, 1979), 2715.

Chapter 6

1. Joshua 11, in *The Holy Bible, New Revised Standard Version* (Nashville, TN: Thomas Nelson Publishers, 1990).

2. Ralph A. Scott, *A New Look at Biblical Crime* (Chicago: Nelson-Hall, 1979), 66, cited in Peter Iadicola and Anson Shupe, *Violence, Inequality, and Human Freedom* (New York: General Hall, 1998), 175

3. Numbers 25:1, 6-8, in *The Holy Bible, New Revised Standard Version* (Nashville, TN: Thomas Nelson Publishers, 1990).

4. Norman Cohn, *The Pursuit of the Millennium* (New York: Oxford University Press, 1971), 61, quoted in Iadicola and Shupe, *Violence, Inequality, and Human Freedom*, 177.

5. For a history of the Assassin movement, see Bernard Lewis, *The Assassins: A Radical Sect in Islam* (New York: Oxford University Press, 1987).

6. In *The Assassins*, Bernard Lewis discounts the assertion that the Assassins drugged themselves and argues that "in all probability it was the name that gave rise to the story, rather than the reverse" (12).

7. James A. Haught, *Holy Horrors* (Buffalo, NY: Prometheus, 1990), 34, cited in Iadicola and Shupe, *Violence, Inequality, and Human Freedom*, 181.

8. Much of the discussion has been adapted from Human Rights Watch, *The Scars of Death: Children Abducted by the Lord's Resistance Army in Uganda* (New York: Human Rights Watch, 1997).

9. *The Protocols of the Learned Elders of Zion* has been extensively published on the Internet. It is readily available from Web sites promoting civil liberties, neo-Nazi propaganda, anti-Semitism, and Islamist extremism.

10. See Michael A. Hiltzik, "Russian Court Rules 'Protocols' an Anti-Semitic Forgery," *Los Angeles Times*, November 28, 1993.

11. Walter Laqueur, *The New Terrorism: Fanaticism and the Arms of Mass Destruction* (New York: Oxford University Press, 1999), 129.

12. Robert J. Kelly and Jess Maghan, eds., *Hate Crime: The Global Politics of Polarization* (Carbondale: Southern Illinois University Press, 1998), 105.

13. See Samuel M. Katz, *The Hunt for the Engineer: How Israeli Agents Tracked the Hamas Master Bomber* (New York: Fromm International, 2001), 97–99.

14. Ibid., 225.

15. "Rabin's Alleged Killer Appears in Court," *CNN World News*, November 7, 1995.

16. Data derived from Central Intelligence Agency, *World Fact Book* (Washington, DC: U.S. Central Intelligence Agency), http://www.cia.gov/cia/publications/factbook.

17. See Monte Morin, "In Iraq, to Be a Hairstylist Is to Risk Death," *Los Angeles Times,* February 22, 2005.

18. A jihadi is one who wages jihad, regardless of whether it is an armed jihad. Mujahideen are jihadis who have taken up arms.

19. David Rohde and C. J. Chivers, "Al Qaeda's Grocery Lists and Manuals of Killing," *New York Times,* March 17, 2002.

20. Robert D. McFadden, "Bin Laden's Journey from Rich Pious Lad to the Mask of Evil," *New York Times*, September 30, 2001.

21. Simon Reeve, *The New Jackals: Ramzi Yousef, Osama bin Laden and the Future of Terrorism* (Boston, MA: Northeastern University Press, 1999), 181.

22. U.S. Department of State, "Aum Supreme Truth (Aum)," in *Patterns of Global Terrorism, 2000* (Washington, DC: U.S. Department of State, April 2001).

23. Andrew Marshall, "It Gassed the Tokyo Subway, Microwaved Its Enemies and Tortured Its Members. So Why Is the Aum Cult Thriving?" *The Guardian*, July 15, 1999.

24. U.S. Department of State, 2001.

25. Marshall, "It Gassed the Tokyo Subway," 1999.

26. Marshall, "It Gassed the Tokyo Subway," 1999; Laqueur, *The New Terrorism*, 54.

27. Initial reports cited this figure. Later studies suggest that physical injuries numbered 1,300 and that the rest were psychological injuries (U.S. Department of State, 2001).

28. For interesting insight about Al Qaeda's use of computers, see Alan Cullison, "Inside Al-Qaeda's Hard Drive," *The Atlantic*, September 2004.

29. Dana Priest, "Report Says Iraq Is New Terrorist Training Ground," *Washington Post*, January 14, 2005.

30. Joash Meyer, "Bin Laden, in Tape, May Have Sights on New Role," *Los Angeles Times*, October 31, 2004.

31. See Alan Cowell, "British Muslims Are Seen Moving Into Mideast Terrorism," *New York Times*, May 1, 2003; Alan Cowell, "Zeal for Suicide Bombing Reaches British Midlands," *New York Times*, May 2, 2003; Craig Whitlock, "Moroccans Gain Prominence in Terror Groups," *Washington Post*, October 14, 2004; Tony Czuczka, "Germans Suspect Terror Pipeline," *Dallas Morning News*, January 9, 2005; Sebastian Rotella, "Europe's Boys of Jihad," *Los Angeles Times*, April 2, 2005.

32. For a discussion of Christian evangelizing in Iraq, see Jane Lampman, "A Crusade After All?" *Christian Science Monitor*, April 17, 2003.

Chapter 7

1. Charles W. Kegley, Jr., "An Introduction," in *International Terrorism: Characteristics, Causes, Controls* (New York: St. Martin's, 1990), 3.

2. Kegley, "An Introduction," 21.

3. See Peter C. Sederberg, *Terrorist Myths: Illusion, Rhetoric, and Reality* (Englewood Cliffs, NJ: Prentice Hall, 1989), 117.

4. Simon Reeve, *The New Jackals: Ramzi Yousef, Osama bin Laden and the Future of Terrorism* (Boston: Northeastern University Press, 1999), 3.

5. James Bruce, "Arab Veterans of the Afghan War," *Jane's Intelligence Review* 7 (April 1, 1995): 175.

6. Donald G. McNeil, "What Will Rise If bin Laden Falls?" *New York Times,* December 2, 2001.

7. MSNBC, "The Lesson of Air France Flight 8969," MSNBC.com, http://www.msnbc.com/news/635213.asp?cp1=1 (accessed: September 30, 2001).

8. From Richard C. Paddock and Bob Drogin, "A Terror Network Unraveled in Singapore," *Los Angeles Times,* January 20, 2002, quoted with permission.

9. Ibid.

10. Keith B. Richburg and Fred Barbash, "Madrid Bombings Kill at Least 190," *Washington Post*, March 11, 2004.

11. Doreen Carvajal, "Spaniards Turn Out in Huge Numbers to Mourn Blast Victims," *New York Times*, March 12, 2004.

12. Glenn Frankel, "London Subway Blasts Almost Simultaneous, Investigators Conclude," *Washington Post*, July 10, 2005.

13. Sarah Lyall, "3 Main British Parties to Back Tougher Antiterrorism Laws," *New York Times*, July 27, 2005.

Chapter 8

1. See Brent L. Smith, *Terrorism in America: Pipe Bombs and Pipe Dreams* (Albany: State University of New York Press, 1994), 35.

2. For good histories of the civil rights movement, see Taylor Branch, *Parting the Waters: American in the King Years, 1954–63* (New York: Simon & Schuster, 1988). See also Henry Hampton and Steve Fayer, *Voices of Freedom: An Oral History of the Civil Rights Movement From the 1950s Through the 1980s* (New York: Bantam Books, 1990).

3. Ted Robert Gurr, "Terrorism in Democracies: Its Social and Political Bases," in Walter Reich, ed., *Origins of Terrorism: Psychologies, Ideologies, Theologies, and States of Mind* (Washington, D.C.: Woodrow Wilson Center, 1998), 89.

4. Bruce Hoffman, *Inside Terrorism* (New York: Columbia University Press, 1998), 107. These numbers declined during the late 1990s and then rebounded after September 11, 2001. For a discussion of the militia movement in retrospect, see Southern Poverty Law Center, *Intelligence Report*, Summer 2001.

5. A copy of *The Turner Diaries* was found with Timothy McVeigh when he was arrested after the Oklahoma City bombing in 1995. See Andrew MacDonald, *The Turner Diaries* (New York: Barricade, 1980).

6. Data are derived from *Riots, Civil and Criminal Disorders—Hearings Before the Permanent Subcommittee on Investigations of the Committee on Government Operations*, United States Senate, Part 25. (Washington, DC: Government Printing Office, 1970), cited in George Prosser, "Terror in the United States: 'An Introduction to Elementary Tactics' and 'Some Questions on Tactics,'" in *Terror and Urban Guerrillas: A Study of Tactics and Documents*, ed. Jay Mallin (Coral Gables, FL: University of Miami Press, 1971), 52.

7. A former professor, Leary was best known for his recreational and spiritual experimentation with the hallucinogenic drug LSD, which he advocated as a consciousness-raising drug.

8. Possibly 26 police officers died in BLA ambushes. Ted Robert Gurr, "Political Terrorism: Historical Antecedents and Contemporary Trends," in *Violence in America: Protest, Rebellion, Reform*, vol. 2, ed. Ted R. Gurr (Newbury Park, CA: Sage, 1989).

9. For sources and further discussion of the FALN, see Brent L. Smith, *Terrorism in America: Pipe Bombs and Pipe Dreams* (Albany: State University of New York Press, 1994).

10. See Smith, *Terrorism in America*, for sources and further discussion of left-wing hard cores.

11. See Smith, *Terrorism in America*, for sources and further discussion of the United Freedom Front.

12. For a discussion of the Aryan Republican Army, see Mark Hamm, *In Bad Company: America's Terrorist Underground* (Boston, MA: Northeastern University Press, 2002).

13. The chronology of The Order's terrorist spree is adapted from the Anti-Defamation League, *Danger: Extremism. The Major Vehicles and Voices on America's Far-Right Fringe* (New York: Anti-Defamation League, 1996), 270–271.

14. Southern Poverty Law Center, *Intelligence Report*, Summer 2001.

15. Ibid.

16. See Southern Poverty Law Center, "Patriot Free Fall," in *Intelligence Report*, Summer 2002.

Chapter 9

1. Brian Jenkins, quoted in Bruce Hoffman, *Inside Terrorism* (New York: Columbia University Press, 1998), 132.

2. Ariana Eunjung Cha, "From a Virtual Shadow, Messages of Terror," *Washington Post*, October 2, 2004.

3. Richard Davis and Diana Owen. *New Media and American Politics*, (New York: Oxford University Press, 1998), 7.

4. Ibid., 9–15.

5. For a discussion of selective reporting, see James Rainey, "Unseen Pictures, Untold Stories," *Los Angeles Times*, May 21, 2005.

6. Hoffman, *Inside Terrorism*, 36–37.

7. See Albert Bandura, "Mechanisms of Moral Disengagement," in *Origins of Terrorism: Psychologies, Ideologies, Theologies, States of Mind*, ed. Walter Reich (Washington, DC: Woodrow Wilson Center, 1998), 169.

8. For a discussion of the participants in media events, see Gabriel Weimann and Conrad Winn, *The Theater of Terror: Mass Media and International Terrorism* (New York: Longman, 1994), 104.

9. Robin P. J. M. Gerrits, "Terrorists' Perspectives: Memoirs," in *Terrorism and the Media*, eds. David L. Paletz

and Alex P. Schmid. (Newbury Park, CA: Sage, 1992), 36.

10. Gerrits, "Terrorists' Perspectives," 39.

11. Ibid.

12. Hoffman, *Inside Terrorism*, 142.

13. Ibid.

14. See Hoffman, *Inside Terrorism*.

15. Steven E. Barkan, and Lynne L. Snowden, *Collective Violence* (Boston, MA: Allyn & Bacon, 2000), 84.

16. Peter Taylor, *States of Terror: Democracy and Political Violence* (London: Penguin, 1993), 8, cited in Hoffman, *Inside Terrorism*, 74.

17. Abu Iyad with Eric Rouleau, *My Home, My Land: A Narrative of the Palestinian Struggle* (New York: Time Books, 1981), 111–112, quoted in Hoffman, *Inside Terrorism*, 73.

18. Hoffman, *Inside Terrorism*, 132.

19. Weimann and Winn, *The Theater of Terror*, 1.

20. Barkan and Snowden, *Collective Violence*.

21. Lawrence Howard, ed., *Terrorism: Roots, Impact, Responses* (New York: Praeger, 1992), 102.

22. For a discussion of the contagion effect and additional discussions, see Weimann and Winn, *Theater of Terror*, 157–160, 211.

23. Weimann and Winn, *Theater of Terror*, 158. See also C. S. Spilerman, "The Causes of Racial Disturbances," *American Psychological Review* 35 (1970): 627–649, and L. Berkowitz and J. Macaulay, "The Contagion of Criminal Violence," *Psychometry* 34 (1971): 238–260.

24. For a discussion of the contagion effect and additional discussions, see Weimann and Winn, *The Theater of Terror*, 157–160, 211.

25. See David L. Paletz and Laura L. Tawney, "Broadcasting Organizations' Perspectives," in *Terrorism and the Media*, eds. David Paletz and Alex P. Schmid (Newbury Park, CA: Sage, 1992), 105.

26. Ibid., 107, 109.

Chapter 10

1. See John Keegan and Richard Holmes, *Soldiers: A History of Men in Battle* (New York: Elisabeth Sifton Books, 1986), 252–253.

2. For a good discussion of the objectives of states vis-à-vis dissidents, see Peter C. Sederberg, *Terrorist Myths: Illusion, Rhetoric, and Reality* (Englewood Cliffs, NJ: Prentice Hall, 1989), 92.

3.   Another inventory of objectives is presented by Paul R. Pillar, *Terrorism and U.S. Foreign Policy* (Washington, DC: Brookings Institution, 2001), 130–131.

4.   Philip B. Heymann, *Terrorism and America: A Commonsense Strategy for a Democratic Society* (Cambridge, MA: MIT Press, 1998), 9.

5.   Regis Debray, *Revolution in the Revolution?* (Westport, CT: Greenwood, 1967), quoted in Christopher Hewitt, "Public's Perspectives," in *Terrorism and the Media*, eds. David L. Paletz and Alex P. Schmid (Newbury Park, CA: Sage, 1992), 189.

6.   See Heymann, *Terrorism and America*, 10–11.

7.   See Jessica Stern, *The Ultimate Terrorists* (Cambridge, MA: Harvard University Press, 1999), 70.

8.   For a discussion of terrorism in the age of globalization, see Gus Martin, "Globalization and International Terrorism," in *The Blackwell Companion to Globalization*, ed. George Ritzer (Malden, MA: Blackwell, 2006).

9.   Pillar, *Terrorism and U.S. Foreign Policy*, 47.

10.   Ibid., 48–49.

11.   Avigdor Haselkorn, "Martyrdom: The Most Powerful Weapon," *Los Angeles Times*, December 3, 2000.

13.   David Ronfeldt, John Arquilla, and Michele Zanini, "Networks, Netwar, and Information-Age Terrorism," in *Countering the New Terrorism*, by Ian Lesser, Bruce Hoffman, John Arquilla, David Ronfeldt, and Michele Zanini (Santa Monica, CA: RAND, 1999), 47.

12.   Ibid., 49.

13.   Haselkorn, "Martyrdom: The Most Powerful Weapon."

14.   See Bruce Hoffman, "Terrorism Trends and Prospects," in *Countering the New Terrorism*, by Ian Lesser, Bruce Hoffman, John Arquilla, David Ronfeldt, and Michele Zanini (Santa Monica, CA: RAND, 1999), 28.

15.   *Report of the Secretary-General on Chemical and Bacteriological (Biological) Weapons and the Effects of Their Possible Use A/7575* (New York: UN General Assembly, 1969), 6, cited in Jessica Eve Stern, "The Covenant, the Sword, and the Arm of the Lord," in *Toxic Terror: Assessing Terrorist Use of Chemical and Biological Weapons*, ed. Jonathan B. Tucker (Cambridge: MIT Press, 2000).

16.   See John Mintz, "Technical Hurdles Separate Terrorists From Biowarfare," *Washington Post*, December 30, 2004.

17.   Stern, *Toxic Terror*, 21–22.

18.   See Joby Warrick, "An Easier, but Less Deadly, Recipe for Terror," *Washington Post*, December 31, 2004.

19. Stern, *Toxic Terror*, 21–22.

20.   See Dafna Linzer, "Nuclear Capabilities May Elude Terrorists, Experts Say," *Washington Post*, December 29, 2004.

21.   Sun Tzu, *The Art of War* (New York: Oxford University Press, 1963), 168.

22.   For a good discussion of Hezbollah's rationale in deploying suicide bombers, see Martin Kramer, "The Moral Logic of Hezbollah," in *Origins of Terrorism: Psychologies, Ideologies, Theologies, States of Mind*, ed. Walter Reich. (Washington, DC: Woodrow Wilson Center, 1998), 131.

23.   Recall that HAMAS means zeal and is an acronym for Harakt al-Muqaqama al-Islamiya.

24.   The 1996 attacks by Hamas led to the election of a hawkish Israeli administration.

25.   See Romesh Ratnesar, "Season of Revenge: The Inside Story of How Israel Imprisoned Arafat—and Why the Rage Keeps Burning," *Time*, April 8, 2002.

26.   For a discussion of several myths about terrorism, including the notion that it is highly effective, see Walter Laqueur, "The Futility of Terrorism," in *International Terrorism: Characteristics, Causes, Controls*, ed. Charles W. Kegley, Jr. (New York: St. Martin's, 1990). 69. Other good discussions of effectiveness are available in Sederberg, *Terrorist Myths* and Heymann, *Terrorism and America*, 12.

27.   Bruce Hoffman, *Inside Terrorism* (New York: Columbia University Press, 1998), 176.

Chapter 11

1.   Only the United States, Great Britain, Russia, and France have the capability to deploy large numbers of seaborne troops.

2.   Muslim commentators criticized the term infinite justice, arguing that only God is capable of infinite justice.

3.   David J. Whittaker, ed., *The Terrorism Reader* (New York: Routledge, 2001), 144–145.

4.   For a history of the Foreign Legion, see Tony Geraghty, *March or Die: A New History of the French Foreign Legion* (New York: Facts on File, 1986).

5.   See Scott Shane and David E. Sanger, "Bush Panel Finds Big Flaws

Remaining U.S. Spy Efforts," *New York Times*, April 30, 2003.

6.   See T. R. Reid, "IRA Issues Apology for All Deaths It Caused," *Washington Post*, July 17, 2002.

7.   Jim Dwyer and Brian Lavery, "I.R.A. to Renounce Violence in Favor of Political Struggle," *New York Times*, July 27, 2005.

8.   Glenn Franke and William Branigin, "IRA Announces End to Armed Campaign," *Washington Post*, July 28, 2005.

9.   See Chapter 2 for a discussion of the Falangist code of self-sacrifice.

10.   For the complete text of the findings of Senate investigations into the scandal, see New York Times, *The Tower Commission Report* (New York: Bantam Books and Times Books, 1987).

11.   See Mike Allen, "House Votes to Curb Patriot Act," *Washington Post*, June 16, 2005. See also Dan Eggen, "Patriot Act Changes to be Proposed," *Washington Post*, April 5, 2005.

12.   Paul Wilkinson, "Fighting the Hydra: Terrorism and the Rule of Law," in *International Terrorism: Characteristics, Causes, Controls*, ed. Charles W. Kegley (New York: St. Martin's, 1990), 255.

13.   Ibid.

14.   Ibid.

15.   Ibid.

16.   Ibid.

Chapter 12

1.   See Mark Ehrman, "Terrorism From a Scholarly Perspective," *Los Angeles Times Magazine*, May 4, 2003.

2.   Samuel P. Huntington, *The Clash of Civilizations and the Remaking of World Order* (New York: Touchstone, 1996).

3.   Andrew MacDonald [William Pierce], *The Turner Diaries* (New York: Barricade Books, 1996).

4.   E. L. Anderson [David L. Hoggan], *The Myth of the Six Million* (Newport Beach, CA: Noontide, 1969).

5.   Adam Cohen, "Following the Money," *Time*, October 8, 2001.

6.   One Hamas leader bragged that suicide bombings were cheap—costing approximately $1,500 each in 2001 dollars.

7.   For a report on charities, see David B. Ottaway, "U.S. Eyes Money Trails of Saudi-Backed Charities," *Washington Post*, August 19, 2004.

8.   Karen DeYoung and Douglas Farah, "Al Qaeda Shifts Assets to Gold," *Washington Post*, June 18, 2002.

9. For online access to EPIC's DCS-1000 FOIA documents, see www.epic.org/privacy/carnivore/foia_ documents.html.

10. The Department of Homeland Security absorbed the Immigration and Naturalization Service, the Coast Guard, the Customs Service, the Federal Emergency Management Agency, and many other smaller bureaus.

11. National Commission on Terrorist Attacks on the United States, *The 9/11 Commission Report: Final Report of the National Commission on Terrorist Attacks Upon the United States* (New York: W. W. Norton, 2004).

12. Commission on the Intelligence Capabilities of the United States Regarding Weapons of Mass Destruction, *Report to the President of the United States* (Washington, DC: 2005).

Index

Abbas, Abu, 100, 188
Abdel-Nasser, President Gamel, 116
Abortion clinic bombings, 131
Absolute deprivation, 46
Absolutes. *See* Absolute deprivation; Moral absolutes
Abu Ghraib prison, 242–243
Abu Hafs al-Masri Brigades, 148
Abu Nidal, 59–60, 62, 93
Abu Nidal Organization (ANO), 59–60, 91, 100, 228
Abu Sayyaf Group (ASG), 91, 128 (table), 135 (table), 212
Achille Lauro cruise ship hijacking, 188, 228
Activist social movements, 44–45, 47, 157–158
Acts of political will, 51–53, 61
Afghan Arabs, 126, 145, 148
Afghanistan, 39, 41, 54, 75, 102, 125, 126, 127, 131, 146, 272 (map)
Africa, 254, 273 (map)
African National Congress (ANC), 72
Air France Airbus hijacking, 228
Air France Flight 8969 hijacking, 146–147
Airliner bombings, 84, 106, 150, 205, 212
Airliner hijackings, 58–60, 137, 143, 146–147, 189–190, 212, 228
   international conventions on, 243–244
   See also Airliner bombings; September 11, 2001 (USA); Targets
AK-47 rifles, 203, 238
Al-Aqsa Martyrs Brigade (Palestinian), 51, 91, 99, 200 (table), 207
Albania, 55
al-Banna, Sabri, 59–60, 100
Albigensian Crusade, 114
Alexander II, Czar of Russia, 26, 30
Alex Boncayo Brigade, 97 (table)
Al Fatah, 59, 60
Alfonso XIII, King of Spain, 34
Algeria, 146, 222, 254
Algerian Armed Islamic Group, 136
Algerian/North African cells, 128 (table)
al-Iraqi, Abu Maysara, 183
Al Jazeera cable news, 183
Al-Jihad (AJ), 91
al-Majid, Ali Hassan, 78

al-Megrahi, Abdel Basset, 84
Al Qaeda Organization for Holy War in Iraq, 148, 186 (table)
Al Qaeda terrorist organization, 2, 91, 93, 97 (table), 128 (table), 130, 150 (table), 253
   activity profile of, 200 (table)
   Air France Flight 8969 hijacking, 146–147
   cell-based network in, 134
   contagion effect and, 190
   Iraq-United States debacle and, 85–86
   Islambouli Brigades, 106
   London transportation system attacks and, 149
   Madrid train bombings and, 148
   Muslim perceptions and, 53–54
   Operation Enduring Freedom and, 221–222
   Reid, Richard and, 107
   religious foundation of, 126–127
   September 11, 2001 attacks, 37–39
   Singapore plot, 147
   suicidal violence and, 207
   terrorist manual of, 208–209
   transnational movement of, 145–148
al-Qassam Brigade cells, 207
al-Sabbah, Hasan ibn, 115
Alternative lifestyles, 158
al-Zarqawi, Abu Musab, 127
Al-Zawahiri, Ayman, 188
Amal movement, 190
American Bar Association, 31
American Center for Law and Justice, 31
American Civil Liberties Union (ACLU), 31
American Federation of Labor and Congress of Industrial Organizations (AFL-CIO), 31
American Federation of State, County and Municipal Employees (AFSCME), 31
American Friends Service Committee (AFSC), 31
Amin, Idi, 73
Amir, Yigal, 124
Ammonium nitrate/fuel oil (ANFO) bombs, 147, 204, 257
Amnesty, 241
Amnesty International, 86
Analyst role, 14, 187
Anarchist philosophy, 25, 26, 30, 57–58, 252
Anfal campaign (Iraq), 77, 78–79, 86

ANFO. *See* Ammonium nitrate/fuel oil (ANFO) bombs
Anglo-Israelism and, 161
Angola, 18–19, 96, 104
Animal Liberation Front (ALF), 168
Animists, 103
Anonymity of banking customers, 263
ANO. *See* Abu Nidal Organization (ANO)
Ansar al-Islam (AI), 91
Anthrax, 38, 127, 202, 205, 256
Antiestablishment radicals, 30
Antimiscegenation, 6
Antiquity, 22–23
Anti-Semitic beliefs, 5, 117–119, 259
Antistate rebels, 36, 96–98
    See also Dissident terrorism; Existing order
Antitechnology activists/terrorists, 25
Antiterrorism, 233
Antiterrorist reprisals, 36–37
Antiwar movement, 158
Apartheid, 19, 72
Appian Way crucifixions, 23
Arab Islamist extremism, 116–117
Arab League, 47
Arab Revolutionary Brigades, 60
Arab Revolutionary Council, 60, 100
Arafat, Yasir, 98, 99, 101, 236
Argentina, 14, 15, 34, 35, 68, 146, 283 (map)
Argentine Anti-Communist Alliance
        (Triple-A), 82 (table)
Aristogeiton, 22
Armed Forces for National Liberation (FALN), 167
Armed Islamic Group (GIA), 91, 146–147
Armed propaganda, 141
Armed Resistance Unit, 168
Armenian Revolutionary Army, 138
Armenian Secret Army for the Liberation of Armenia
        (ASALA), 138
Army of God, 31, 174
Aryan Nations, 160, 169
Aryan Republican Army (ARA), 31, 172
Aryan revolution (United States), 160, 169, 173
Asahara, Shoko, 127
ASALA. *See* Armenian Secret Army for the Liberation of
        Armenia (ASALA)
Ásatrú, 161
Asbat al-Ansar, 91
Asia, 254, 282 (map)
Askaris, 72
Assassination, 9, 22, 23, 24, 30, 52, 55, 60, 67, 72, 101, 142
    counterterrorist option of, 224
    Order of Assassins and, 115
    Rabin assassination, 124
Assault rifles, 203
Assistance model for terrorism, 68 (table), 69
    domestic policy and, 70–71
    foreign policy and, 69–70

Asymmetrical warfare, 133, 201, 258
    appeal of, 201–202
    high-tech terrorism and, 255–257
    martyr nation strategy and, 202
    netwar, new organizational structures and, 202
Atheism, 30
Atrocities, 7, 68
Atta, Mohammed, 262
AUC. *See* United Self-Defense Forces of Colombia (AUC)
Augustine, 36, 37 (table)
Aum Shinrikyō (AUM), 91, 97 (table), 127–128,
        128 (table), 156, 201
Austria, 28, 60, 134, 236
Authoritarianism, 3, 14, 72, 73 (table), 85–86, 87–88,
        103, 181, 193, 251, 254
Auto-genocide, 79
Autonomy, 94, 107, 136, 202
Axis armies, 103
Axis of evil, 85
Ayyash, Yehiya "the Engineer," 207
Azerbaijan, 102

Baader-Meinhof Gang, 94
Ba'ath Party (Iraq), 78
Babylonian Exile, 23
Baghdad. *See* Iraq
Bakunin, Mikhail, 30, 37 (table)
Bali (Indonesia) bombing, 41
Bangladesh, 123
Banking anonymity, 263
Basque Fatherland and Liberty (ETA), 47 (table),
        50, 94, 97 (table), 98, 134, 236–237
Batista, Fulgencio, 18, 93
Battlefield detainees, 10
Beheading, 24, 214
Beirut bombings, 216–217
Beka'a Valley, 122
Belgium, 28, 50, 134
Belief systems, 2
    anti-Semitic beliefs, 5
    chosen people belief, 5
    conspiratorial beliefs, 5
    dogmatic style and, 3
    extremism and, 3–6
    ideologies, 26, 41
    mystical belief system, 157
    religious beliefs, 2, 5
    simplified goals/generalizations and, 4–5
    violent extremists, characteristics of, 4–5
    See also Terrorism
Berlin Wall, 17
Berri, Nabih, 190
Bethlehem, 115
Bhagwan Shree Rajneesh, 157
Big Brother atmosphere, 270
bin Laden, Osama, 2, 5, 53, 93, 126–127, 130, 188, 207

Biological weapons, 38, 105, 127, 157, 202, 205
Birmingham Six case, 240
Black Death, 205
Black Hebrew Israelites, 156
Black Hundreds group, 118
Black June group, 60
Black Liberation Army (BLA), 31, 165–166, 167
Black Liberation Movement, 165–166, 166 (table)
Black Nationalism, 162
Black Panther Party for Self-Defense, 31, 162, 165
Black Power movement, 157–158, 165
Black Pride movement, 158
Black Repartition society, 26
Black September group, 60, 62, 97 (table), 100, 135 (table),
    137, 224, 226, 227, 253
Black Tigers (Sri Lanka), 108
Black Widows, 50, 51
Blacklisting, 87
Bloody Sunday (Ireland), 54, 61–62
Boland Amendment, 70
Bolshevik Revolution (Russia), 32, 37 (table), 50
Bombs. See Weaponry
Booth, John Wilkes, 90
Bosnians, 12, 16, 76, 81 (table),
    83 (table), 102, 145, 146
Botulinum toxin/botulism, 205
Bourgeoisie, 33 (figure), 192
Brazil, 76, 146, 182, 283 (map)
Bribery, 236
Broad conclusions, 4–5
Bubonic plague, 205
Buddhist Sinhalese, 102, 108–109
Building/site targets, 211, 212
Bullying, 3
Bureau for the Protection of the Constitution
    (Germany), 231
Burke, Edmund, 24
Burma, 73
Burundi, 95, 102
Bush, President George W., 39, 85, 241
Bush wives, 117
The Butcher of Kurdistan, 78
Butler, "Reverend" Richard, 169, 170
Byzantine Empire, 114

Cadre of fighters, 145, 162
Caligula, 23
Cambodia, 75, 282 (map)
    Khmer Rouge, 79–80, 81 (table)
    Killing Fields, 77, 80, 81 (table)
Cambodian Communist Party, 79
Campus disturbances, 158, 163
Canaan conquest, 113–114
Canada:
    Front de Liberation du Québec,
        47 (table), 94, 186 (table)

Capitalism:
    anti-imperialist umbrella group and, 134
    capitalist society, 25, 103
    collapse of, 52
    nihilist activism and, 98
    opposition to, 30, 150 (table)
CAP. See Civil Patrols (CAP)
Caracalla, 23
Carlos the Jackal, 100, 134, 188, 236
Carmichael, Stokely, 158
Carnivore (DCS–1000) surveillance system, 264
Carthage, 23
Castro, Fidel, 18, 93
Castro, Raul, 18
Catholic Worker movement, 31
Caucasus, 102, 254
Causes of terrorism, 44–45
    absolute deprivation, 46
    activist social movements and, 44–45, 47
    acts of political will and, 51–53
    adversaries, freedom fighter-terrorist dilemma and, 53–54
    codes of self-sacrifice and, 57–60
    contextual influence on theory development and, 61
    counterrevolutionary elements, suppression of, 55–56
    cultural perspectives and, 53–54
    dramatic events and, 44, 45
    gender and terrorism, 50–51
    good/evil, simplified definitions of, 56–57
    group level and, 44–45, 48
    group membership drive, group dynamics and, 48
    individual level and, 44, 48
    initiation decisions/perpetuation processes and, 48–49
    injustice, violent response to, 45–51
    intergroup conflict, collective violence and, 45–47
    international unrest, 46–47, 47 (table)
    mental illness/lunatic fringe and, 47–48
    moral convictions of terrorists and, 54–56
    moral justification of violence, 54–60
    people's war strategy, 52
    psychological explanations, 47–51
    relative deprivation theory and, 46
    sociological explanations, 45–46
    Stockholm Syndrome and, 49
    strategic choice of violence, 51–54
    structural theory and, 45–46
    utopia, seeking of, 57
    See also Future trends/projections in terrorism;
        International terrorism; Political violence;
        Terrorism; Violence
Cave of the Patriarchs, 207
Cell-based terrorist environment, 144, 144 (table)
Cells of terrorists, 26, 30, 39, 41, 99, 100, 107, 127, 134, 136,
    146, 173, 174, 202, 261
Central Intelligence Agency (CIA), 19, 53, 54, 129, 231, 238–239
    See also Homeland security reorganization
        (United States)

Centralized economic planning, 93
Chains of command, 8
Champions of an ideal, 28, 30
Change process:
    activist social movements and, 44–45
    acts of political will and, 51–53
    dramatic events and, 44, 45
    fear, exploitation of, 8
    government policy, terrorist pressures and, 8, 9
    See also Revolution
Charismatic leadership, 33, 156
Chechen Republic of Ichkeria, 105
Chechen separatists, 50, 51, 105–106, 206
Chechnya, 105, 146
Chemical Ali, 78
Chemical weapons, 78–79, 86, 105,
    127, 128, 202, 205–206
Chiang Kai-shek, 52
Chiapas paramilitaries, 82 (table)
Child soldiers, 94, 95, 96
Chile, 14, 34, 35, 75, 137 (table), 283 (map)
Chile Security Service (DINA), 137 (table)
China:
    Angolan civil war and, 19
    Chinese Revolution, 52
    communist insurgents and, 103
    Communist Red Army, 52, 71
    Cultural Revolution, 70–71
    Four Olds and, 71
    Long March, 52
    Nationalists and, 52
    People's Liberation Army, 71
    struggle meetings and, 24
    Tibet and, 94
    totalitarianism in, 73
Chinese Communist Party, 70–71
Chinese Revolution, 52
Chlorine gas, 206
Chosen people belief, 5
Christian Crusades, 114–115, 131
Christian Identity creation myth, 161, 174
Christian Right religious politics, 158–159
Church of England, 24
Cienfuegos, Camilo, 18
Civil disobedience, 157, 161
Civil disorder, 165–167, 165 (table)
Civilians:
    citizen militias, 160
    cleansing society, 75
    collateral costs of war and, 12, 16, 188
    combatant status, 15–16
    Communist insurgent activities and, 103–104
    disposables, 76
    ethnic cleansing, 12, 15, 76
    genocide and, 12, 15, 42, 67
    innocents as targets, 2, 8, 9, 12, 23, 98, 105
    intimidation of, 9
    noncombatants status, 15–16, 15 (figure)
    onlookers, 13–14
    oppressed masses, 4
    rebellious/resisting masses, 7
    scapegoat groups, 67, 77
    social cleansing, 76
    sympathy among, 99
    total war and, 41–42
    See also Governments; Military campaigns
Civil liberties, 10, 38, 240–241, 263, 270
Civil Patrols (CAP), 82 (table)
Civil rights movement, 157, 171
Civil wars, 16, 18, 19, 103–104
Clandestine agents, 9
    See also Covert operations
Class pyramid of the industrial age, 33 (figure)
Class warfare, 28, 30, 52
Classless society, 93
Cleansing campaigns, 75–76
Cleansing society, 75
Clear vision, 92–93, 96
Cleaver, Eldridge, 163
Clinton, President Bill, 236
Codes of self-sacrifice, 57–60
Coercive covert operations, 224, 229 (table), 262
Cold War, 4, 12, 17–18, 19, 116, 159, 204, 254
Collateral damage, 12, 16, 188
Collective nonviolence, 157, 158, 159
Collective violence, 45–47
Colombia, 51, 76, 82 (table),
    95, 255 (table), 283 (map)
Colonialism, 18–19, 32, 71–72
    Arab world and, 116
    cost of war and, 141
    freedom fighters and, 141
    India/Pakistan and, 123
    neocolonialism, 138–139
    postcolonial developing world and, 103–104
    revolutionary solidarity and, 138
Columbia University takeover, 163
Combatants, 15–16, 15 (figure)
Commodus, 23
Common Law Courts and Constitutionalists, 159
Communal violence, 16, 18, 19, 101
    ethno-nationalist groups and, 102
    genocide and, 76
    ideological terrorism, Communist insurgents, 103–104
    Palestinian resistance and, 98
    Reign of Terror and, 24
    religious terrorism/sectarian violence, 102–103
    ritualistic communal lynchings, 172
    See also Dissident terrorism
Communication technologies, 140, 141, 181
    digital/video/audio terrorism and, 213–214
    high-tech terrorism and, 255

surveillance of, 261, 263–264
  See also Information technologies; Internet resources;
      Mass media; War on terrorism
Communism, 12, 18, 25, 28, 32
  Bolshevik Revolution, 32
  capitalism, collapse of, 52
  insurgent groups and, 103–104
  May 19 Communist Organization, 167–168
  Red Scare, 87
  United States, Communist Menace and, 159
Communist Combat Cells (Belgium), 50, 134
Communist Menace, 159
Communist Organization, 31
Communist Party of Nepal, 51
Communist Party of Philippines/New People's Army
      (CPP/NPA), 91
Communist Party, USA, 31, 87
Communist Red Army (China), 52–53, 71
Community Alert Patrol (Los Angeles), 162
Composite-4 (C–4), 204
Concessions, 214, 220, 237, 238 (table)
Confrontational style of activism, 159, 164
Congo, 102
Conservatism, 29 (table), 31 (table), 33
Conspiratorial beliefs, 5, 118
  American right and, 159–160
  conspiracy theories, 5
  Creativity Movement and, 161
  extremists' worldview, 5–6
  international terrorist networks and, 143–144
  Jones, Jim and, 156
  Ku Klux Klan and, 160
  Patriot movement and, 159
  Universal Product Code and, 160
  See also Protocols of the Learned Elders of Zion
Conspiratorial organizational structure, 8
Constantinople, 114
Contagion effect, 190
Continuity Irish Republican Army (CIRA), 91
Contra fighters, 19, 69, 70
Convention of the Prevention and Punishment of the Crime
      of Genocide (UN), 77
Convention to Prevent and Punish Acts of Terrorism Taking
      the Form of Crimes Against Persons and Related
      Extortion That Are of International Significance, 244
Counterculture of the New Left, 158, 164
Counterterrorism, 219, 220 (table), 234 (table)
  antiterrorism and, 233
  coercive covert operations, 219, 224
  concessionary options, 214, 220, 237
  conciliatory options, 219–220, 235–237, 238 (table)
  covert operations, 219, 224, 230
  diplomatic efforts, 220
  domestic laws and, 240–243
  economic sanctions, 219, 233
  elite counterterrorist units, utility of, 246–247

enhanced security, 219, 233
extraordinary renditions, 219
force, use of, 219, 221–229, 229 (table)
intelligence gathering, 219, 230–233
international law and, 243–244
law enforcement and, 220, 238–240
legalistic options, 220, 237–244, 245 (table)
nonviolent covert operations, 219, 230
other-than-war operations, 219–220, 229–237
paramilitary repressive options for, 219
repressive options, 219, 233, 234 (table)
social reform, 220, 236–237
special operations forces, 224–229
suppression campaigns, 219, 221–223
  See also Counterterrorist policies; War on terrorism
Counterterrorist policies, 7–8, 9
  American security environment, 38–39
  authoritarian measures, 14
  security environment, state participation in, 67
  terrorist suspects, labeling of, 9–10
  war on terror and, 39–40, 41
  See also Counterterrorism; War on terrorism
Covert official state terrorism, 75
Covert operations, 219
  coercive covert operations, 224
  nonviolent covert operations, 230
  See also Counterterrorism; Suppression campaigns
Crazies (New Left terrorism), 164
Crazy states, 73, 73 (table)
Creation myth, 161
Creativity Movement, 161
Criminal profiling, 240
Criminal terrorism, 137 (table)
Croats, 12, 76, 81 (table), 102, 145
Crucifixion, 23
Crusades, 114–115, 131
Cuba:
  Angolan civil war and, 19, 104
  Cuban Revolution, 18, 93, 158
  Ethiopian unrest and, 83 (table)
  Guantanamo Bay detentions, 10, 39
  Nicaragua and, 19
Cuban Missile Crisis of 1962, 18
Cult of personality, 193
Cults:
  American cults/terrorist violence, 156–157
  Aum Shinriky?, 91, 97 (table), 127–128, 128 (table)
  Bhagwan Shree Rajneesh, 157
  Black Hebrew Israelites, 156
  Holy Spirit Mobile Force, 117
  Lord's Resistance Army, 117
  People's Temple of the disciples of Christ, 156
  racial mysticism, neofascist movements and, 160
  Thuggee cult, 116
Cultural conservatism, 33
Cultural destruction, 79, 81 (table)

Cultural Revolution (China), 70–71
Cyberwar, 219, 230, 257
Cyprus Airways hijacking, 228

Dacian nation, 23
Darfur region, 103
Days of Rage action, 164
DCS–1000 (Carnivore surveillance system), 264
Death Angels, 156
Death Night, 228
Death squads, 72, 74, 156
Decentralization of power, 30
Decommissioning of paramilitaries, 235
Defense of faith. *See* Religious extremism
Defense Intelligence Agency (DIA), 231
DeFreeze, Donald, 164, 165
Delta Force, 224, 225, 228–229
Democratic Front for the Liberation of Palestine, 100
Democratic Kampuchea, 80
Democratic processes, 28
    authoritarian methods and, 87–88
    fascism and, 33
    freedom of reporting, security issues and, 194–195
    nihilist activism and, 98
    state authority, legitimization of, 72, 73 (table)
    terrorism/subversive movements and, 85
    Zimbabwe and, 71
    See also Capitalism
Democratic Socialist Republic of Sri Lanka, 108
Department of Homeland Security, 267
Deprivation. *See* Relative deprivation theory
Destroying the town. *See* It became necessary to destroy the town to save it
Destruction of existing order, 28, 30, 41, 81 (table)
Detention facilities, 10, 39
Developing world, 103–104, 141–142
Dictatorship of the proletariat, 32, 37 (table)
Dictatorships, 34
Dien Bien Phua (Vietnam), 79
Digital fingerprinting, 263
DINA. *See* Chile Security Service (DINA)
Diplomacy, 220, 235, 238 (table)
Diplomatic personnel, 209–210, 215 (table), 244
Direct Action (France), 50, 133–134, 138
Direct action strategy, 158
Directorate for Inter-Services Intelligence (ISI), 123
Dirty bomb, 206
Dirty War, 14, 15, 68, 75
The Disappeared, 14
Discriminate force, 15 (figure), 16
Disinformation campaigns, 219, 230
Disposables, 76
Dissident terrorism, 11, 13, 22, 37 (table), 90–91, 137 (table)
    antistate dissident terrorism, 96–98
    child soldiers and, 94, 95, 96
    communal terrorism and, 98, 101–104

criminal dissidents, 96
Cuban Revolution and, 93
defeat/victory attitudes, 98–101
genocide, communal dissident violence, 76
intensities of conflict, antistate terrorist environments and, 96–98
international terrorism and, 135, 135 (table)
just war doctrine and, 36–37
Marxist revolutions and, 32, 93
nationalist dissident terrorism, 94–95, 96
New Terrorism, new morality and, 104–107
nihilist dissidents, 57, 93, 96
Palestinian nationalist movement, coalitional features of, 99–101
practices of dissident terrorism, 96–104
revolutionary dissident terrorism, clear world vision and, 92–93, 96
unconventional war/destabilized central authority and, 93
utopian vision and, 99
violent dissent, 92–96
    See also Future trends/projections in terrorism
Domestic law and counterterrorism, 240–243, 245 (table)
Domestic policy:
    assistance model for terrorism and, 68 (table), 70–71
    patronage model for terrorism and, 68–69, 68 (table)
    See also Domestic terrorism practices; Foreign policy; Foreign policy terrorism; State terrorists
Domestic terrorism (United States), 154, 155 (figure, table), 175
    Animal Liberation Front, 168
    Aryan Nations, 169
    Aryan revolution and, 160
    Black Liberation Army and, 165–166
    Black Liberation Movement and, 165–166, 166 (table)
    Black Panther Party for Self-Defense, 162
    Black Power movement, 157–158, 165
    Christian Identity creation myth, 161
    Christian Right religious politics and, 158–159
    civil rights movement and, 157–158
    Cold War, Communist Menace and, 159
    concentration camps and, 159
    conspiracy theories, American right and, 159–160
    cults/terrorist violence, 156–157
    Days of Rage action, 164
    direct action strategy and, 158
    Earth Liberation Front, 168
    empowerment ideology and, 158
    extremism in America, 157–161
    FALN activities, 167
    Free Speech Movement (Berkeley) and, 158
    hostile un-American interests and, 159
    Ku Klux Klan, 112, 160, 170–172, 171 (table)
    leftist ethno-nationalist terrorism/civil strife, 165–167
    leftist hard core groups, 167–168
    leftist terrorism, 161–168
    left-wing extremism, 157–158

May 19 Communist Organization, 167–168
moralist terrorism, 174
moral suasion/nonviolence and, 157, 158
National Alliance, 169–170
neo-Nazi terrorism, 172–173
New Left, counterculture of, 158, 163
New Left terrorism, generational rebellion, 164–165
New World Liberation Front, 165
New World Order, 159, 160
The Order, 160, 169, 173
Patriot movement, 97, 159, 160, 173–174
Progressive Labor Party, 163
Puerto Rican Independencistas, 167
racial mysticism, 160, 172–173
racial supremacy activity and, 160, 169–172
radicalized resistance, civil
    disobedience/confrontation, 161–162
reactionary tendencies and, 159
Revolutionary Youth Movement, 163
right-wing extremism, 158–160
right-wing terrorism, 168–174
single-issue leftist violence, 168
Student Nonviolent Coordinating Committee, 158
Students for a Democratic Society and, 158, 162–164
Symbionese Liberation Army, 164–165
Unabomber activities, 168
United Freedom Front, 168
voter registration and, 158
Weatherman group, 163, 164
Weather Underground Organization, 93, 164, 165
White Aryan Resistance, 169
See also Counterterrorism; United States; War on terrorism
Domestic terrorism practices, 72
antiapartheid reformist agitation and, 72
cleansing society and, 75
covert official state terrorism, 75
death squads, 72
ethnic cleansing, 76, 81 (table)
genocide, 76–80, 81 (table)
official domestic state terrorism, 74–76
overt official state terrorism, 75
paramilitary groups, 72
secret police services and, 75
social cleansing, 76
state authority, legitimization of, 72–73, 73 (table)
state domestic authority, 73–81
vigilante domestic state terrorism, 74
See also Domestic policy; Domestic terrorism (United
    States); Foreign policy; Foreign policy terrorism
    practices; State terrorists
Domination, 23.37
Dramatic events, 44, 45
Drug trafficking, 128
Druze, 103
Duelfer, C. A., 86
Dynamite, 203

Earth Liberation Front (ELF), 168
Eastern Bloc, 97, 98, 223
Ebola virus, 127
Echelon satellite surveillance network, 264
Economic sanctions, 219, 233, 234 (table)
Eelam Tamil Association (ETA), 47 (table)
Egalitarian social order, 32, 93
Egypt, 47, 254
    democracy in, 73
    Force 777, 228
    Nasserism and, 116
    Sharm el Sheikh bombing, 26, 41, 105, 148
Egyptair Flight 648, 228
Elagabalus, 23
el-Assad, Bashar, 78
el–Assad, President Hafez, 75
Elders of Zion. See Protocols of the Learned Elders of Zion
Electromagnetic pulse (EMP) technologies, 257
Electronic Privacy Information Center (EPIC), 264
Electronic surveillance, 264, 270
Electronic triggers, 204
Elite counterterrorist units, 246–247
    See also Counterterrorism; Special operations forces
ELN. See National Liberation Army (ELN) of Colombia
El Salvador, 34, 68
Embassies, 209–210, 212
Emergency Search Team, 226
Empowerment ideology, 158
The end justifies the means, 57, 74
Enemies:
    demonization of, 5
    just war doctrine and, 36
    labeling of, 4, 5, 9–10, 184–185
    prisoners of war, international law and, 9
    scapegoat groups, 67
    terrorist suspects, 9–10
    See also Causes of terrorism; Moral justification;
        Targets; Victims of terrorism
Enemy combatant status, 9, 10, 67
Engels, Friedrich, 32, 37 (table)
England. See Great Britain
Enhanced security, 219, 233, 234 (table)
Ensslin, Gudrun, 28, 50, 94
Episode-specific sponsorship of terrorism, 83 (table), 84
Equatorial Guinea, 18
The Establishment, 161
Estimative language, 289–290
    See also National Intelligence Estimate report
ETA. See Basque Fatherland and Liberty (ETA)
Ethical issues:
    Abu Ghraib prison abuse, 242–243
    combatants vs. noncombatants and, 15–16, 15 (figure)
    extraordinary renditions, 219
    Geneva Convention protocols and, 9–10
    Guantanamo Bay detentions, 10, 39
    indiscriminate vs. discriminate force and, 15 (figure), 16

privacy protections, 242, 263, 264, 270
secret detentions, 39
See also Genocide; Human rights issues; Torture
Ethiopia, 18, 83
Ethnic cleansing, 12, 15, 76, 81 (table), 102
Ethno-nationalist groups, 2, 5, 11, 16, 28, 33, 37
civil strife, leftist ethno-nationalist terrorism, 165–167
code of self-sacrifice and, 57
communal terrorism and, 102
ethnic cleansing and, 76
international unrest and, 46–47, 47 (table)
Middle Eastern politics and, 37–38
moral superiority and, 55
Serbs in Kosovo, 55
traumatic events and, 45
Euphemistic language, 184–185
Europe, 7, 24, 25–26, 28, 73, 254, 275 (map)
European Police Office (EUROPOL), 238
Evil. *See* Good-evil dichotomy
Executions, 22, 23, 24, 25, 80
Exile, 75, 80, 81 (table), 117
Existing order, 28, 30, 41, 74, 92, 93, 198
Exploitation, 37, 54, 90, 116, 138–139
Extradition treaties, 244
Extraordinary renditions, 219
Extremism, 3
broad conclusions/incontrovertible
generalizations and, 4–5
conspiratorial beliefs/language usage and, 5
definition of, 3
elite grouping and, 5
enemies, labeling of, 4, 5
Great Proletarian Cultural Revolution in China, 70–71
intolerance and, 4
moral absolutes and, 4
rationalization/justification for violence, 3
religious belief systems and, 5
terrorism/terrorists and, 3, 41
violent extremists, characteristics of, 4–5
worldview of extremists, 5–6
Extremism in defense of liberty is no vice, 12

Facial imaging, 263
Falange party, 34–35
Falklands War, 194
FALN. *See* Armed Forces for National Liberation (FALN)
Fanon, Frantz, 162
FARC. *See* Revolutionary Armed Forces of Colombia (FARC)
Far left ideology, 28, 29 (table), 31 (table)
Farmland seizure, 71
Far right conspiracy proponents, 5
Far right ideology, 28, 29 (table), 30, 31 (table)
Fascism, 32–35, 75
Fatah Revolutionary Council, 100
Fawkes, Guy, 24–25
Fear, 8, 9

February 1993 World Trade Center attack, 136, 201
Federal Bureau of Investigation (FBI), 9, 162–163,
166, 226, 231, 264
See also Homeland security reorganization
(United States)
Fhima, Lamen Khalifa, 84
Financial operations, 261, 262–263
Fingerprinting technology, 263
First Navy Parachute Infantry Regiment (1RPIMa), 225
Five Pillars of Islam, 124
Flavius Josephus, 23
FLQ. *See* Front de Liberation du Quebec (FLQ)
FNLA. *See* Front for the Liberation of Angola (FNLA)
Force:
antiterrorist reprisals, 36–37
counterterrorism activities and, 219
fear and, 8
illegitimate use of, 6, 8, 9
indiscriminate vs. discriminate force, 15 (figure), 16
lethal force, selfless application of, 12
warfare environments and, 16
war on terrorism and, 262
See also Counterterrorism; Military campaigns;
New Terrorism; Terrorism
Force 17, 99
Force 777, 228
Forced labor, 80
Foreign policy:
assistance model for terrorism and, 68 (table), 69–70
patronage model for terrorism and, 68–69, 68 (table)
See also Domestic policy; State terrorists
Foreign policy terrorism practices, 81–83, 83 (table)
joint operations/active participation, 83 (table), 84–85
logistical/technical support for, 83 (table), 84
politically sympathetic sponsorship/moral support
for, 83–84
selective/episode-specific participation, 83 (table), 84
state terrorism, advantages of, 82
See also Domestic policy; Domestic terrorism practices;
Foreign policy; International terrorism;
State terrorists
Foreign terrorist organizations, 90, 91 (table)
Forrest, Nathan Bedford, 171
Four Olds, 71
The Fourteen Words, 160
France, 7, 28, 87, 146
Abu Nidal and, 60
Direct Action, 50, 133–134, 138
National Front, 28
special operations forces, 225, 226
See also French Revolution
Franco, Francisco, 33, 35, 98
Free press, 181, 191–193
Free Speech Movement (University of
California/Berkeley), 158
Free world, 4

Freedom Birds (Sri Lanka), 51
Freedom fighters, 2, 7
    anticolonial extremists, 141
    nihilists/nationalists/revolutionaries and, 96, 97 (table)
    terrorist or freedom fighter, dilemma of, 11–14, 53–54
    See also National liberation conflicts; One person's
        terrorist is another person's freedom fighter
Freedom of Information Act (FOIA), 264
Freemasons, 35, 118
Freemen, 31
French Canadians, 47 (table), 94, 186 (table)
French Foreign Legion, 225
French Navy Special Assault Units, 225
French Revolution, 24, 26, 27–28, 27 (table)
Front de Liberation du Québec (FLQ), 47 (table),
        94, 186 (table)
Front for the Liberation of Angola (FNLA), 19
Fund for the Martyrs (Iran), 122–123
Furrow, Buford Oneal, 169
Future trends/projections in terrorism, 249
    Africa and, 254
    Asia and, 254
    dissident terrorism, 252
    electromagnetic pulse technologies and, 257
    Europe and, 254
    exotic technologies and, 256–257
    high-tech terrorism and, 255–257
    historic waves of terrorism and, 249–250
    Holocaust denial and, 259
    ideological terrorism, 252
    information technologies and, 256
    international terrorism, 258
    Latin America and, 253–254
    Middle East and, 253
    Palestine/Israel and, 253
    religious terrorism, 252
    soft targets and, 257
    sources of terrorism and, 253–255, 255 (table)
    state-sponsored terrorism, 251–252
    symbolism and, 257
    traditional vs. new terrorism, characteristics of,
        250–251, 251 (table)
    twenty-first century terrorist environments and, 251–252
    United States, international terrorism in, 258
    United States, violent left in, 258
    United States, violent right in, 259
    United States, Western allies and, 254–255
    weapons of mass destruction and, 256
    See also New Terrorism; War on terrorism

Galba, 23
Gama'a al-Islamiyya (IG), 91
Gandhi, Prime Minister Rajiv, 109
Gasoline bombs, 204
Gender. *See* Women terrorists
Generalizations, 4–5

Generational rebellion (United States), 164–165
Geneva Convention protocols, 9–10
Genocidal state terrorism, 79
Genocide, 12, 15, 42, 67, 76–77, 81 (table)
    auto-genocide, 79
    definition of, 77
    genocidal state terrorism, 79, 103
    Holocaust denial and, 259
    rationale for, 77
    scapegoat groups and, 77
    state resources for, 77
    See also Domestic terrorism practices
Germany, 7, 37 (table), 87, 146
    Gastarbeiters in, 98
    intelligence agencies in, 231
    Munich Olympics massacre, 60
    Nazi Party, 33, 34, 35, 81 (table)
    neo-Nazi violence and, 98
    People's Union, 28
    Red Army Faction, 28, 50, 94, 97–98, 134, 138
    socialism and, 32, 33
    special operations forces, 226
    total war doctrine and, 42
Gibbs, J. P., 8
GIGN (Groupe d'Intervention Gendarmerie
        Nationale), 226
Global Islamists, 116–117, 130
Globalized terrorism, 2, 40–41
    cycles of violence, global awareness and, 40–41
    terrorist methodology and, 201
    United States, reputation of, 53–54
    See also International terrorism; War on terrorism
Goldman, Emma, 50
Goldstein, Baruch, 124, 207
Goldwater, Senator Barry M., 12
Good-evil dichotomy, 4, 5, 12, 47, 56–57
Good Friday Agreement, 235
Governments:
    antiterrorist policy and, 7–8, 9–10, 14
    domestic/foreign policy, terrorist practices in, 7
    intimidation of, 8, 9
    policy, terrorist pressures and, 8, 9
    proportional response to threats, 14
    secret policies, 67–68
    separation of church and state, 5–6
    soft targets of, 6
    subnational/non-state entities, 8, 9
    terrorism, unacceptability of, 141–142
    total war and, 41–42
    See also Political violence; State terrorists
Grail missiles, 203
Grand Mosque Incident, 124
Grand vision, 92–93, 96
Great Britain, 2, 7, 14, 26, 28, 87, 146
    Abu Nidal and, 60
    Angolan civil war, mercenaries in, 19

civil liberties, suspension of, 240–241
Guy Fawkes, gunpowder plot of, 24–25
intelligence agency/MI-5 and, 231
London bombing, 2, 26, 41, 105, 134,
    137 (table), 148, 149, 254
Luddite movement, 25
Northern Ireland battles, 97
special operations forces, 224–225
Great Depression, 32
Great Proletarian Cultural Revolution (China), 70–71
Greater good justification, 12, 22
Greater jihad, 120
Greece, 23, 28, 104
Greek Communist Party, 104
Green Police, 226
Grenada invasion, 168
Grey Wolves, 129
Ground Zero site. *See* September 11, 2001 (USA)
Group-level causes of terrorism, 44–45
group membership drive, rationalizations of
        violence and, 48
intergroup conflict, collective violence and, 45–47
See also Causes of terrorism
GSG-9 (Grenzschutzgruppe 9), 226, 228
Guantanamo Bay (Cuba), 10, 39
The Guardia (El Salvador), 74
Guatemala, 76, 82 (table)
Guerrilla warfare, 7, 12, 15, 67
Afghan Arabs, 145
Angolan raids, 18–19
anti-Sandinista guerrillas, 19
Chechen separatists, 105–106
China's Red Army, 52–53
Communist insurgents and, 103–104
Cuban Revolution, 18, 93
Iraq-United States conflict and, 39–40
Khmer Rouge, 79–80, 81 (table)
Vietnam War, 13, 18, 84–85
Guevara, Ernesto "Che," 18, 58, 93
Guildford Four case, 240
Guinea-Bissau, 18
Gulf War of 1990–1991, 82, 127, 194
Gurr, T. R., 7
Guyana, 156

Habash, George, 100
Hague Convention, 36, 243
Hamas (Islamic Resistance Movement), 14, 47, 84, 91,
    97 (table), 100, 122–123, 128 (table), 131, 207
Harakat ul-Mujahidin (HUM), 91
Harakt al-Muqaqama al-Islamiya. *See* Hamas
    (Islamic Resistance Movement)
Hardening the target, 219, 233
Harmodius, 22
Hate crimes, 170, 252
Hawala practices, 263

Hearst, Patricia, 165
Hebron Mosque massacre, 124, 207
Heinzen, Karl, 25
Henry VIII, King of England, 24
Heritage Foundation, 31
Hezbollah, 83 (table), 84, 93, 122, 131, 136,
    186, 189–190, 207, 217
Hidden agendas, 5
High-tech terrorism, 255
electromagnetic pulse technologies, 257
exotic technologies, 256–257
information technologies and, 256
liquid metal embrittlement, 257
plastics, 257
soft targets, symbolism and, 257
tech-terror, feasibility/likelihood of, 257
weapons of mass destruction and, 256
See also Future trends/projections in terrorism;
        War on terrorism
High-yield explosives, 25
Hijackings, 58–60, 136–137, 143,
    146–147, 188, 189–190, 228
international conventions on, 243–244
See also September 11, 2001 (USA)
Hindus (India), 123
Hindu Tamils, 102, 108–109
Hindu Thuggee cult, 116
Hipparchus, 22
Hiss, Alger, 87
Historical perspectives on terrorism, 22
anarchist philosophy and, 25, 26
ancient/medieval Middle East, 23–24
antiquity, 22–23
Babylonian Exile, 23
Carthage, destruction of, 23
communist philosophy and, 25
crucifixion and, 23
French Revolution, 24
Guy Fawkes, gunpowder plot of, 24–25
ideas, danger of, 23
lower classes, rights of, 25
Luddite movement, 25
Marxist philosophy and, 25, 26
Masada, 24
modern era, 26
nineteenth century Europe, 25–26
People's Will movement, 25–26
populist revolution, 26
regicide and, 23
Reign of Terror, 24
religious terrorism, 111–119
Revolutionary Tribunal, 24
revolutionary vanguard strategy and, 26
Roman age, 23
sicarii rebels, 23, 24
tyrannicide and, 22

Zealots, 23–24
    See also Causes of terrorism; Ideological origins of
        terrorism; International terrorism
Hitler, Adolf, 33, 34, 35, 37 (table), 118
Hizballah, 91
Ho Chi Minh, 79, 167
Hoffman, B., 8, 112
Holocaust (Germany), 81 (table)
Holocaust denial, 259
Holy cause, 54, 55, 102
Holy Spirit Mobile Force (Uganda), 117
Holy wars, 53–54, 120–121, 123, 125–126, 148, 169
Homeland security reorganization (United States), 265–269,
        266 (table), 268–269 (tables)
Honorable causes, 56–57
Hoover, J. Edgar, 162–163
Hostage Rescue Team, 226
Hostages, 137 (table), 213–214, 226, 227–229
House Un-American Activities Committee (USA), 87
Human intelligence/HUMINT, 230–231
Humanitarian protections, 10
Human progress, 32
Human rights issues:
    child soldiers and, 95
    Sandinistas and, 69
    terrorist suspects, incarceration/treatment of, 10
    See also Ethical issues
Human Rights Watch, 86
Humiliation, 24, 67
HUMINT. *See* Human intelligence/HUMINT
Hussein, King of Jordan, 62
Hussein, Saddam, 39–40, 67, 75, 78,
        85–86, 124, 149, 193, 251

Ibrahim Mosque, 207
Ideas, expression of, 23
Ideological objectives, 8, 16, 18, 22, 252
Ideological origins of terrorism, 26
    anarchism, 25, 26, 30
    classical ideological continuum and, 26, 27–28, 27 (table)
    codes of self-sacrifice and, 57
    communal terrorism and, 103–104
    destruction of existing order and, 28
    far left ideology, 28, 29 (table), 31 (table)
    far right ideology, 28, 29 (table), 31 (table)
    fascism, 32–35
    fringe left ideology, 28, 29 (table), 31 (table)
    fringe right ideology, 28, 29 (table), 30, 31 (table)
    ideologies and, 26, 37 (table)
    just war doctrine and, 36–37
    left/center/right designations, 27, 29, 31
    Marxism, 4, 10, 19, 25, 26, 32, 33 (figure)
    modern political environments, classical ideological
        continuum and, 28, 29 (table)
    See also Causes of terrorism; Historical perspectives
        on terrorism

Imperialism, 54–55, 134, 138, 150 (table)
Improvised explosive devices (IEDs), 203
Indefinite detention, 10
India, 51, 123, 254, 255 (table), 281 (map)
Indiscriminate force, 15 (figure), 16
Individualism, 30
Individual-level causes of terrorism:
    psychological theory and, 48
    sociological theory and, 44
    See also Causes of terrorism
Individuals with symbolic value, 211, 212
Indochinese Communist Party, 79
Indonesia, 41, 102, 104, 146, 282 (map)
Indonesian Communist Party (PKI), 104
Industrial age class pyramid, 33 (figure)
Industrial Revolution, 25
Industrial working class, 32
Infidels, 5
Infiltration, 230
Information is power, 180, 189–190
Information technologies, 40–41, 141, 183, 201
    disinformation campaigns and, 219, 230
    high-tech terrorism and, 256
    See also Communication technologies; Internet
        resources; Mass media
Ingram, J., 106
Inkatha Freedom Party (South Africa), 72
INLA. *See* Irish National Liberation Army (INLA)
Institute for Historical Review, 259
Insurgent groups, 18, 19, 39–40, 51, 90, 103–104,
        145, 149–150, 200 (table)
Intelligence community, 231
Intelligence gathering, 219, 230, 234 (table)
    human intelligence/HUMINT, 230–231
    intelligence agencies and, 231
    intelligence community and, 231, 232 (figure)
    National Intelligence Estimate report, 285–291
    signal intelligence/SIGINT, 230
    war on terrorism, international cooperation and, 264–265
    See also Counterterrorism; Other-than-war operations
Intergroup conflict, 45–47, 61
International Court of Justice, 244
International Criminal Court (ICC), 244
International Criminal Court for Rwanda (ICTR), 244
International Criminal Police Organization
        (INTERPOL), 238
International Criminal Tribunal for the Former Yugoslavia
        (ICTY), 244
International Justice Group, 97 (table)
International law:
    counterterrorism and, 220, 243, 245 (table)
    diplomat protections, 244
    extradition treaties, 244
    hijacking offenses, international conventions on, 243–244
    international courts/tribunals, 244
    prisoners of war, 9

terminal institutions and, 265
terrorism, unacceptability of, 141–142
war on terrorism, international law enforcement
    cooperation and, 264–265
International mujahideen, 125–126, 151
International symbols, 211, 212
International terrorism, 11, 133
    Afghan Arabs at war, 145, 148
    airline/ship hijackings/bombings, 136–137
    Al Qaeda transnational movement, 145–148
    anti-imperialist umbrella group, 134
    asymmetrical warfare and, 133
    Carlos the Jackal and, 100
    cases of, 46–47, 47 (table)
    cells of terrorists and, 134, 136, 144, 144 (table), 146
    cooperation, European connection and, 133–134
    domestic-oriented dissident terrorist groups and,
        135, 135 (table)
    domestic victims with international profile and, 136
    foreign country-based operations and, 136
    hijackings and, 136–137
    historical reasons for, 141–142
    ideological reasons for, 138–139
    imperialism and, 138
    Iraq, terrorist violence in, 149–150
    Madrid train bombings, 2, 26, 41, 105, 134, 148
    mechanisms of, 134–142
    mujahideen warriors, 54, 145
    neocolonialism and, 138–139
    networks in, 142–144
    New Terrorism, international dimension of,
        145–150, 150 (table)
    Palestinian terrorists and, 58–60
    patronage model for terrorism and, 68–69, 68 (table)
    practical reasons/perceived efficiency and, 140
    psychological anxiety and, 140
    publicizing a cause, 137, 140
    qualities of, 134
    rationale for, 138–142
    Singapore plot, 147
    spillover effect of, 135–137, 151–152
    tactical reasons for, 140–141
    terrorist environments and, 137, 137 (table),
        139–140, 143–144
    unacceptability of, 141–142
    unambiguous international implications of, 136–137
    United States as target of, 38
    Zionist movement and, 139
    See also Future trends/projections in terrorism;
        War on terrorism
Internationalism, 35
Internet resources:
    communication technologies, 140, 141, 181
    cyberwar, 219, 230, 257
    digital/video/audio terrorism and, 213–214
    netwar/internetted movements and, 202

    newscasts, audience to terrorism and, 14
    student study site, xv
    surveillance, Carnivore system and, 264
    terrorist activities and, 40–41, 130, 183, 213–214
Interrogation, 10, 67
Intifada, 45, 50, 98, 105, 207, 209, 210 (table)
Intimidation, 8, 9, 67, 72, 76
Intolerance, 3, 4
    one true faith concept and, 130–131
    See also Extremism
Inversion process, 164
Iran:
    Anfal campaign, Kurdish population and, 77, 78–79
    axis of evil member, 85
    Fund for the Martyrs, 122–123
    Hezbollah movement and, 122
    Iranian Revolution, 50, 121, 228
    komitehs and, 24
    Mujahideen-e Khalq Organization, protected persons
        status and, 10
    Palestinian Islamists and, 122–123
    Qods Force, 121
    Revolutionary Guards Corps, 121
    Revolutionary Justice Organization and, 122
    state-sponsored religious terrorism, 121–122
    terrorism, moral support for, 83–84, 83 (table)
Iran-Contra scandal, 70, 237
Iran-Iraq War of 1980–1988, 86
Iraq, 2, 17, 39, 146, 279–280 (maps)
    Abu Nidal and, 60
    Al Qaeda Organization for Holy War in Iraq, 148
    Anfal campaign, Kurdish genocide, 77, 78–79
    axis of evil member, 85
    Ba'ath Party, 78
    improvised explosive devices/roadside bombs, 203
    Marsh Arabs, 78
    secret police in, 75
    sectarian violence in, 124–125, 125 (table)
    suicide bombings and, 40, 41
    terrorist violence in, 149–150
    United States conflict with, 39–40, 41, 85–86,
        149–150, 179, 195
    United States private contractors, 105
IRA. See Provisional Irish Republican Army (PROVOS);
        Provisional Irish Republican Army (Provos)
Ireland, 276 (map)
    Bloody Sunday, 54, 61–62
    communal violence, religious justification for, 102–103
    Irish Catholic civil rights movement, 44, 47
    Northern Ireland Emergency Provisions Act and, 87–88
    Northern Ireland peace process, 235
    the Troubles, Northern Ireland and, 44, 45, 46–47, 235
    See also Provisional Irish Republican Army (Provos)
Irish National Liberation Army (INLA), 241
Irish Republican Army. See Provisional Irish Republican
        Army (Provos)

ISI. *See* Directorate for Inter-Services Intelligence (ISI)

Islambouli Brigades (Al Qaeda), 106

Islamic extremists, 5, 54, 69, 93, 116–117, 120, 129, 131

Islamic Group, 147

Islamic holy wars, 53–54, 120–121

Islamic Movement of Uzbekistan (IMU), 91

Islamic Resistance Movement. *See* Hamas
    (Islamic Resistance Movement)

Islamic world, 53–54, 93, 116–117

Israel, 146, 253, 255 (table), 277 (map)

    Abu Nidal and, 60

    Cain, descendents of, 161

    communal violence, religious justification of, 102

    Gaza/West Bank occupation, 126

    Halacha/Jewish Code, 124

    Hebron Mosque massacre, 124, 207

    Judaism and, 131

    Judeo-Christian antiquity and, 113–114

    Kach movement and, 91, 102, 131

    King David Hotel bombing, 216

    Lod Airport massacre, 143

    Munich Olympics massacre,
        60, 135 (table), 147, 224, 227

    Palestinians and, 46, 47, 50, 53, 98, 100

    September 11th attack and, 5

    special operations forces, 225, 226

    suicide bombings and, 207

    terrorist suicide campaign in, 105

    war of statehood, 47

    Wrath of God unit, 224

    Zionist movement and, 139

    See also Hamas (Islamic Resistance Movement); Protocols
        of the Learned Elders of Zion

Israel Defense Force (IDF), 225

Italy, 28, 37 (table), 87

    Abu Nidal and, 60

    cleansing society and, 75

    fascism in, 33, 35

    Red Brigades, 15, 50, 93, 98, 134, 135 (table), 138, 241

    repentance laws and, 241

It became necessary to destroy the town to save it, 13

Izzedine al-Qassam Brigade, 100, 207

Jacobin dictatorship, 24

Jaish-e-Mohammed (JEM), 91

Jamahiriya Security Organization (JSO), 84

James I, King of England, 24, 25, 46

Jammu-Kashmir groups, 128 (table)

Jammu Kashmir Liberation Front, 68

Japan, 38

    Aum Shinrikyō and, 91, 97 (table), 127–128, 128 (table)

    democracy in, 73

    hijackings, airline crimes' treaty and, 243

    invasion of China, 52

    Japanese Red Army, 50, 134, 142–143, 150 (table)

    nerve gas attack, 128

    total war doctrine and, 42

Japanese Red Army (JRA), 134, 142–143, 150 (table)

Jemaah Islamiya Organization (JI), 91, 147

Jenkins, B., 8

Jerusalem, 98, 121

Jewish Defense League (JDL), 97 (table), 131

Jewish rebellion, 23–24

Jibril, Ahmid, 100

Jihad, 120–121, 122, 127, 129–130

Johnson, President Lyndon B., 165

Joint operations of terrorism, 83 (table), 84–85

Jones, Jim, 156

Jonestown, 156

Jordan, 47, 60, 62, 83 (table), 137

Joyu, Fumihiro, 128

JSO. *See* Jamahiriya Security Organization (JSO)

Judaism, 131

Judeo-Christian antiquity, 113–114

Julius Caesar, 23

July 23, 2005 bombing (Egypt), 2, 26, 41, 105, 148

July 7, 2005 bombing (Great Britain), 2, 26, 41, 105,
        134, 137 (table), 148, 149, 254

Jus ad bellum criterion, 36

Jus in bello criterion, 36

Just war doctrine, 36–37

Justice Commandos of the Armenian Genocide, 138

Justification of violence, 3, 11–13, 14, 22, 34

        See also Causes of terrorism; Moral justification

Kach movement, 91, 102, 131

Kaczynski, Theodore "Ted," 25, 168, 181

Kahane Chai (Kach), 91, 102, 131

Kahane, Rabbi Meir, 131

Kalashnikov, Mikhail, 203

Kalashnikov rifles, 203

Kampuchea, 79, 80

Kansi, Mir Aimal, 238–239

Kashmir, 254

Kerner Commission, 166

Khaled, Leila, 50, 58–59, 62, 137

Khmer Rouge (Cambodia), 75, 79–80, 81 (table)

Kidnapping, 9, 55, 105, 117, 122, 128, 135, 143

        See also Hostages

Killing Fields, 77, 80, 81 (table)

Kill one man, terrorize a thousand, 12

Kim Il Sung, 193

Kim Jong Il, 193

King David Hotel bombing, 216

King, Reverend Martin Luther, Jr., 157, 158

KLA. *See* Kosovo Liberation Army (KLA)

Komitehs (Iran), 24

Kongra-Gel (KGK), 91

Korean War, 15

Kosovo, 55, 146

Kosovo Liberation Army (KLA), 55

Kropotkin, Petr, 30

Kuclos, 170
Ku Klux Klan (KKK), 5–6, 112, 160, 170, 171 (table)
  fifth-era Klan, 172
  first-era Klan, 170–171
  fourth-era Klan, 172
  Holocaust denial and, 259
  kuclos and, 170
  moderate Klan, 172
  purist Klan, 172
  second-era Klan, 171
  third-era Klan, 171–172
Kumpulan Militan Malaysia, 147
Kuomintang, 52
Kurdish genocide, 77, 78–79, 86, 251
Kurdistan, 78, 94, 124
Kuwait, 82, 149

Labeling choices, 4, 5, 9–10, 184–185
Labor unions, 28, 31
Lakwena, Alice Auma, 117
Land and Liberty society, 26
Land mines, 203
Land redistribution, 71
Lane, David, 160
Language usage:
  coded references and, 5
  conspiratorial beliefs and, 5
  enemies, labeling of, 4, 5, 9–10, 184–185
  euphemistic language, 184–185
  labeling decisions in news reporting, 184–185
  terror estimate, estimative language and, 289–290
Laos, 79
Laqueur, W., 6, 8, 76
Lashkar i Jhangvi (LJ), 91
Lashkar e-Tayyiba (LT), 91
Laskar Jihad, 128 (table)
Latin America, 253–254, 274 (map), 283 (map)
  dictatorships in, 34
  Falangist ideology and, 35
  far left dissent in, 28
  Marxists in, 4, 93
  narcotraficantes and, 253
  reoccupation of, 35
  social cleansing, 76
  United States as imperialist enemy and, 54
Law enforcement approach, 8, 38, 220, 238–240,
    245 (table), 264–265
Lazar, Prince (Serb hero), 55
Leaderless resistance, 173
Leary, Timothy, 164
Lebanon, 47, 83 (table), 84, 93, 100, 103, 121,
    122, 136, 146, 206–207, 216–217
Left fringe ideology, 28, 29 (table), 31 (table), 57–58
Left-wing movements. See Ideological
    origins of terrorism
Left-wing terrorism, 137 (table)

Legal issues:
  counterterrorism and, 220
  humanitarian protections of prisoners, 10
  prisoners of war, Geneva Convention protocols and, 9–10
  See also Legalistic options in counterterrorism
Legalistic options in counterterrorism, 220, 237, 245 (table)
  civil liberties, suspension of, 240–242
  domestic law, counterterrorism and, 240–243
  European Police Office and, 238
  International Criminal Police Organization and, 238
  law enforcement, counterterrorism and, 238–240
  law enforcement, international cooperation and, 264–265
  miscarriage of justice and, 240
  police repression, security activities and, 239–240
  privacy protections, 242, 263
  profiling of terrorists and, 240
  qualified amnesty, 241
  repentance laws, 241
  terrorist environments, police and, 239
  USA PATRIOT Act and, 38, 241–242
  See also International law
Legitimized conflict, 37 (table)
Lemkin, Raphael, 76
Leninism, 104
Lenin, Vladimir Ilich, 52
Lesser jihad, 120
Liberalism, 29 (table), 31 (table), 35
Liberation fighters. See Dissident terrorism; Freedom
    fighters; National liberation conflicts
Liberation theology, 31
Liberation Tigers of Tamil Eelam (LTTE), 14, 27, 45, 50, 51,
    91, 95, 97 (table), 108–109, 186 (table), 200 (table)
Liberia, 73, 95, 102, 254
Liberty. See Extremism in defense of liberty is
    no vice; Liberty Lobby
Liberty Lobby, 259
Libya, 18, 84, 100, 150, 204, 223
Libyan Islamic Fighting Group (LIFG), 91
Linzer, D., 86
Lipstadt, D. E., 259
Liquid metal embrittlement, 257
Lockerbie airliner bombing, 84, 150
Lod Airport massacre, 143
Logistical/technical support for terrorism, 83 (table), 84
London (England) bombing, 2, 26, 41, 105,
    134, 137 (table), 148, 149, 254
Lone wolf model, 107, 172, 174
Long Hot Summer, 166 (table)
Long live death slogan, 35
Long March, 52
Lord's Resistance Army (Uganda), 117, 128 (table)
Lower classes, 25
Lowndes County Freedom Organization, 162
LTTE. See Liberation Tigers of Tamil Eelam (LTTE)
Luddite movement, 25
Lufthansa airliner hijacking, 228

Lumpenproletariat, 33 (figure)
Lunatic fringe, 47–48
Lynchings, 172
Lyndon Larouche groups, 31

M-16 rifles, 203
Madrid (Spain) bombing, 2, 26, 41, 105, 134, 148, 202, 254
Mala prohibita, 41
Mala in se, 41
Malaysia, 146, 147, 254, 282 (map)
Malcolm X., 162, 167
Mallin, J., 56
Malum in se, 2
Managua harbor, 19
Manaysia, 103
Manifesto of the Communist Party, 32
Mao Zedong, 12, 52, 70–71, 140, 162, 202
Map resources, 272–284
March 11, 2004 (Spain), 2, 26, 41, 105, 134, 148, 202, 254
Marighella, Carlos, 57, 99, 141
Maronite Christians, 103
Marsden, V. E., 119
Martial law, 62
Martyr nation strategy, 202
Martyrdom, 114, 121, 122–123, 207
Marx, Karl, 25, 30, 32, 37 (table), 52
Marxism, 4, 10, 19, 25, 26, 32, 33 (figure), 52, 56, 93, 104, 252
Marxist-Leninist Party, USA, 192
Masada, 24
Mass communication. *See* Communication technologies;
    Internet resources; Mass media
Mass destruction, 26, 40–41
    genocide, 12, 15, 42, 67, 76–80, 81 (table)
    total war doctrine and, 41–42
    See also Weapons of mass destruction
Mass media, 2, 180, 213
    Al Jazeera cable news, 183, 188
    analysts/interpreters, 14
    backlash, risk of, 190–191
    coded language/secret messages and, 188, 192
    contagion effect and, 190
    content decisions and, 184
    criticisms of news reporting, 188–189
    cult of personality and, 193
    digital/video/audio terrorist tactics, 213–214
    euphemistic language and, 184–185
    free press and, 181, 191–193
    gatekeeping function and, 192–193
    information is power, media as weapon and, 180, 189–190
    Internet use and, 183
    journalistic self-regulation/media gatekeeping and, 192
    labeling of terrorists and, 184–185, 186 (table)
    manipulation of news by terrorists and, 188
    market competition and, 183–184
    mass communications, 181
    media-oriented terrorism and, 191

media scooping and, 188
    media spin and, 181, 187
    national security issues and, 194–195
    New Media and, 183
    onlookers/terrorist audience, 13
    participant roles in terrorism and, 185–187
    practical considerations, using the media, 187–189
    print media, 181
    propaganda and, 180, 185, 189
    quality of reporting, 184
    radio broadcasts, 182
    reporting on terrorist activities, 183–185
    roles of media, 180–185
    selective release of information and, 193
    state-regulated press and, 193
    television, 182
    terrorists' messages and, 180–183
    TWA Flight 847 hijacking, Hezbollah and, 189–190
    war for information and, 185–193
Mass oppression, 4, 28, 30
    absolute deprivation and, 46
    relative deprivation theory and, 46
    structural theory of revolution and, 45–46
    See also Domestic terrorism practices; Mass destruction;
        Oppressed masses
Mass revolution, 100
Matak, Prince Sirik, 80
Mathews, Robert Jay, 173
May 19 Communist Organization (M19CO), 167–168
McCarthy, Senator Joseph, 87
McVeigh, Timothy, 147, 170, 173
Means of production, 32
Media coverage. *See* Mass media
Media scooping, 188
Media spin, 181, 187
Media-oriented terrorism, 191
Medieval Middle East, 23–24
Meinhof, Ulrike, 50, 94
Mental illness, 47–48
Mercenaries, 19, 70, 105
Meredith, James, 158
Methods. *See* Terrorist tactics
Metzger, Tom, 169
Mexico, 51, 82 (table), 90
MI-6 (Great Britain), 231
Middle Core Faction, 97 (table)
Middle East, 23–24, 37–38, 58–60, 83,
        123, 125, 253, 278 (map)
Middle East Media Research Institute (MEMRI), 183
Military campaigns:
    chains of command, 8
    combatants vs. noncombatants and, 15–16, 15 (figure)
    conventional fighting, rules of engagement and,
        7, 16, 36–37, 67
    ethnic cleansing and, 12
    guerrilla warfare, 7

indiscriminate vs. discriminate force and, 15 (figure), 16
    just war doctrine and, 36–37
    occupation forces, 13
    passive military as targets, 8, 16
    total war and, 41–42
    unconventional warfare, 8
    warfare environments and, 16
Military campaigns, *See also* Prisoner
        of war status; Terrorism
Military Intelligence Service (Germany), 231
Military as national savior, 35
Militias. *See* Paramilitary groups
Mines, 203
Mini-Manual of the Urban Guerrilla, 57
Miscarriage of justice, 240
Missiles. *See* Weaponry
Mitchell, Hulon, Jr., 156
Mob brawls, 252
Moderate center ideology, 29 (table), 31 (table)
Modern era. *See* French Revolution; Historical perspectives
        on terrorism; New Terrorism
Molotov cocktails, 204
Molotov, Vyacheslav, 204
Monolithic terrorist environment, 143, 144 (table)
Montoneros, 215
Montreal Convention of 1971, 244
Moral absolutes, 4, 11
Moral conviction, 54–56, 131
    See also Torture
Moral justification, 14, 44–45, 54
    codes of self-sacrifice and, 57–60
    modern dissident terrorists, new morality and, 104–107
    moral convictions of terrorists, 54–56
    political violence and, 54–60
    simplified definitions of good/evil and, 56–57
    utopia, seeking of, 57
    See also Causes of terrorism
Moral Majority, 31
Moral suasion strategy, 157
Moral support for terrorism, 83–84, 83 (table)
Moralistic philosophy, 36
Moralist terrorism, 174, 215 (table)
More, Sir Thomas, 57
Moroccan Islamic Combatant Group, 148
Morocco, 254
Moussaoui, Zacaria, 147
Movement for Democratic Change (MDU)
        of Zimbabwe, 71–72
Movement for the Liberation of Angola (MPLA), 19, 104
Movimiento Nacional (Spain), 35
Mozambique, 18, 95
MPLA. *See* Movement for the Liberation of Angola (MPLA)
MR–8 (Brazil), 182
Mud People, 5, 161
Mugabe, Robert, 71, 72
Mujahedin-e Khalq Organization (MEK), 91

Mujahideen-e Khalq Organization (MKO), 10
Mujahideen warriors, 54, 125–126, 145
Multinational Force (MNF), 217
Munich Olympics massacre, 60, 135 (table),
        147, 189, 224, 226, 227
Muslim Brotherhood, 69, 93, 100, 122
Muslims, 12, 37, 53–54
    Albanian Muslims, 55
    Bosnian Muslims, 76, 81 (table)
    Chechen separatists, 105–106
    Five Pillars of Islam, 124
    Lebanon, 103
    Pakistan, 123
    Sudan region, 103
    See also Shi'a Muslims; Sunni Muslims
Mussolini, Benito, 33, 34, 37 (table), 75
Mustard gas, 206
Myanmar, 73, 282 (map)

Napoleonic campaigns, 7
Narcotraficantes, 253
Nasserism, 116
National Advisory Commission on Civil Disorders, 166
National Alliance, 169–170
National Association for the Advancement of Colored
        People (NAACP), 31, 157
National Association for the Advancement
        of White People, 31
National autonomy, 94
National Bar Association, 31
National Commission on Terrorist Attacks on the United
        States, 267
National Conference of Christians and Jews, 31
National Council for the Defense of Democracy-Forces
        for the Defense of Democracy (Burundi), 95
National Council of Resistance, 97 (table)
National independence, 94
National Intelligence Council (NIC), 54, 287–288
National Intelligence Estimate report, 285–291
National Lawyers Guild, 31
National Liberation Action (Brazil), 182
National Liberation Army (ELN) of Colombia, 51, 91
National liberation conflicts, 2, 18
    China, 70–71
    Nicaragua, 19, 69, 70
    Zimbabwe, 71–72
National Security Agency (NSA), 231, 264
National Security Council, 70
National security threats, 163
    See also National Intelligence Estimate report
National Socialist German Worker's (Nazi)
        Party, 33, 34, 35, 118
National Union for the Total Independence of Angola
        (UNITA), 19, 104
Nationalism, 32–33, 35, 47, 50, 51, 94–95, 145
Nationalist dissident terrorism, 94–95

Nationalist precepts, 26
Nationalist terrorism, 27
Nationalists (China), 52
Native populations, 54–55, 81 (table), 160
Nativism, 170
NATO. *See* North Atlantic Treaty Organization (NATO)
Natural law, 131
Navy Sea Air Land Forces (SEALs), 225
Naxalites (India), 51
Nazi Germany, 26, 33, 34, 35, 75, 81 (table), 118
Nechayev, Sergei, 30, 59
Negotiations, 235–236
Neocolonialism, 138–139
Neo-Confederate movement, 160, 172
Neofascist movements, 98, 160, 252
Neofascist youths, 98
Neo-Nazi rightists, 5, 98, 118, 137 (table), 156, 160, 168,
        169, 172–173, 259
Neopagan movement, 161
Nepal, 51, 254
Nerve gas, 127, 128, 201, 206
Netanyahu, Prime Minister Benjamin, 236
Netwar, 202
Networks of terrorists. *See* Al Qaeda terrorist organization;
        Cells of terrorists; Netwar; New Terrorism
New Africa Freedom Fighters, 167
New Left counterculture, 158, 163
New Left terrorism, 164–165
New order, 37 (table), 96
New People's Army, 97, 186 (table)
New Terrorism, 2, 26, 40
    asymmetrical warfare and, 133
    cells of terrorists and, 26, 30, 39, 41, 99, 100, 107
    characteristics of, 40
    contagion theory and, 190
    enemies, labeling of, 9–10
    horizontal organizational structure, 26
    independent autonomous terrorist cells
            and, 26, 30, 39, 41
    information technologies/Internet access and, 40–41
    international dimension of, 145–150, 150 (table)
    lone wolf model and, 107
    mass-casualty attacks and, 40–41
    netwar and, 202
    new morality/lethal weapons and, 104–107
    objectives, vague articulation of, 200–201
    religious terrorism and, 129–130
    suicide bombings, 40, 41, 101, 106, 108
    tech-terrorism, feasibility/likelihood of, 257
    traditional vs. New Terrorism, characteristics of,
            250–251, 251 (table)
    violent extremism and, 40
    waging war in era of, 39–40
    See also Future trends/projections in terrorism;
            September 11, 2001 (USA); State terrorists;
            Terrorism; Terrorist suspects; Terrorist tactics

New World Liberation Front, 165
New World Order, 159–160
Newton, Huey, 163
Nicaragua:
    Contra fighters, 19, 69, 70
    Guardia of, 19
    Iran-Contra scandal, 70
    mining of Managua harbor, 19
    Sandinista National Liberation Front, 19, 68–69, 70
    United States influence in, 19
Nicholas II, Czar, 118
Nihilism, 93
Nihilist dissidents, 57, 93
    See also Dissident terrorism
9/11 Commission, 267
Nineteenth century Europe, 25–26
Nol, Prime Minister Lon, 80
Noncombatants, 9, 15–16, 15 (figure)
Non-European races, 5
    antimiscegination beliefs, 5–6
    See also Racism
Non-state entities, 8, 10, 11, 36
    See also Dissident terrorism
Nonviolence, 157, 158, 159, 163
Nonviolent covert operations, 219, 230, 234 (table)
North Atlantic Treaty Organization (NATO), 55, 134, 254
North Korea, 73, 85
North, Marine Lieutenant Colonel Oliver, 19, 70
North Vietnamese Army, 55
Northern Ireland Act of 1993, 240–241
Northern Ireland Civil Rights Association, 61
Northern Ireland Emergency Provisions Act of 1973, 88
Nuclear weapons, 105, 123, 202, 206, 257

Oakland Black Panthers, 162
Oakland School District assassination, 164
Objective data, 5
Objectives. *See* Terrorist tactics
Occupation forces, 13
Office of the Director of National Intelligence (ODNI),
        231, 232 (figure)
Official Secrets Act, 194
Official state terrorism, 74–75
    See also Domestic terrorism practices; State terrorists
Okhrana, 118
Oklahoma City attack, 168, 173, 174, 201, 233
Olds. *See* Four Olds
One man willing to throw away his life is enough to terrorize
        a thousand, 12, 206
One person's terrorist is another person's freedom fighter,
        2, 11–12, 53
One-seedline Christian Identity, 161
One true faith. *See* Religious extremism; Religious terrorism
Onlooker role, 13–14, 187
OPEC. *See* Organization of Petroleum Exporting
        Countries (OPEC)

Operation Eagle Claw, 228–229
Operation El Dorado Canyon, 223
Operation Enduring Freedom, 221–222
Operation Infinite Justice, 222
Operation Iraqi Freedom, 39, 149–150
Oppressed masses, 4, 28, 30, 44–45, 90
    assistance in domestic policy and, 70–71
    imperialistic approach and, 54
    indigenous populations and, 54–55
    Mini-Manual of the Urban Guerrilla and, 57
    patronage in domestic policy and, 69
    See also Mass oppression; State terrorists
Orange Volunteers, 97 (table)
The Order, 160, 169, 173
Order of Assassins, 115
Organization of American States, 244
Organization of Petroleum Exporting Countries
    (OPEC), 134, 188, 236
Organizations of terrorists. *See* Foreign terrorist
    organizations
Osawatomie periodical, 164
Other-than-war operations, 219–220, 229
    concessions and, 214, 220, 237
    conciliatory options, 235–237, 238 (table)
    cyberwar, 219, 230
    disinformation, 219, 230
    economic sanctions, 219, 233
    hardening the targets/enhanced security, 219, 233
    infiltration, 230
    intelligence gathering, 230–233
    negotiations, 235–236
    nonviolent covert operations, 230
    peace processes, 235
    reasoned dialogue/diplomatic options, 235
    repressive responses and, 229, 233, 234 (table)
    social reform and, 236–237
    See also Counterterrorism; Legalistic options in
        counterterrorism; Suppression campaigns
Ottoman Turks, 55, 116
Overt official state terrorism, 75

Pahlavi, Shah Muhammed Reza, 121
Pakistan, 68, 123, 146, 254, 255 (table)
Palestine, 23, 27, 45, 253, 255 (table)
    Abu Nidal and, 59–60, 62, 93
    Abu Nidal Organization, 59–60, 91, 100
    Al-Aqsa Martyr Brigades, 51, 91, 99
    Black September group, 60, 62, 100
    Democratic Front for the Liberation
        of Palestine, 100
    Hamas, 14, 47, 84, 91, 97 (table), 100, 122–123
    intifada, 45, 50, 98, 100, 105
    Iranian support for Islamists, 122–123
    Izzedine al-Qassam Brigade and, 100
    Palestine Islamic Jihad, 84, 91, 97 (table), 101, 122
    Palestine Liberation Front, 91, 100

Palestinian nationalist movement, coalitional
    features of, 99–101
Palestinians in Israel, 46, 47, 53, 98
Popular Front for the Liberation of Palestine, 58–59, 100
Popular Front for the Liberation of Palestine-General
    Command, 91, 100
secular/religious Palestinian groups, 47 (table)
violent extremists, 58–60
Palestine Islamic Jihad (PIJ), 84, 91, 97 (table),
    101, 122, 128 (table)
Palestine Liberation Front (PLF), 91, 100, 188
Palestine Liberation Organization (PLO), 47, 51, 60, 83
    (table), 98, 99, 101, 134, 217, 253
Palestinian National Council, 59
Palestinian Revenge Organization, 97 (table)
Palmer Raids, 87
Pan Am Flight 103, 84, 150
Pan Am Flight 830, 212
Pan-Arabist ideology, 59–60, 116
Pan-Islamic revival, 93, 145
Paraguay, 146, 283 (map)
Paramilitary groups, 72, 73 (table), 76, 77,
    80, 82 (table), 95, 99
    Black Panther Party for Self-Defense, 162–163
    citizen militias, 160
    decommissioning of, 235
    Patriot movement, 97, 159, 160
    shadow wars and, 224, 261
Paramilitary repressive options, 219, 222
Participants in terrorism, 7, 13
    analysts, 14
    emotional responses of, 14
    labeling of, 4, 5, 9–10, 14
    onlookers, 13–14
    political associations of, 14
    supporters, 13
    symbolism, target selection and, 14
    terrorists, 13
    victims, 13, 14
    See also Terrorist environment
Party of God, 93, 186 (table)
Passenger carrier targets, 149, 188, 212, 228
    See also Achille Lauro cruise ship hijacking; Airliner
        bombings; Airliner hijackings; Lod Airport
        massacre; London (England) bombing; Madrid
        (Spain) bombing
Patriot movement (United States), 97, 159–160,
    168, 173–174, 259
Patronage model for terrorism, 68, 68 (table)
    domestic policy and, 69
    foreign policy and, 68–69
    See also State terrorists
Peace processes, 235
Pearl, Daniel, 214
Pearl Harbor attack, 38
People's Liberation Army (China), 71

People's Revolutionary Party of Kampuchea, 79
People's rights movements (United States), 97
People's Temple of the Disciples of Christ, 156
People's war strategy, 52, 202
People's Will movement (Russia), 25–26, 30, 37 (table), 50
Perceptions, 53–54, 61, 141–142
Perfect society, 57
Peru, 51, 135 (table), 137 (table), 283 (map)
PFLP. *See* Popular Front for the Liberation of
          Palestine (PFLP)
PGMs, 203
Phalangists, 217
Phansi, 116
Phantom cells, 173
Philippines, 35, 91, 125 (table), 128 (table),
          146, 147, 254, 282 (map)
Phineas Priesthood, 174
Phoenix Program collaboration (Vietnam War),
          83 (table), 84–85
Phosgene gas, 206
Pierce, William, 169, 170
PIJ. *See* Palestine Islamic Jihad (PIJ)
Pinochet, General Agusto, 75
Pipe bombs, 205
PKI. *See* Indonesian Communist Party (PKI)
Plastic explosives, 204, 257
PLO. *See* Palestine Liberation Organization (PLO)
Pogroms, 118
Poison gas, 25
Police. *See* Law enforcement approach; Legalistic options in
          counterterrorism; Police Border Guards
Police Border Guards, 226
Political beliefs, 2
     activist social movements and, 44–45, 47
     antimiscegination beliefs, 5–6
     classical ideological continuum and, 28, 29 (table)
     political extremism, 3–6
Political objectives, 8, 9
Political violence, 2, 6, 22
     ideologies and, 41
     injustice as precipitating cause, 45–51
     Iraqi sectarian violence, 40
     justification of, 11, 14
     left fringe ideology and, 28
     mala in se/mala prohibita and, 41
     morality of, 54–60
     participants' perspectives on, 7, 13–14
     political violence matrix, 15–16, 15 (figure)
     strategic choice of, 51–54
     terrorism, definition of, 7, 8, 11
     terrorists or freedom fighters, dilemma of, 12–14
     women in, 50–51
     See also Causes of terrorism; Domestic terrorism (United
          States); Extremism; State terrorists; Terrorism
Political will, 51–53
Pol Pot, 79, 80

Popular Front for the Liberation of Palestine (PFLP),
          58–59, 100, 134, 137, 143, 186 (table), 228
Popular Front for the Liberation of Palestine-General
          Command (PFLP-GC), 91, 100
Populations. *See* Civilians; Exile
Port Huron Statement, 160
Portugal, 18–19
Posse Comitatus, 31
Postrevolutionary society, 30, 70–71, 93
Praetorian Guard, 23
Prairie Fire manifesto, 164
Precision-guided munitions (PGMs), 203
Preemptive strikes, 222, 223
Preemptive war, 85
Pressure triggers, 204
Prevention and Punishment of Crimes Against
          Internationally Protected Persons, Including
          Diplomatic Agents, 244
Print media, 181
Prisoner of war status, 9–10
Privacy protections, 242, 263, 264, 270
Private contractors in wars, 105
Private property, 30
Proactive policies. *See* Counterterrorism; Counterterrorist
          policies; War on terrorism
Profiling, 240
Profit motive, 96
     mercenaries, 19, 70, 105
     private contractors in wars, 105
Progressive Labor party, 163
Projections. *See* Future trends/projections in terrorism
Proletariat, 32, 33 (figure)
Propaganda, 35, 52–53, 75, 77, 82, 129, 141, 180, 189, 219
Propaganda by the deed, 30
Property is theft, 30
Property targets, 9, 14, 97
Protected persons status, 10
Protocols of the Learned Elders of Zion, 117–119, 259
Proudhon, Pierre-Joseph, 30, 37 (table)
Provisional Irish Republican Army (Provos),
          4, 14, 47 (table), 51, 54, 94, 97, 135 (table),
          186 (table), 200 (table), 239–240, 241, 255 (table)
Provos. *See* Provisional Irish Republican Army (Provos)
Proxy wars, 18, 69–70, 83 (table), 252
Psychological anxiety, 8, 140, 198–199, 256
Psychological theory, 47
     gender and terrorism, 50–51
     good-evil dichotomy and, 47
     individual motivation/group dynamics
          explanations, 48–49
     mental illness/lunatic fringe and, 47–48
     moral convictions and, 47
     psychopathology, terrorists and, 49
     Stockholm Syndrome and, 49
     strategic choice of violence and, 51–54, 61
     See also Causes of terrorism; Sociological theory

Public criticism sessions, 24
Public humiliation/confessions, 24
Publicizing a cause, 137, 140
Puerto Rican Independencistas, 167
Punitive strikes, 222–223
Purity notion, 28, 55, 76, 93, 160
Putin, President Vladimir, 106

Qaddafi, Muammar el-, 84, 223
Qatar, 183
Q-fever, 127
Qods Force (Iran), 121
Qualified amnesty, 241
Quebec Liberation Front, 47 (table), 94, 186 (table)

Rabin, Prime Minister Yitzhak, 124
Racial Holy War, 169, 172
Racial mysticism, 160, 172–173
Racial profiling, 240
Racial supremacists, 5–6, 31, 97, 156, 160, 169–172
Racism:
    antimiscegination beliefs, 5–6
    apartheid, 19, 72
    Ásatrú adherents and, 161
    biblical interpretation and, 161
    Black Liberation Army and, 31, 165–166
    Black Liberation Movement and, 165–166, 166 (table)
    Black Power movement, 157–158
    chauvinistic racial dimension, 28
    ethnic cleansing, 12
    Ku Klux Klan and, 5–6, 112, 170–172
    Nazi Germany and, 81 (table)
    racial mysticism, 160, 172–173
    racial polarization, 166
    racial supremacists, 5–6, 31, 97, 156, 160, 169–172
    segregation, 160, 166
Radical decentralization of power, 30
Radical opinion. See Extremism
Radical socialism, 32
Radio broadcasts, 182
Radiological weapons, 105, 202, 206, 257
Rahowa, 172
Rajneeshees, 157
Rapoport, D. C., 249, 250
Rationality and terrorism. See Psychological theory
Rationalization of violence, 3, 11–13, 14, 22, 36, 68, 74
    See also Causes of terrorism; Moral justification
RDX compound, 204
Reactionary tendency, 159
Reagan, President Ronald, 70, 188
Real IRA (RIRA), 91
Rebel movements, 95
Red Army (Japan), 50, 97 (table)
Red Army Faction (Germany), 28, 50, 94, 97–98,
    134, 138, 227
Red Brigades (Italy), 15, 50, 93, 98, 134, 135 (table), 138, 241

Red Guards (China), 71
Red Guerrilla Resistance, 168
Red Scare, 87
Refugees, 117
    See also Exile
Regicide, 23
Regimes and terrorism, 67–68, 72–73
    See also State terrorists
Regional self-governance, 94
Reich, W., 8
Reid, Richard C., 107
Reign of Terror (1793–1794), 24
Relative deprivation theory, 46
Religious extremism, 2, 5
    Al Qaeda organization and, 147–148
    chosen vs. not-chosen ones and, 5
    Christian Right religious politics, 158–159
    codes of self-sacrifice and, 57
    faith-based natural law and, 131
    gender and, 51
    holy causes, 54, 55
    infidels, 5
    just war doctrine and, 36
    separation of church and state, 5–6
    United States and, 97
    See also Extremism; Religious terrorism; Terrorism
Religious objectives, 8, 16, 26, 37
Religious terrorism, 4, 11, 26, 27, 111, 137 (table)
    Afghan Arabs and, 126
    Al Qaeda's religious foundation and, 126–127, 130
    Aum Shinrikyō, 127–128
    Canaan, conquest of, 113–114
    Christian Crusades and, 114–115, 131
    communal terrorism/sectarian violence, 102–103
    cults and, 116, 117
    dissident religious terrorism,
        123–128, 125 (table), 128 (table)
    Grand Mosque incident, 124
    Grey Wolves and, 129
    historical perspective on, 111–119
    India vs. Pakistan, religious animosity and, 123
    Iranian support of, 121–123
    Iraqi sectarian violence, 124–125, 125 (table)
    Islamist extremists and, 116–117, 120, 121–123
    jihad and, 120–121, 122, 127, 129–130
    Judeo-Christian antiquity and, 113–114
    martyrdom and, 114, 121
    mujahideen warriors, 54, 125–126
    one true faith concept and, 130–131
    Order of Assassins and, 115
    Palestinian Islamists and, 122–123
    practices of, 120–128, 128 (table)
    promise of paradise and, 115, 121
    Protocols of the Learned Elders of Zion, 117–119
    Rabin assassination, 124
    religious vs. secular terrorism, 111, 112 (table)

Revolutionary Justice Organization and, 122
   September 11, 2001 attacks and, 37–39
   state-sponsored religious terrorism, 121–123
   Thuggee cult, 116
   trends/projections for, 129–130
   Ugandan mysticism/rebellion and, 117
   See also Future trends/projections in terrorism
Renamo rebel movement (Mozambique), 95
Repentance laws, 241
Repression. *See* Counterterrorism;
   Domestic terrorism practices; Oppressed masses;
   Other-than-war operations; State terrorists;
   Suppression campaigns
Republic of New Africa, 167
Republic of South Vietnam, 55
Resources:
   map resources, 272–284
   National Intelligence Estimate report, 285–291
   See also Internet resources
Revolution:
   acts of political will and, 51–53
   anarchist philosophy and, 25, 26, 30, 57–58
   structural theories of, 45–46
   See also Causes of terrorism; Moral justification
Revolutionary Armed Forces of Colombia (FARC),
   51, 82 (table), 91, 139
Revolutionary Catechism, 58
Revolutionary dissident terrorism, 92–93
Revolutionary Fighting Group, 168
Revolutionary Guards Corps (Iran), 121
Revolutionary Justice Organization, 97 (table), 122
Revolutionary Nuclei (RN), 91
Revolutionary Organization of Socialist Muslims, 60
Revolutionary People's Liberation Party/Front (DHKP/C),
   91, 97 (table)
Revolutionary People's Struggle, 97 (table)
The Revolutionary Tribunal, 24
Revolutionary United Front (Sierra Leone), 95
Revolutionary vanguard strategy, 26
Revolutionary Youth Movement, II, 163
Rhodesia. *See* Zimbabwe
Righteousness. *See* Extremism; Good-evil dichotomy;
   Moral justification; Religious extremism; Religious
   terrorism; Terrorism
Right-wing movements. *See* Ideological
   origins of terrorism
Right-wing terrorism, 137 (table)
Roadside bombs, 203
Robin Hood image, 91
Rocket-propelled grenades (RPGs), 203
Rockwell, George Lincoln, 169
Roman age, 23
Roman Catholicism, 5–6, 24, 114–115
Romania, 23
Royal Marine Commandos, 225
RPG-7, 203

Rules of engagement, 7, 36–37
Ruling class, 32, 33 (figure)
Russia:
   Black Widows, 50, 51
   Bolshevik Revolution, 32, 37 (table), 50
   Chechen separatists, 50, 51, 105–106
   nihilists, 93
   People's Will movement,
      25–26, 30, 37 (table), 50
   Revolutionary Catechism, 58
   revolutionary vanguard strategy and, 26
   Social Revolutionary Party, 50, 105
   Stalinist Russia, 75
Rwanda, 15, 81 (table), 102
Rwandan genocide, 15, 81 (table)

SA-7 missiles, 203
Salafist Group for Call and Combat (GSPC), 91
Salmonella poisoning, 157
Sam Melville-Jonathan Jackson Unit, 168
Samoza, Anistasio, 19
Sanchez, Ilich Ramirez, 134, 188, 236
Sandinista National Liberation Front (Nicaragua),
   19, 68–69, 70
Sargent, L. T., 6
Sarin gas. *See* Aum Shinrikyō (AUM); Nerve gas
Satellite surveillance network, 264
Saudia Arabia, 2, 124, 131, 142
Savimbi, Jonas, 19
Scapegoat groups, 67, 77, 117–119
Scheuer, M., 53
Scooping, 188
Scottish Protestants, 46–47
SEALs. *See* Navy Sea Air Land Forces (SEALs)
Searches, 88
2nd of June Movement, 134
Secret cult of murder, 116
Secret detentions, 10
Secret police services, 75
Secret policies, 67–68, 75
Sectarian violence, 102–103, 124–125
Security services, 72, 77, 82 (table), 219
Sederberg, P. C., 73
Self-sacrifice, 57–60, 126
Semtex, 204
Separation of church and state, 5–6
September 11, 2001 (USA), 1, 5, 26, 37–38,
   136, 137 (table), 174, 212
   counterterrorist policies and, 9–10, 38–39
   effectiveness of attacks, 215 (table)
   financial operations and, 262–263
   homeland security reorganization and,
      265–269, 266 (table), 268–269 (tables)
   media coverage of, 13–14, 188
   Muslim perceptions and, 53–54
   New Terrorism, objectives of, 201

war on terror and, 39–40, 41, 118
    See also February 1993 World Trade Center attack;
        United States
Serbs, 12, 55, 76, 81 (table), 102, 145, 200 (table)
Servile War of 73–71 B.C.E., 23
17 November, 91
Sex slavery, 117
Shadow wars, 224, 261
Shari'a law, 93
Sharm el Sheikh (Egypt) bombing, 26, 41, 105
Sharon, General Ariel, 98
Shi'a Muslims, 103, 117, 122, 124–125, 125 (table),
        136, 149, 189, 190, 217, 236, 237
Shigenobu, Fusako, 50
Shining Path (SL), 51, 91, 135 (table)
Sicarii rebels, 23, 24
Sierra Leone, 95, 254
SIGINT. *See* Signal intelligence/SIGINT
Signal intelligence/SIGINT, 230
Signature method, 206
Sihanouk, Prince Norodom, 79, 80
Sikh groups, 128 (table)
Singapore plot, 147
Sinhalese. *See* Buddhist Sinhalese
Sinn Féin, 186
Skinheads, 98, 160, 169, 172, 252
Slavery, 160, 164
Sleeper cells, 146, 148, 188, 202
Slovenia, 81 (table)
Small Boy Unit (Liberia), 95
Small Boy Units (Sierra Leone), 95
Small Girl Units (Sierra Leone), 95
Smallpox, 205
Smith Act of 1940, 87
Social cleansing, 76
Social evolution, 32, 52
Social objectives, 9, 199
Social reform, 220, 236–237, 238 (table)
Social Revolutionary Party (Russia), 50, 105
Socialist states, 30, 32
Sociological theory, 45
    intergroup conflict, collective violence and, 45–47, 61
    relative deprivation theory, 46
    structural theory of revolution, 45–46
    See also Causes of terrorism; Psychological theory
Socrates, 23
Soft targets, 6, 8, 136–137, 257
Somalia, 73, 102, 146, 226, 228, 254
South Africa:
    Angolan civil war and, 19, 104
    anti-apartheid supporters, 72
    apartheid, 19, 72
South America. *See* Latin America
Southern Baptist Convention, 31
Southern Poverty Law Center, 169
Southern Rhodesia. *See* Zimbabwe

Soviet Union, 12, 17–18
    Afghanistan war, 125–126
    Angolan civil war and, 19, 104
    Chechen separatists and, 105
    Cuban troops, Ethiopian unrest and, 83 (table)
    Nicaragua and, 19
    post-World War II era and, 103–104, 105
    repressive police state in, 26
    Social Revolutionary Party and, 105
    See also Russia
Spain, 2, 7, 18, 26, 28, 146
    Basque ETA, 47 (table), 50, 94, 98, 134, 236–237
    civil war in, 33, 34–35
    Falange party, 34–35
    Madrid bombing, 2, 26, 41, 105, 134, 148, 202, 254
    Movimiento Nacional, 35
Spanish-American War of 1898, 18
Spanish Civil War, 33, 34–35
Spanish Falange, 34–35
Spanish National Action, 98
Spartacus, 23
Special Air Service (SAS), 224–225
Special Boat Service (SBS), 225
Special operations forces, 224, 229 (table)
    Delta Force, 224, 225, 228–229
    elite counterterrorist units, utility of, 246–247
    France and, 225, 226
    Germany, 226
    Great Britain and, 224–225
    hostage incident cases, 227–229
    Israel and, 225, 226
    military units, 224–225
    police units, 226–227
    SEALs, 225
    United States, 225, 226–227
    See also Counterterrorism
Special Weapons and Tactics (SWAT) units, 227
Spillover effect, 135–137, 151–152
Spin, 181, 187
Sri Lanka, 14, 45, 50, 51, 95, 102, 108–109, 254
Stalinist Russia, 75
Stalin, Joseph, 75, 193
Starvation, 80
State assistance. *See* Assistance model for terrorism
State authority. *See* Domestic terrorism practices; State
        terrorists
Stateless revolutions, 150 (table), 255 (table)
Stateless society, 37 (table)
State patronage. *See* Patronage model for terrorism
State-regulated presses, 193
State terrorists, 8, 10, 22, 37 (table), 66, 137 (table)
    assassinations, 67
    assistance model for terrorism and, 68 (table), 69–71
    domestic policy and, 69, 70–71
    domestic terrorism practices, 72–81
    ethnic cleansing, 12, 15, 76, 81 (table)

foreign policy and, 68–70
foreign policy terrorism practices, 81–86, 83 (table)
genocide, 12, 15, 42, 67, 76–80, 81 (table)
just war doctrine and, 36–37
Marxist revolutions and, 32
official domestic state terrorism, 74–76
patronage model for terrorism, 68–69, 68 (table)
religious terrorism and, 121–123
Revolutionary Tribunal and, 24
scales of violence in, 67
security environment, state participation in, 67
state authority, legitimization of, 72–73, 73 (table)
state domestic authority, 73–81
state sponsorship, understanding of, 67–72, 68 (table)
structural theory of revolution and, 45–46
vigilante domestic state terrorism, 74
See also Future trends/projections in terrorism;
        Governments; Political violence
Stinger missiles, 203
Stockholm Syndrome, 49, 165
Street gangs, 159–160
Strong multipolar terrorist environment,
        143–144, 144 (table)
Structural theory of revolution, 45–46
Struggle meetings (revolutionary China), 24
Student dissidents, 25–26
Student Nonviolent Coordinating
        Committee (SNCC), 158
Students for a Democratic Society (SDS), 158, 162–164
Submachine guns, 203
Subnational entities, 8, 9
Sudan, 100, 103, 121, 146, 254
Suicide bombings, 10, 40, 41, 50, 51, 98, 99, 101,
        106, 108, 122, 149, 206
    Al Qaeda-related suicidal violence, 207
    intifada-motivated suicide/Israel, 207, 209, 210 (table)
    maximizing casualties and, 257
    religion-motivated suicide/Lebanon model, 206–207
    signature method of, 206
Sunni Muslims, 103, 117, 124–125, 125 (table), 149, 217
Sun Tzu, 12
Supergrass program, 241
Superiority notion, 28, 32–33, 55, 57, 118, 160
Supporter role, 13, 186
Supporters of terrorism, 13
Suppression campaigns, 219, 221, 229 (table)
    military suppression campaigns, 221–222
    paramilitary suppression campaigns, 222
    punitive/preemptive strikes, 222–223
    See also Counterterrorism; Other-than-war operations;
        Special operations forces
Supreme Truth. See Aum Shinrikyō (AUM)
Surveillance activities, 38, 242, 261, 263–264
Symbionese Liberation Army (SLA), 164–165
Symbolism, 14, 30, 39, 137, 160, 209–212, 257
Syria, 47, 59, 62, 69, 73, 75, 78, 83 (table), 100, 122

Tactics. See Terrorist tactics
Taliban regime, 73, 75, 131, 141
Tamil Tigers. See Liberation Tigers of
        Tamil Eelam (LTTE)
Tanzim Qa'idat al-Jihad fi Bilad al-Rafidayn (QJBR), 91
Targets, 6, 8, 9
    buildings/sites, 211, 212
    communal violence and, 16, 18
    counterrevolutionaries, suppression of, 55–56
    embassies/diplomatic personnel, 209–210
    global audience and, 40
    hardening targets, 219, 233
    individuals with symbolic value, 211, 212
    international symbols, 211, 212
    international terrorism and, 11, 136
    off-duty military personnel, 16
    passenger carriers, 212
    perspective on terrorism, 13
    soft targets, 6, 8, 136–137, 257
    symbolism of, 14, 30, 39, 137, 209–212, 257
    total war and, 41–42
    United States symbolic targets, 212
    See also Civilians; Terrorist tactics;
        Victims of terrorism
Taylor, Charles, 95
Teamsters Union, 31
Technology. See Communication technologies; High-tech
        terrorism; Information technologies; Internet; Mass
        media; Weaponry
Television broadcasts, 182
Temple Mount (Jerusalem), 98
Terminal institutions, 265
Terrorism, 2
    antiterrorist policies and, 7–8, 9–10, 14
    conspiratorial organizational structure/secretive features
        and, 8
    definitions of, 6–10
    enemies, labeling of, 4, 5
    fear, exploitation of, 8
    force, illegitimate use of, 6, 8
    formal definitions of, 7–8
    freedom fighters or terrorists, dilemma of, 11–14
    government domestic/foreign policy and, 7
    guerrilla warfare and, 7
    informal definitions of, 6–7
    interpretation/analysis of, 14
    mala in se/mala prohibita and, 41
    mass media, globalized terrorism and, 2
    perspective of participants in, 7, 13–14
    political extremism, 3–6
    rationalization/justification of violence, 3, 11–13
    subnational/non-state entities and, 8
    terrorists groups vs. terrorist states, 8
    types of terrorism, 10–11
    unacceptability of, 141–142
    United States, government policy of, 8–10

violent extremists, characteristics of, 4–5
women terrorists, 50–51, 106, 109, 117
See also Causes of terrorism; Counterterrorist
    policies; Dissident terrorism; Freedom fighters;
    Future trends/projections in terrorism;
    Historical perspectives on terrorism; Ideological
    origins of terrorism; International terrorism; New
    Terrorism; Political violence; Religious extremism;
    Religious terrorism; State terrorists; Terrorist
    suspects; Torture
Terrorist environments, 137, 137 (table),
    139–140, 143, 144 (table)
    analyst role and, 187
    cell-based terrorist environment, 144
    globalized environment and, 201
    monolithic terrorist environment, 143
    onlooker role and, 187
    participant roles in, 185–187
    police agencies and, 239
    strong multipolar terrorist environment, 143–144
    supporter role and, 186
    target role and, 187
    terrorist role and, 185–186
    twenty-first century and, 251–252
    victim role and, 187
    weak multipolar terrorist environment, 144
    See also Participants in terrorism; Terrorist tactics
Terrorist profiling, 240
Terrorist role, 13, 185–186
Terrorist suspects, 9
    enemy combatant status and, 9, 10
    Geneva Convention protocols and, 9–10
    Guantanamo Bay detentions, 10, 39
    innocent prisoners, reclassification of, 10
    labeling of, 9–10
    profiling of, 240
    protected persons status and, 10
    secret detentions, 39
    special status labeling of, 10
    See also Freedom fighters; New Terrorism; Participants
        in terrorism; Terrorism; Terrorist tactics
Terrorist tactics, 197–198
    adaptation, methods/targets/constituencies and,
        199–200, 200 (table)
    asymmetrical warfare concept and, 201–202
    audience impact and, 213
    concessions from enemies, 214, 220
    digital/video/audio terrorism, 213–214
    effectiveness analysis, 213–214, 215 (table)
    existing order, disruption of, 198
    globalized environment and, 201
    hostages and, 213–214
    information technology and, 201
    martyr nation strategy and, 202
    media/political attention and, 213
    netwar, new organizational structure and, 202

New Terrorism, vague objectives of, 200–201
    normal routines, disruption of, 214
    objectives of terrorists, 198–201, 200 (table)
    provoked overreactions, 215
    psychological anxiety and, 198–199
    senselessness/randomness and, 197
    social disruption and, 199
    United States events, 209, 211 (figure)
    weaponry and, 202–209
    weapons of mass destruction and, 201, 205–206
    See also Suicide bombings; Targets
The Terrorist Threat to the US Homeland (National
        Intelligence Estimate report), 285–291
Terrorizing a thousand. *See* Kill one man, terrorize a
        thousand; One man willing to throw away his life
        is enough to terrorize a thousand
Third World groups, 27
    postcolonial developing world, 103–104
    See also Colonialism
Thomas More Law Center, 31
Threat estimate. *See* National Intelligence
        Estimate report
Thuggee cult, 116
Tibet, 94
Tigers of Tamil. *See* Liberation Tigers of
        Tamil Eelam (LTTE)
TNT, 203
Tokyo Convention on Offences and Certain Other Acts
        Committed on Board Aircraft, 243
Torture:
    Abu Ghraib prison abuse and, 242–243
    coercive interrogations, 10, 242
    extraordinary renditions and, 219
    government sponsored torture, 2, 10, 67, 219, 242–243
    Guantanamo Bay detentions and, 10, 39
    See also Legalistic options in counterterrorism
Total war doctrine, 41–42
Totalitarianism, 72, 73 (table), 75, 193
Trends. *See* Future trends/projections in terrorism
Triple-A. *See* Argentine Anti-Communist Alliance (Triple-A)
The Troubles, 44, 45, 46–47, 235
Truth:
    codes of self-sacrifice and, 57
    cults and, 157
    propaganda and, 180
    protectors of, extremists and, 5–6
    religious extremists and, 5
Tupac Amaru Revolutionary Movement (Peru), 137 (table)
Tupamaros, 215
Turkey, 138
The Turner Diaries, 160, 170, 173
TWA Flight 840 bombing, 212
TWA Flight 847 hijacking, 189–190, 215 (table), 236
Twin Towers. *See* September 11, 2001 (USA)
Two-seedline Christian Identity, 161
Tyrannicide, 22

Uganda, 73, 117, 228
Unabomber, 168, 181
Unconventional warfare, 8
UNITA. *See* National Union for the Total Independence of
    Angola (UNITA)
United Arab Republic, 116
United Freedom Front (UFF), 168
United Kingdom. *See* Great Britain
United Nations, 71, 77, 85, 159, 244, 254
United Self-Defense Forces of Colombia (AUC), 82 (table), 91
United States, 2, 146, 254–255, 284 (map)
    Abu Nidal and, 60
    anarchist assassination in, 30
    Angolan civil war and, 19, 104
    civil liberties, suspension of, 241–242
    civil rights movement, 44
    classical ideological continuum and, 27, 31 (table)
    Cold War and, 4, 12, 17–18
    counterterrorist policies of, 9, 38–40
    Cuban Revolution and, 18
    democracy in, 73
    Department of Homeland Security, 38, 40 (figure), 267
    detention of suspects, 10, 39
    dissident terrorists and, 93
    estimative language, terror threat evaluation and, 289–290
    far right movements, 28, 30, 31 (table)
    Geneva Convention protocols and, 9–10
    global reputation of, 53–54
    homeland security community, reorganization of,
        265–269, 266 (table), 268–269 (tables)
    imperialist tendencies in, 54–55
    intelligence agencies in, 231
    intelligence community in, 231, 232 (figure)
    Iran-Contra scandal, 70, 237
    Iraq war and, 39–40, 41, 85–86, 149–150, 179, 195
    law enforcement approach and, 8, 38
    leftist terrorism in, 97
    National Intelligence Council and, 287–288
    National Intelligence Estimate report, 285–291
    national security, authoritarian methods and, 87–88
    native populations, 81 (table)
    Nicaragua and, 19
    Patriot movement, 97
    post-World War II global rivalries and, 103–104
    privacy rights, limitations on, 242
    private contractors in Iraq, 105
    profiling of terrorists, 240
    protected persons status, Mujahideen-e Khalq
        Organization and, 10
    Red Scares and, 87
    religious extremists and, 97
    right-wing terrorism in, 97
    special operations forces, 225, 226–227
    surveillance activities in, 38
    terrorism, definitions of, 7, 8–10
    terrorist activities in, 209, 211 (figure)
    terrorist suspects, labeling of, 9–10
    torture debate within, 242–243
    USA PATRIOT Act of 2001, 38, 241–242
    war protesters, 97
    See also Central Intelligence Agency (CIA); Domestic
        terrorism (United States); Federal Bureau of
        Investigation (FBI); Future trends/projections in
        terrorism; September 11, 2001 (USA); Vietnam War
Unlawful combatants, 10
Urban II, Pope, 114
Uruguay, 34, 283 (map)
USA PATRIOT Act of 2001, 38, 241–242
USS Cole bombing, 147
Utopia, 57
Utopian vision, 99

Vandalism sprees, 168
Vanguard strategy, 26
Vehicular bombs, 205, 207
Victims of terrorism, 13, 14, 187
    civilians, 2, 8, 9, 12
    perspectives on terrorism of, 13
    selection of, global attention and, 40
    See also Civilians; Targets; Terrorist tactics
Viet Cong, 55–56, 84–85, 159
Vietnam, 79, 158, 282 (map)
Vietnamese Communists, 55
Vietnam War, 13, 18, 55–56, 83 (table), 84–85, 97, 158, 194
Vigilante domestic state terrorism, 74
Violence:
    civilian targets and, 2, 8, 9, 12
    collective violence, intergroup conflict and, 45–47
    communal violence/civil war, 16
    fear, exploitation of, 8
    rationalization/justification of, 3
    self-sacrifice and, 12
    violent extremists, characteristics of, 4–5
    See also New Terrorism; Political violence; Terrorism
Violent extremism. *See* Extremism
Vision. *See* Grand vision; Utopian vision
Voter registration, 158

War, 67
    asymmetrical warfare, 133
    cyberwar, 219, 230, 257
    Islamic holy wars, 53–54, 120–121
    people's war strategy, 52, 202
    postcolonial civil wars, 103–104
    preemptive war, 85
    proxy wars, 18, 69–70, 83 (table), 252
    shadow wars, 224, 261
    total war doctrine, 41–42
    war on terrorism, 9, 39–40, 41
    See also Military campaigns; New Terrorism;
        Other-than-war operations; War on terrorism;
        specific names of wars

Warfare environments, 16, 67
Warlords, 73
Warrantless searches, 88
  See also War on terrorism
War on terrorism, 9, 39–40, 41, 106, 260
  adaptation, resilient adversaries and, 263
  American homeland security community, reorganization
      of, 265–269, 266 (table), 268–269 (tables)
  banking systems, customer anonymity and, 263
  Carnivore Internet surveillance system, 264
  digital fingerprinting and, 263
  Echelon satellite surveillance network, 264
  electronic surveillance and, 264, 270
  extremist ideologies and, 261
  facial imaging and, 263
  financial operations and, 261, 262–263
  force, utility of, 262
  global surveillance of communications technologies, 261
  government responses and, 260
  international cooperation with, 264–265
  law enforcement, intelligence and, 264–265
  new fronts/new approaches in, 261
  shadow wars and, 224, 261
  societal responses and, 260–261
  surveillance technologies and, 263–264
  terrorist cells, disruption of, 261
  See also Counterterrorism; Counterterrorist policies;
      Future trends/projections in terrorism
Watts riots (Los Angeles), 162
Weak multipolar terrorist environment, 144, 144 (table)
Weaponry, 202
  ammonium nitrate/fuel oil bombs, 147, 204
  assault rifles, 203
  barometric bombs, 205
  biological weapons, 38, 105, 127, 157, 202, 205
  black market in, 202–203
  bomb typology, 204–209
  chemical weapons, 78–79, 86, 105,
      127, 128, 202, 205–206
  common explosives, 147, 203–204
  composite-4, 204
  dynamite/TNT, 203
  electronic triggers and, 204
  firearms, 203
  gasoline bombs, 204
  high range weapons, 202
  improvised explosive devices/roadside bombs, 203
  infrared-targeted missiles, 203
  land mines, 203
  low range weapons, 203
  medium range weapons, 202–203
  Molotov cocktails, 204
  nuclear weapons, 105, 123, 202, 206
  pipe bombs, 205
  plastic explosives, 204
  precision-guided munitions, 203

  pressure triggers and, 204
  radiological weapons, 105, 202, 206
  rocket-propelled grenades, 203
  Semtex, 204
  submachine guns, 203
  tech-terror and, 257
  triggering devices, 204
  vehicular bombs, 205
  See also Suicide bombings; Targets; Terrorist tactics;
      Weapons of mass destruction (WMDs)
Weapons of mass destruction (WMDs),
      2, 25, 39, 85–86, 251
  biological agents, 38, 105, 127, 157, 202, 205
  chemical agents, 78–79, 86, 105, 127, 128, 202, 205–206
  dissident terrorists, new morality of, 105
  high-tech terrorism and, 256
  New Terrorism, objectives of, 201
  nuclear weapons, 105, 123, 202, 206
  radiological agents, 105, 202, 206
  See also Terrorist tactics; Weapons
Weather Bureau, 164
Weather Underground Organization (United States),
      93, 164, 165, 167
Weathermen, 163, 164
Weinrich, Johannes, 236
West Bank, 126, 146
White Army, 118
White Aryan Resistance (WAR), 169
White supremacy, 5–6, 31, 97, 160
Wilcox, Laird, 3
Wilson, President Woodrow, 87, 171
Women as second class, 131
Women terrorists, 50–51, 106, 109, 117
Worker-Student Alliance, 163
Working class struggles, 32, 52
World Trade Center (New York City). *See* February 1993
      World Trade Center attack; September 11, 2001
World War, II, 15, 34, 35, 37 (table)
  casualty count for, 42
  post-war global rivalries, 103–104
  total war doctrine and, 41–42
Wrath of God unit, 224, 227
Wu Ch'i, 12, 206

YAMAM group, 226
YAMAS group, 226
Yankee imperialism, 4
Yemen, 146

Zapatista National Liberation Front (Mexico), 51, 90–91
Zealots, 23–24
Zimbabwe, 71–72
Zimbabwe African National Union-Patriotic Front
      (ZANU-PF), 71–72
Zionist movement, 139, 150 (table)
Zionist Occupation Government (ZOG), 5, 159, 173